P9-CKX-109

Integral Methods in Science and Engineering

Theoretical and Practical Aspects

C. Constanda
Z. Nashed
D. Rollins
Editors

Birkhäuser
Boston • Basel • Berlin

C. Constanda
University of Tulsa
Department of Mathematical
and Computer Sciences
600 South College Avenue
Tulsa, OK 74104
USA

Z. Nashed
University of Central Florida
Department of Mathematics
4000 Central Florida Blvd.
Orlando, FL 32816
USA

D. Rollins
University of Central Florida
Department of Mathematics
4000 Central Florida Blvd.
Orlando, FL 32816
USA

Cover design by Alex Gerasev.
AMS Subject Classification: 45-06, 65-06, 74-06, 76-06

Library of Congress Cataloging-in-Publication Data
Integral methods in science and engineering : theoretical and practical aspects / C. Constanda, Z. Nashed, D. Rollins (editors).
p. cm.
Includes bibliographical references and index.
ISBN 0-8176-4377-X (alk. paper)
1. Integral equations–Numerical solutions–Congresses. 2. Mathematical analysis–Congresses. 3. Science–Mathematics–Congresses. 4. Engineering mathematics–Congresses. I. Constanda, C. (Christian) II. Nashed, Z. (Zuhair) III. Rollins, D. (David), 1955-
QA431.I49 2005
518′.66–dc22 2005053047

ISBN 0-8176-4377-X eISBN 0-8176-4450-4 Printed on acid-free paper.
ISBN-13 978-0-8176-4377-5

 Birkhäuser 

Printed in the United States of America. (IBT)

9 8 7 6 5 4 3 2 1

www.birkhauser.com

Contents

Preface

The purpose of the international conferences on Integral Methods in Science and Engineering (IMSE) is to bring together researchers who make use of analytic or numerical integration methods as a major tool in their work.

The first two such conferences, IMSE1985 and IMSE1990, were held at the University of Texas at Arlington under the chairmanship of Fred Payne. At the 1990 meeting, the IMSE consortium was created, charged with organizing these conferences under the guidance of an International Steering Committee. Thus, IMSE1993 took place at Tohoku University, Sendai, Japan, IMSE1996 at the University of Oulu, Finland, IMSE1998 at Michigan Technological University, Houghton, MI, USA, IMSE2000 in Banff, AB, Canada, IMSE2002 at the University of Saint-Étienne, France, and IMSE2004 at the University of Central Florida, Orlando, FL, USA. The IMSE conferences have now become established as a forum where scientists and engineers working with integral methods discuss and disseminate their latest results concerning the development and applications of a powerful class of mathematical procedures.

An additional, and quite rare, characteristic of all IMSE conferences is their very friendly and socially enjoyable professional atmosphere. As expected, IMSE2004, organized at the University of Central Florida in Orlando, FL, continued that tradition, for which the participants wish to express their thanks to the Local Organizing Committee:

David Rollins, *Chairman;*

Zuhair Nashed, *Chairman of the Program Committee;*

Ziad Musslimani;

Alain Kassab;

Jamal Nayfeh.

The organizers and the participants also wish to acknowledge the support received from

The Department of Mathematics, UCF,

The College of Engineering, UCF,

and the University of Central Florida itself for the excellent facilities placed at our disposal.

The next IMSE conference will be held in July 2006 in Niagara Falls, Canada. Details concerning this event are posted on the conference web page, http://www.civil.uwaterloo.ca/imse2006.

This volume contains eight invited papers and nineteen contributed papers accepted after peer review. The papers are arranged in alphabetical order by (first) author's name.

The editors would like to record their thanks to the referees for their willingness to review the papers, and to the staff at Birkhäuser Boston, who have handled the publication process with their customary patience and efficiency.

Tulsa, Oklahoma, USA *Christian Constanda, IMSE Chairman*

Contributors

Mario Ahues: Équipe d'Analyse Numérique, Université Jean Monnet de Saint-Étienne, 23 rue Dr. Paul Michelon, F-42023 Saint-Étienne, France

mario.ahues@univ-st-etienne.fr

Ivanilda B. Aseka: UFSM–CCNE, Departamento de Matematica, Campus Universitario, Santa Maria (RS) 971-5-900, Brazil

iaseka@smail.ufsm.br

Florin Badiu: Department of Mathematics, University of Texas at Arlington, P.O. Box 19408, Arlington, TX 76019-0408, USA

fvbadiu@uta.edu

Barbara S. Bertram: Department of Mathematical Sciences, Michigan Technological University, 1400 Townsend Drive, Houghton, MI 49931-1295, USA

bertram@mtu.edu

Haroldo F. Campos Velho: Laboratório de Computação e Matemática Aplicada, Instituto Nacional de Pesquisas Espaciais, Av. dos Astronautas 1758, P.O. Box 515, 12245-970 São José dos Campos (SP), Brazil

haroldo@lac.inpe.br

Ke Chen: Department of Mathematical Sciences, University of Liverpool, Peach Street, Liverpool L69 7ZL, UK

k.chen@liv.ac.uk

Jin Cheng: Department of Mathematics, Fudan University, Shanghai 200433, China

jcheng@fudan.edu.cn

Igor Chudinovich: Department of Mechanical Engineering, University of Guanajuato, Salamanca, Mexico

chudynovich@salamanca.ugto.mx

Christian Constanda: Department of Mathematical and Computer Sciences, University of Tulsa, 600 S. College Avenue, Tulsa, OK 74104-3189, USA

christian-constanda@utulsa.edu

Eduardo A. Divo: Engineering Technology Department, University of Central Florida, Orlando, FL 32816-2450, USA

edivo@mail.ucf.edu

Delfina Gómez: Departamento de Matematicas, Estadistica y Computación, Universidad de Cantabria, Av. de los Castros s.n., 39005 Santander, Spain

gomezdel@unican.es

Charles W. Groetsch: Department of Mathematics, University of Cincinnati, P.O. Box 210025, Cincinnati, OH 45221-0025.

groetsch@uc.edu

Paul J. Harris: School of Computational Mathematics and Informational Sciences, University of Brighton, Lewes Road, Brighton BN2 4GJ, UK

p.j.harris@brighton.ac.uk

Markku Hihnala: Mathematics Division, Department of Electrical Engineering, Faculty of Technology, University of Oulu, 90570 Oulu, Finland

markku.hihnala@ee.oulu.fi

John W. Hilgers: Signature Research Inc., 56905 Calumet Avenue, Calumet, MI 49913, USA

jwhilger@mtu.edu

Hiroshi Hirayama: Department of System Design Engineering, Kanagawa Institute of Technology, 1030 Shimo-Ogino, Atsugi-Shi, Kanagawa-Ken, 243-0292, Japan

hirayama@sd.kanagawa-it.ac.jp

Alexander O. Ignatyev: Institute for Applied Mathematics and Mechanics, R. Luxemburg Street 74, Donetsk-83111, Ukraine

mila@budinf.donetsk.ua, ignat@iamm.ac.donetsk.ua

Oleksiy A. Ignatyev: Department of Mathematical Sciences, Kent State University, Kent, OH 44242, USA

aignatye@kent.edu

Alain J. Kassab: Mechanical, Materials, and Aerospace Engineering, University of Central Florida, Orlando, FL 32816-2450, USA

kassab@mail.ucf.edu

Alain Largillier: Équipe d'Analyse Numérique, Université Jean Monnet de Saint-Étienne, 23 rue Dr. Paul Michelon, F-42023 Saint-Étienne, France

larg@anum.univ-st-etienne.fr

Gilbert Lewis: Department of Mathematical Sciences, Michigan Technological University, 1400 Townsend Drive, Houghton, MI 49931–1295, USA

lewis@mtu.edu

Miguel Lobo: Departamento de Matematicas, Estadistica y Computación, Universidad de Cantabria, Av. de los Castros s.n., 39005 Santander, Spain

miguel.lobo@unican.es

Svitlana Mayboroda: Department of Mathematics, University of Missouri-Columbia, Mathematical Sciences Building, Columbia, MO 65211, USA

svitlana@math.missouri.edu

Sergey E. Mikhailov: Department of Mathematics, Glasgow Caledonian University, Cowcaddens Road, Glasgow G4 0BA, UK

s.mikhailov@gcal.ac.uk

Dorina Mitrea: Department of Mathematics, University of Missouri-Columbia, 202 Mathematical Sciences Building, Columbia, MO 65211, USA

dorina@math.missouri.edu

Marius Mitrea: Department of Mathematics, University of Missouri-Columbia, 305 Mathematical Sciences Building, Columbia, MO 65211, USA

marius@math.missouri.edu

Zuhair Nashed: Department of Mathematics, University of Central Florida, P.O. Box 161364, Orlando, FL 32816, USA

znashed@mail.ucf.edu

Adriana Nastase: Aerodynamik des Fluges, Rhein.-Westf. Technische Hochschule, Templergraben 55, 52062 Aachen, Germany

nastase@lafaero.rwth-aachen.de

Fred R. Payne: 1003 Shelley Court, Arlington, TX 76012, USA

frpdfi@airmail.net

Eugenia Pérez: Departamento de Matematica Aplicada y Ciencia de la Computación, E.T.S.I. Caminos, Canales y Puertos, Universidad de Cantabria, Av. de los Castros s.n., 39005 Santander, Spain

meperez@unican.es

Shirley Pomeranz: Department of Mathematical and Computer Sciences, University of Tulsa, 600 S. College Avenue, Tulsa, OK 74104-3189, USA

pomeranz@utulsa.edu

Stanislav Potapenko: Department of Civil Engineering, University of Waterloo, 200 University Avenue West, Waterloo, Ontario N2L 3G1, Canada

spotapen@uwaterloo.ca

Cynthia F. Segatto: Departamento de Matematica Pura e Aplicada, Av. Bento Gonçalves, 9500, Predio 43-111, Agronomia, Porto Alegre (RS) 91509-900, Brazil

csegatto@mat.ufrgs.br

Seppo Seikkala: Division of Mathematics, Department of Electrical Engineering, Faculty of Technology, University of Oulu, 90570 Oulu, Finland

seppo.seikkala@ee.oulu.fi

Hua Shan: Department of Mathematics, University of Texas at Arlington, P.O. Box 19408, Arlington, TX 76019-0408, USA

hshan@uta.edu

Jianzhong Su: Department of Mathematics, University of Texas at Arlington, P.O. Box 19408, Arlington, TX 76019-0408, USA

su@uta.edu

Tadie: Matematisk Institut, Universitetsparken 5, 2100 Copenhagen, Denmark

tad@math.ku.dk

Johannes Tausch: Department of Mathematics, Southern Methodist University, P.O. Box 750156, Dallas, TX 75275-0156, USA

Tausch@mail.smu.edu

Marco T. Vilhena: Departamento de Matematica Pura e Aplicada, Av. Bento Gonçalves, 9500, Predio 43-111, Agronomia, Porto Alegre (RS) 91509-900, Brazil

vilhena@mat.ufrgs.br

Sergio Wortmann: Departamento de Matematica Pura e Aplicada, Av. Bento Gonçalves, 9500, Predio 43-111, Agronomia, Porto Alegre (RS) 91509-900, Brazil

wortmann@mat.ufrgs.br

Leon Xu: Nokia Research Center, 6000 Connection Drive, Irving, TX 75039, USA

leon.xu@nokia.com

Jiansen Zhu: Nokia Research Center, 6000 Connection Drive, Irving, TX 75039, USA

jiansen.zhu@nokia.com

1 Newton-type Methods for Some Nonlinear Differential Problems

Mario Ahues and Alain Largillier

1.1 The General Framework

The goals of this paper are to show how to formulate different kinds of nonlinear differential problems in such a way that a Newton–Kantorovich-like method may be used to compute an approximate solution, and to establish the rates of convergence corresponding to a Hölder continuity assumption on the derivative of the associated nonlinear operator. This generalizes the classical convergence results (see [6] and [7]).

The abstract general framework is a complex Banach space and applications include evolution equations and spectral problems. Computations have been done on standard model problems to illustrate practical convergence. A perturbed fixed-slope inexact variant is proposed and studied.

Let X be a complex Banach space, $\mathcal{O}_r(\varphi)$ the open disk centered at φ with radius $r > 0$, $\mathcal{L}(X)$ the Banach algebra of all linear bounded operators in X, I the identity operator, and $\mathcal{A}(X)$ the open subset of automorphisms.

Let $\ell > 0$ and $\alpha \in]0, 1]$. We recall that an operator $P : \mathcal{D} \subseteq X \to X$ is (ℓ, α)-Hölder continuous on $\mathcal{D}$ if for all x and y in $\mathcal{D}$,

$$\|P(y) - P(x)\| \leq \ell \|y - x\|^{\alpha}.$$

Let $\mathcal{O}$ be an open set in X, $F : \mathcal{O} \to X$ a nonlinear Fréchet differentiable operator, and $(B_k)_{k \geq 0}$ a sequence in $\mathcal{A}(X)$. A Newton-type iterative process reads as

$$\varphi_0 \in \mathcal{O}, \quad \varphi_{k+1} := \varphi_k - B_k^{-1} F(\varphi_k), \quad k \geq 0. \tag{1.1}$$

Obviously, if the sequence of operators $(B_k)_{k \geq 0}$ is bounded in $\mathcal{L}(X)$, and the sequence $(\varphi_k)_{k \geq 0}$ is convergent in X, then the limit of the latter is a zero of F.

The Newton–Kantorovich method corresponds to the choice

$$B_k := F'(\varphi_k). \tag{1.2}$$

Theorem 1. (A priori convergence of (1.1)–(1.2) with Hölder derivative) *Let $\varphi_\infty \in \mathcal{O}$ be a zero of F. Suppose that*

(1) $F'(\varphi_\infty) \in \mathcal{A}(X)$;
(2) *there is* $r > 0$ *such that* $F' : \mathcal{O}_r(\varphi_\infty) \subseteq \mathcal{O} \to \mathcal{L}(X)$ *is* (ℓ, α)*-Hölder continuous.*

Then there exists $\varrho \in]0, r]$ *such that, for all* $\varphi_0 \in \mathcal{O}_\varrho(\varphi_\infty)$, *the sequence* (1.1)–(1.2) *is well defined, and there exists* $C > 0$ *such that for all* $k \geq 0$,

$$\|\varphi_{k+1} - \varphi_\infty\| \leq C\|\varphi_k - \varphi_\infty\|^{\alpha+1}.$$

Proof. From (2) there follows the existence of $r_1 \in]0, r[$ such that, for all $\varphi \in \mathcal{O}_r(\varphi_\infty)$, if $\|\varphi - \varphi_\infty\| < r_1$, then $\|F'(\varphi) - F'(\varphi_\infty)\| < 1/\|F'(\varphi_\infty)^{-1}\|$. Hence, $F'(\varphi) \in \mathcal{A}(X)$ for all $\varphi \in \mathcal{O}_{r_1}(\varphi_\infty)$. Since $\varphi \mapsto F'(\varphi)^{-1}$ is continuous on $\mathcal{O}_{r_1}$ there exist $r_2 \in]0, r_1[$ and $\mu > 0$ such that, for all $\varphi \in \mathcal{O}_{r_2}(\varphi_\infty)$, $\|F'(\varphi)^{-1}\| \leq \mu$. The Hölder continuity of F' on $\mathcal{O}_{r_2}(\varphi_\infty)$ implies its uniform continuity and, in particular, the existence of $\varrho \in]0, r_2[$ such that, for all φ, ψ in $\mathcal{O}_\varrho(\varphi_\infty)$, $\|F'(\varphi) - F'(\psi)\| < 1/(2\mu)$. Take $\varphi_0 \in \mathcal{O}_\varrho(\varphi_\infty)$ and suppose that $\varphi_k \in \mathcal{O}_\varrho(\varphi_\infty)$. For $t \in [0, 1]$, define

$$\varphi_k(t) := (1 - t)\varphi_\infty + t\varphi_k \in \mathcal{O}_\varrho(\varphi_\infty).$$

Then

$$F(\varphi_k) = F(\varphi_k) - F(\varphi_\infty) = \int_0^1 F'(\varphi_k(t))(\varphi_k - \varphi_\infty)\, dt,$$

$$\varphi_{k+1} - \varphi_\infty = F'(\varphi_k)^{-1} \int_0^1 (F'(\varphi_k) - F'(\varphi_k(t)))(\varphi_k - \varphi_\infty)\, dt,$$

and hence,

$$\|\varphi_{k+1} - \varphi_\infty\| \leq \tfrac{1}{2}\|\varphi_k - \varphi_\infty\| < \tfrac{1}{2}\varrho.$$

This proves that $\varphi_{k+1} \in \mathcal{O}_\varrho(\varphi_\infty)$ and that φ_k converges to φ_∞. We can estimate the rate of convergence more precisely by means of the (ℓ, α)-Hölder continuity:

$$\|\varphi_{k+1} - \varphi_\infty\| \leq \mu\ell\|\varphi_k - \varphi_\infty\|^{\alpha+1} \int_0^1 (1 - t)^\alpha\, dt = \frac{\mu\ell}{1 + \alpha}\|\varphi_k - \varphi_\infty\|^{\alpha+1}.$$

Theorem 2. (A posteriori convergence of (1.1)–(1.2) with Hölder derivative) *Suppose that* $\mathcal{O}$, F, $\varphi_0 \in \mathcal{O}$, $c_0 > 0$, $\ell > 0$, $\alpha > 0$, *and* $m_0 > 0$ *satisfy*

(1) $F'(\varphi_0) \in \mathcal{A}(X)$ *and* $\|F'(\varphi_0)^{-1}\| \leq m_0$;
(2) $\|F'(\varphi_0)^{-1}F(\varphi_0)\| \leq c_0$;
(3) $\mathcal{D}_0 := \{\varphi \in X : \|\varphi - \varphi_0\| \leq 2c_0\}$ *is included in* $\mathcal{O}$;
(4) F' *is* (ℓ, α)*-Hölder continuous on* $\mathcal{D}_0$;
(5) $h_0 := m_0\ell c_0^\alpha < \alpha/(1 + \alpha)$.

Then F *has a unique zero* $\varphi_\infty \in \mathcal{D}_0$, *and for all* $k \geq 0$,

$$\|\varphi_{k+1} - \varphi_\infty\| \leq \frac{m_0\ell}{(1 - 2^\alpha h_0)(1 + \alpha)}\|\varphi_k - \varphi_\infty\|^{1+\alpha}.$$

Proof. For all $\varphi \in \mathcal{D}_0$,

$$\|I - F'(\varphi_0)^{-1}F'(\varphi)\| \leq \|F'(\varphi_0)^{-1}\|\|F'(\varphi_0) - F'(\varphi)\| \leq m_0\ell(2c_0)^\alpha = 2^\alpha h_0.$$

But for all $\alpha \in]0,1]$, $2^\alpha \leq 1+\alpha$ so with hypothesis (5) we get

$$\|I - F'(\varphi_0)^{-1}F'(\varphi)\| < \alpha \leq 1.$$

We conclude that $I - (I - F'(\varphi_0)^{-1}F'(\varphi)) = F'(\varphi_0)^{-1}F'(\varphi)$ is an automorphism of X, and hence so is $F'(\varphi)$ for all $\varphi \in \mathcal{D}_0$. Also, the family of inverses $\{F'(\varphi)^{-1} : \varphi \in \mathcal{D}_0\}$ is bounded in $\mathcal{L}(X)$: for all $\varphi \in \mathcal{D}_0$,

$$\begin{aligned}\|F'(\varphi)^{-1}\| &= \|(F'(\varphi_0)^{-1}F'(\varphi))^{-1}F'(\varphi_0)^{-1}\| \\ &\leq \|(I - (I - F'(\varphi_0)^{-1}F'(\varphi)))^{-1}F'(\varphi_0)^{-1}\| \leq \mu_0 := \frac{m_0}{1-2^\alpha h_0}.\end{aligned}$$

Let us prove that

$$\|F'(\varphi_1)^{-1}\| \leq m_1 := \frac{m_0}{1-h_0}.$$

We consider the auxiliary operator $A := F'(\varphi_0)^{-1}F'(\varphi_1)$. Then

$$\|I - A\| = \|F'(\varphi_0)^{-1}(F'(\varphi_0) - F'(\varphi_1))\| \leq m_0 \ell c_0^\alpha = h_0 < \frac{\alpha}{1+\alpha},$$

and hence

$$\|A^{-1}\| = \|(I-(I-A))^{-1}\| \leq \frac{1}{1-\|I-A\|} \leq \frac{1}{1-h_0} < 1+\alpha \leq 2.$$

So,

$$\|F'(\varphi_1)^{-1}\| = \|A^{-1}F'(\varphi_0)^{-1}\| \leq \|F'(\varphi_0)^{-1}\|\|A^{-1}\| \leq m_1 := \frac{m_0}{1-h_0}.$$

Set $G(\varphi) := \varphi - F'(\varphi_0)^{-1}F(\varphi)$ for $\varphi \in \mathcal{D}_0$. Then

$$G'(\varphi) = I - F'(\varphi_0)^{-1}F'(\varphi) \quad \text{for all } \varphi \in \mathcal{D}_0, \quad G'(\varphi_0) = O,$$

and hence

$$\begin{aligned}&\|G(\varphi_1) - G(\varphi_0)\| \\ &\quad = \left\|\int_0^1 G'(\varphi_0 + t(\varphi_1 - \varphi_0))(\varphi_1 - \varphi_0)\,dt\right\| \\ &\quad \leq \int_0^1 \|G'(\varphi_0 + t(\varphi_1 - \varphi_0)) - G'(\varphi_0)\|\|\varphi_1 - \varphi_0\|\,dt \\ &\quad \leq \int_0^1 \|F'(\varphi_0)^{-1}\|\,\|F'(\varphi_0 + t(\varphi_1 - \varphi_0)) - F'(\varphi_0)\|\,\|\varphi_1 - \varphi_0\|\,dt \\ &\quad \leq m_0 \ell \|\varphi_1 - \varphi_0\|^{1+\alpha} \int_0^1 t^\alpha\,dt \leq \frac{c_0}{1+\alpha} h_0.\end{aligned}$$

It follows that

$$\|\varphi_2 - \varphi_1\| = \|A^{-1}(G(\varphi_1) - G(\varphi_0))\| \leq c_1 := \frac{h_0}{1-h_0} \cdot \frac{c_0}{1+\alpha} < \frac{\alpha c_0}{1+\alpha}.$$

We remark that

$$h_1 := m_1 \ell c_1^\alpha = \Big(\frac{h_0}{1-h_0}\Big)^{1+\alpha} \frac{1}{(1+\alpha)^\alpha} \leq \alpha^\alpha (1+\alpha)^{1-\alpha} \cdot \frac{\alpha}{1+\alpha} \leq \frac{\alpha}{1+\alpha}$$

because $\alpha \mapsto \alpha^\alpha(1+\alpha)^{1-\alpha}$ is a convex function with limit 1 at both endpoints of the interval $]0,1]$.

A recursive argument shows that there exist real sequences $(m_k)_{k\geq 0}$ and $(c_k)_{k\geq 0}$, $(h_k)_{k\geq 0}$ such that for all $k \geq 0$,

$$\|F'(\varphi_k)^{-1}\| \leq m_k, \quad \|\varphi_{k+1} - \varphi_k\| \leq c_k,$$
$$m_{k+1} := \frac{m_k}{1-h_k}, \quad h_k := m_k \ell c_k^\alpha < \frac{\alpha}{1+\alpha}, \quad c_{k+1} < \frac{\alpha c_k}{1+\alpha}.$$

and for all $n \geq 1$,

$$\sum_{k=0}^{n-1} \|\varphi_{k+1} - \varphi_k\| \leq c_0 \sum_{k=0}^{n-1} \big(\frac{\alpha}{1+\alpha}\big)^k < 2c_0.$$

But $\varphi_n = \varphi_0 + \sum_{k=0}^{n-1}(\varphi_{k+1} - \varphi_k)$, the series with general term $\varphi_{k+1} - \varphi_k$ is normally convergent, and X is complete, so $(\varphi_n)_{n\geq 0}$ is convergent. Let $\varphi_\infty \in \mathcal{D}_0$ be its limit. Since F' and F are continuous at φ_∞ and for all $k \geq 0$, $F'(\varphi_k)(\varphi_{k+1} - \varphi_k) = -F(\varphi_k)$, we get $F'(\varphi_\infty)(\varphi_\infty - \varphi_\infty) = -F(\varphi_\infty)$, that is $F(\varphi_\infty) = 0$. Now, for all $k \geq 0$,

$$\begin{aligned}\varphi_{k+1} - \varphi_\infty &= -F'(\varphi_k)^{-1}\big(-F'(\varphi_k)(\varphi_k - \varphi_\infty) + F(\varphi_k) - F(\varphi_\infty)\big)\\ &= -F'(\varphi_k)^{-1}\int_0^1 [F'((1-t)\varphi_\infty + t\varphi_k) - F'(\varphi_k)](\varphi_k - \varphi_\infty)\,dt,\end{aligned}$$

hence

$$\|\varphi_{k+1} - \varphi_\infty\| \leq \frac{\mu_0 \ell}{1+\alpha}\|\varphi_k - \varphi_\infty\|^{1+\alpha}.$$

If $\alpha = 1$ in Theorem 2, then we get the classical convergence theorem for the case of a Lipschitz Fréchet derivative [7].

Theorem 3. (A posteriori convergence of (1.1)–(1.2) with Lipschitz derivative) *Suppose that* $\mathcal{O}$, F, $\varphi_0 \in \mathcal{O}$, $c_0 > 0$, $\ell > 0$, *and* $m_0 > 0$ *satisfy*

(1) $F'(\varphi_0) \in \mathcal{A}(X)$ *and* $\|F'(\varphi_0)^{-1}\| \leq m_0$;
(2) $\|F'(\varphi_0)^{-1}F(\varphi_0)\| \leq c_0$;
(3) $\mathcal{D}_0 := \{\varphi \in X : \|\varphi - \varphi_0\| \leq 2c_0\}$ *is included in* $\mathcal{O}$;

(4) ℓ *is a Lipschitz constant for* F' *on* $\mathcal{D}_0$;
(5) $h_0 := m_0 \ell c_0 < 1/2$.
Then F *has a unique zero* $\varphi_\infty \in \mathcal{D}_0$ *and for all* $k \geq 0$,

$$\|\varphi_{k+1} - \varphi_\infty\| \leq \frac{m_0 \ell}{1 - 2h_0} \|\varphi_k - \varphi_\infty\|^2.$$

A fixed-slope iteration is defined as

$$\varphi_0 \in \mathcal{O}, \quad \varphi_{k+1} := \varphi_k - B^{-1} F(\varphi_k), \quad k \geq 0, \tag{1.3}$$

where $B \in \mathcal{A}(X)$. The authors have proved in [1] the following a posteriori convergence result for this kind of method.

Theorem 4. (A posteriori convergence of (1.3) with Lipschitz derivative) *Suppose that* $\mathcal{O}$, F, B, $\varphi_0 \in \mathcal{O}$, $\delta \geq 0$, $\ell > 0$, $m > 0$, *and* $c > 0$ *satisfy*
(1) $m\delta < 1$ *and* $4m\ell c \leq (1 - m\delta)^2$;
(2) $\|B^{-1} F(\varphi_0)\| \leq c$;
(3) $\mathcal{O}$ *includes the closed disk* $\mathcal{D}_0 := \{\varphi \in X : \|\varphi - \varphi_0\| \leq \varrho_0\}$, *where*

$$\varrho_0 := \frac{1 - m\delta - \sqrt{(1 - m\delta)^2 - 4m\ell c}}{2m\ell};$$

(4) $F' : \mathcal{D}_0 \to \mathcal{L}(X)$ *exists and is* ℓ*-Lipschitz;*
(5) $\|F'(\varphi_0) - B\| \leq \delta$, $B \in \mathcal{A}(X)$ *and* $\|B^{-1}\| \leq m$.
Then F *has a unique zero* φ_∞ *in* $\mathcal{D}_0$ *and for all* $k \geq 0$,

$$\|\varphi_k - \varphi_\infty\| \leq c \frac{\gamma^k}{1 - \gamma},$$

where

$$\gamma := \tfrac{1}{2}\left(1 + m\delta - \sqrt{(1 - m\delta)^2 - 4m\ell c}\right) \in [0, 1[.$$

If we choose

$$B := F'(\varphi_0) \tag{1.4}$$

in Theorem 4, then we get the following classical result.

Theorem 5. (A posteriori convergence of (1.1)–(1.4) with Lipschitz derivative) *Suppose that* $\mathcal{O}$, F, $\varphi_0 \in \mathcal{O}$, $\ell > 0$, $m > 0$, *and* $c > 0$ *satisfy*
(1) $4m\ell c \leq 1$;
(2) $\|F'(\varphi_0)^{-1} F(\varphi_0)\| \leq c$;
(3) $\mathcal{O}$ *includes the closed disk* $\mathcal{D}_0 := \{\varphi \in X : \|\varphi - \varphi_0\| \leq \varrho_0\}$, *where*

$$\varrho_0 := \frac{1 - \sqrt{1 - 4m\ell c}}{2m\ell};$$

(4) $F' : \mathcal{D}_0 \to \mathcal{L}(X)$ *exists and is* ℓ*-Lipschitz;*
(5) $\|F'(\varphi_0)^{-1}\| \leq m$.

Then F has a unique zero φ_∞ in $\mathcal{D}_0$ and for all $k \geq 0$,

$$\|\varphi_k - \varphi_\infty\| \leq c\frac{\gamma^k}{1-\gamma},$$

where

$$\gamma := \tfrac{1}{2}(1 - \sqrt{1 - 4m\ell c}) \in \left[0\,,\,\tfrac{1}{2}\right[.$$

An extension to the case of a Hölder-continuous derivative is in preparation [4].

1.2 Nonlinear Boundary Value Problems

We are interested in applying Newton-type methods to solve nonlinear differential problems like

$$-\varphi'' + \alpha\,\varphi\,\varphi' = \psi \text{ in }]0,1[, \quad \varphi(0) = \varphi(1) = 0, \tag{1.5}$$

$$\varphi''' + \varphi^2 = \psi \text{ in }]0,1[, \quad \varphi(0) = \varphi'(0) = \varphi'(1) = 0, \tag{1.6}$$

$$-\Delta\varphi + \frac{\partial}{\partial x}f(\varphi) + \frac{\partial}{\partial y}g(\varphi) = \psi \text{ in } \Omega :=]0,1[\times]0,1[, \quad \varphi\big|_{\partial\Omega} = 0, \tag{1.7}$$

$$\frac{\partial\varphi}{\partial t} - \Delta\varphi + \frac{\partial}{\partial x}f(\varphi) + \frac{\partial}{\partial y}g(\varphi) = \psi \text{ in } \Omega :=]0,1[\times]0,1[,$$
$$\varphi(t,\cdot)\big|_{\partial\Omega} = 0, \quad \varphi(0,\cdot) = \phi^0. \tag{1.8}$$

In abstract terms, these problems enter the following setting. Let X and Y be complex Banach spaces (both norms being denoted by $\|\cdot\|$), let $L : \mathcal{D}(L) \subseteq X \to X$ be a linear operator with bounded inverse $L^{-1} : X \to \mathcal{D}(L)$ (hence, L is closed), and let $M : \mathcal{D}(M) \subseteq Y \to X$ be a closed linear operator and $N : X \to Y$ a nonlinear Fréchet differentiable operator such that $N(\mathcal{D}(L)) \subseteq \mathcal{D}(M)$. We assume that $L^{-1}M$ admits a continuous extension $T : Y \to X$. Given $\psi \in X$, we are interested in solving the problem

$$\text{Find } \varphi_\infty \in D \text{ such that } L\,\varphi_\infty + MN(\varphi_\infty) = \psi. \tag{1.9}$$

In the case of (1.5), we identify $X = Y = C^0([0,1])$, $\mathcal{D}(L) = \{\varphi \in C^2([0,1]) : \varphi(0) = \varphi(1) = 0\}$, $L = -\dfrac{d^2}{ds^2}$, $\mathcal{D}(M) = C^1([0,1])$, $M = \dfrac{d}{ds}$, $N(\varphi) = \dfrac{\alpha}{2}\varphi^2$, and in (1.7), $X := L^2(\Omega)$, $Y := X\times X$, $\mathcal{D}(L) := H_0^2(\Omega)$, $\mathcal{D}(M) := H^1(\Omega)\times H^1(\Omega)$, $L := -\Delta$, $M := \operatorname{div}$, $N(\varphi) := (f(\varphi), g(\varphi))$.

Applying L^{-1} to both sides in (1.9) we are led to find the zeros of

$$F(\varphi) := \varphi + TN(\varphi) - L^{-1}\psi.$$

We remark that

$$F'(\varphi) = I + TN'(\varphi).$$

In the present application, (1.1)–(1.2) amounts to solving for φ_{k+1} the linear problem

$$(I + TN'(\varphi_k))\varphi_{k+1} = T(N'(\varphi_k)\varphi_k - N(\varphi_k)) + L^{-1}\psi, \tag{1.10}$$

or equivalently,

$$(L + MN'(\varphi_k))\varphi_{k+1} = M(N'(\varphi_k)\varphi_k - N(\varphi_k)) + \psi \tag{1.11}$$

and, if φ_0 is chosen such that $N'(\varphi_0) = O$, then the computation of φ_{k+1} using (1.1) and (1.4) may be either explicit:

$$\varphi_{k+1} := L^{-1}\psi - TN(\varphi_k), \tag{1.12}$$

or implicit: $L\varphi_{k+1} := \psi - MN(\varphi_k)$.

For problem (1.5), $(L^{-1}\varphi)(s) := \int_0^1 \kappa(s,t)\varphi(t)\,dt$, where

$$\kappa(s,t) := \begin{cases} s(1-t) & \text{if } 0 \le s \le t \le 1, \\ t(1-s) & \text{if } 0 \le t < s \le 1. \end{cases}$$

Integrating by parts, we find that for $\varphi \in X$ and $s \in [0,1]$,

$$(T\varphi)(s) := (s-1)\int_0^s \varphi(t)\,dt + s\int_s^1 \varphi(t)\,dt.$$

It follows that we may choose $\varphi_0 := 0$, $\psi(s) := 1$, $m_0 = 1$, $\ell = |\alpha|/2$, $c_0 = \|L^{-1}\| = 0.125$. The iterative process defined by equation (1.10) amounts to a Fredholm integral equation of the second kind to be solved for φ_{k+1}.

For problem (1.6), L^{-1} is as in the case of (1.5) but the kernel is now

$$\kappa(s,t) := \begin{cases} \dfrac{s^2}{2}(t-1), & \text{if } s \le t, \\ \dfrac{t^2}{2} - st + \dfrac{s^2 t}{2}, & \text{if } t < s. \end{cases}$$

Since $M = I$, $T = L^{-1}$. Choose

$$\varphi_0 := 0, \quad \psi(s) := \beta, \quad m_0 = 1, \quad \ell = 2\|L^{-1}\| = 1/6,$$
$$c_0 = |\beta|\|L^{-1}\| = |\beta|/12.$$

Following Theorems 1 and 2, sufficient conditions on the data for the convergence of the iterative methods in the case of problems (1.5) and (1.6) are found to be $|\alpha| < 8$ and $|\beta| \le 18$, but in practice convergence still holds in less restrictive situations (see Table 1). The trapezoidal composite rule with 100 subintervals has been used for computational purposes. The integral equation (1.10) has been solved using the Fredholm approximation as it is described in [2]. Another application can be found in [1].

In the case of problem (1.7), we have taken

$$\alpha = 10,000, \qquad f(\varphi) := \frac{\alpha}{2}\varphi^2, \qquad g(\varphi) := 0,$$

$$\psi(x,y) := \begin{cases} +1 & \text{if } \max\{|x-0.5|, |y-0.5|\} \le 0.25, \\ -1 & \text{otherwise,} \end{cases}$$

and chosen the Newton–Kantorovich method (1.1)–(1.2).

The computation of φ_{k+1} amounts to the resolution of the partial differential equation (1.11):

$$-\Delta\varphi_{k+1} + \alpha\frac{\partial}{\partial x}\left(\varphi_k\,\varphi_{k+1}\right) = \alpha\varphi_k\frac{\partial\varphi_k}{\partial x} + \psi \quad \text{in } \Omega, \quad \varphi_{k+1}\big|_{\partial\Omega} = 0.$$

This problem has been solved numerically using central second-order finite differences with constant step 0.04 both in x and in y (see Table 1).

Table 1.

Problem	(1.5)	(1.6)	(1.7)
Data	$\alpha = 5\,000$	$\beta = 20$	$\alpha = 10,000$
Method	(1.10)	(1.12)	(1.11)
k	$\\|F(\varphi_k)\\|$	$\\|F(\varphi_k)\\|$	$\\|\varphi_{k+1} - \varphi_k\\|$
0	1.3×10^{-1}	1.7×10^{-1}	1.9×10^{-2}
3	4.1×10^{-2}	4.7×10^{-4}	2.3×10^{-2}
6	7.2×10^{-8}	2.1×10^{-7}	7.1×10^{-4}
8	2.1×10^{-16}	1.3×10^{-9}	2.2×10^{-7}

The nonstationary problem (1.8) can be treated in the preceding framework when the derivative with respect to time is approximated by a finite difference in an implicit way. For example, with the functions f and g defined before, using (1.1)–(1.2), and setting $W := I - \tau\Delta$, we get

$$W\varphi_{k+1}^{[m+1]} + \alpha\tau\frac{\partial}{\partial x}\left(\varphi_k^{[m+1]}\,\varphi_{k+1}^{[m+1]}\right) = \alpha\tau\varphi_k^{[m+1]}\frac{\partial\varphi_k^{[m+1]}}{\partial x} + \varphi_{\text{Last}}^{[m]} + \tau\psi^{[m+1]},$$

$$\varphi_{k+1}^{[m+1]}\Big|_{\partial\Omega} = 0, \; \varphi^{[0]} = \phi^0, \; \varphi_0^{[m+1]} = \varphi_{\text{Last}}^{[m]},$$

where τ is the time mesh size, $\varphi_{\text{Last}}^{[m]}$ denotes the last Newton–Kantorovich iterate at instant number m, and $\psi^{[m+1]}(x,y) = \psi((m+1)\tau, x, y)$.

Satisfactory numerical computations have been done with $\alpha = 1\,000$,

$$\psi(t,x,y) := \begin{cases} -10 & \text{if } |x-0.5| \le 0.1 \text{ and } |y-0.5| \le 0.1, \\ +10 & \text{otherwise,} \end{cases}$$

$$\phi^0(x,y) := \begin{cases} +1 & \text{if } |x-0.5| \le 0.1 \text{ and } |y-0.5| \le 0.1, \\ +0 & \text{otherwise,} \end{cases}$$

a time step $\tau = 0.001$, $m \in [\![0, 10]\!]$, and $n = 19 \times 19 = 361$ points in Ω.

1.3 Spectral Differential Problems

We consider here the application of Newton-type methods to solve a differential spectral problem. For this purpose, let X be a complex Hilbert space, $L : \mathcal{D}(L) \subseteq X \to X$ a linear operator with compact inverse $T : X \to \mathcal{D}(L)$ and domain $\mathcal{D}(L)$ dense in X, in order to ensure the existence and uniqueness of the adjoint operator L^*. For integers $p \ge 1$ and $m \ge 1$, $\mathbf{I}_m$ denotes the identity matrix of order m, for $x := [x_1, \dots, x_m] \in X^m$, $y := [y_1, \dots, y_p] \in X^p$, $< x|y > (i,j) :=< x_j, y_i >$ defines a Gram matrix, the natural extension of L to X^m is $Lx := [Lx_1, \dots, Lx_m]$ and, for $\Theta \in \mathbb{C}^{m \times p}$, $x\Theta := \left[\sum\limits_{i=1}^{m} \Theta(i,1)x_i, \dots, \sum\limits_{i=1}^{m} \Theta(i,p)x_i \right] \in X^p$.

We state the m-dimensional spectral problem for L as follows: Given $\Psi \in X^m$, find $\Phi \in X^m$ such that

$$L\Phi - \Phi < L\Phi|\Psi > = 0, \quad < \Phi|\Psi > = \mathbf{I}_m. \tag{1.13}$$

Let $M := \text{Span}(\Phi) \subseteq X$. Equations (1.13) translate the fact that M is invariant under L and imply that the m elements in Φ are linearly independent, as well as those in Ψ. Moreover, the matrix $< L\Phi|\Psi > \in \mathbb{C}^{m \times m}$ represents the restricted operator $L\big|_{M,M} : M \to M, \quad \varphi \mapsto L\varphi$ with respect to the ordered basis Φ of M, independently of the choice of the adjoint basis Ψ. Obviously, $\Lambda := \text{sp}\left(L\big|_{M,M}\right) = \text{sp}(< L\Phi|\Psi >)$. In many applications, Λ is a singleton containing a multiple possibly defective eigenvalue of L or a cluster of such eigenvalues, which will be approximated by a cluster of eigenvalues of an approximation of L (see [2]).

The product space X^m is a Hilbert space with the Hilbert–Schmidt inner product $< x, y > := \text{tr} < x|y >$. For more details on notation and properties of this kind of formulation, the reader is referred to [2].

Applying T to both sides in (1.13) we are led to compute the zeros of the nonlinear operator $F : X^m \to X^m$ defined by $F(x) := x - Tx < x|L^*\Psi >$ whose Fréchet derivative is $F'(x)h = h - Th < x|L^*\Psi > -Tx < h|L^*\Psi >$. It follows that $\ell := 2\|T\|\,\|L^*\Psi\|$ is a Lipschitz constant of F' over X^m. The iterative process defined by (1.1)–(1.2) amounts to solving for Φ_{k+1}:

$$\Phi_{k+1} - T\Phi_{k+1} < \Phi_k|L^*\Psi > -T\Phi_k < \Phi_{k+1}|L^*\Psi > = -T\Phi_k < \Phi_k|L^*\Psi > . \tag{1.14}$$

If T is not available for computational purposes, we apply L to both sides of equation (1.14) and we get the Sylvester equation

$$(I - P_k)L\Phi_{k+1} - \Phi_{k+1} < L\Phi_k|\Psi> = -P_k L\Phi_k, \tag{1.15}$$

where $P_k x := \Phi_k < x|\Psi>$. If Φ_0 is chosen so that $<\Phi_0|\Psi> = \mathbf{I}_m$, then $(P_k)_{k\geq 0}$ is a sequence of projections along $\Psi^\perp$.

The iterative process defined by (1.1)–(1.4) amounts to solving for Φ_{k+1}:

$$\begin{aligned}\Phi_{k+1} &- T\Phi_0 < \Phi_{k+1}|L^*\Psi> - T\Phi_{k+1} < \Phi_0|L^*\Psi> \\ &= T\Phi_k(<\Phi_k|L^*\Psi> - <\Phi_0|L^*\Psi>) - T\Phi_0 < \Phi_k|L^*\Psi>. \end{aligned} \tag{1.16}$$

The choice of Ψ and Φ_0 may involve spectral computations on an approximation $\widetilde{T}$ of T. For instance, Ψ and Φ_0 may be chosen to be the exact solutions of the approximate problem

$$\widetilde{T}\Phi_0 = \Phi_0 < \widetilde{T}\Phi_0|\Psi>, \quad \widetilde{T}^*\Psi = \Psi < \widetilde{T}\Phi_0|\Psi>^*, \quad <\Phi_0|\Psi> = \mathbf{I}_m.$$

Then $<\widetilde{T}^{-1}\Phi_0|\Psi> = <\widetilde{T}\Phi_0|\Psi>^{-1}$, since both of them are equal to the unique matrix representing in the basis Φ_0 the inverse of the restricted operator $\widetilde{T}|_{M_0,M_0} : M_0 \to M_0$, where $M_0 := \text{Span}(\Phi_0)$ is invariant under $\widetilde{T}$. We may interpret $\widetilde{T}^{-1}$ as an approximation of L and replace equation (1.16) with

$$\begin{aligned}\Phi_{k+1} &- T\Phi_0 < \widetilde{T}\Phi_{k+1}|\Psi>^{-1} - T\Phi_{k+1} < \widetilde{T}\Phi_0|\Psi>^{-1} \\ &= T\Phi_k(<\widetilde{T}\Phi_k|\Psi>^{-1} - <\widetilde{T}\Phi_0|\Psi>^{-1}) - T\Phi_0 < \widetilde{T}\Phi_k|\Psi>^{-1}. \end{aligned} \tag{1.17}$$

If the global multiplicity of the spectral set Λ is $m = 1$, then all the Gram matrices are scalars, and equation (1.17) reduces to one of the algorithms presented and studied in [5].

Again, if T is not available for computational purposes, apply L to both sides of equation (1.16) and get the Sylvester equation

$$(I - P_0)L\Phi_{k+1} - \Phi_{k+1} < L\Phi_0|\Psi> = P_k L(\Phi_k - \Phi_0) - P_0 L\Phi_k. \tag{1.18}$$

Both (1.15) and (1.18) can be solved numerically through a weak formulation method:

Suppose there exist a Hilbert space H with inner product $<\cdot,\cdot>_{\mathbf{H}}$, forms $a : H\times H \to \mathbb{C}$ and $b : H\times H \to \mathbb{C}$, and linear bounded operators $A \in \mathcal{L}(H)$ and $B \in \mathcal{L}(H)$ satisfying $a(\varphi, v) = <L\varphi, v> = <A\varphi, v>_{\mathbf{H}}$ for all $(\varphi, v) \in \mathcal{D}(L)\times H$, and $b(u, v) = <u, v> = <Bu, v>_{\mathbf{H}}$ for all $(u, v) \in H\times H$. Then this leads to the weak problem: Find $\Phi_{k+1} \in H^m$ such that, for all $v \in H^m$,

$$a(\Phi_{k+1}|v) - b(\Phi_k|v)a(\Phi_{k+1}|\Psi) - b(\Phi_{k+1}|v)a(\Phi_k|\Psi) = -b(\Phi_k|v)a(\Phi_k|\Psi), \tag{1.19}$$

where the extension of a (and b) to $H^m \times H^m$ is defined in the obvious way: For any positive integers p and q, if $u := [u_1, \ldots, u_p] \in H^p$ and $v := [v_1, \ldots, v_q] \in H^q$, $a(u|v)(i,j) := a(u_j, v_i)$.

In terms of operators A and B, the weak problem reads: Find $\Phi_{k+1} \in H^m$ such that

$$(I - Q_k)A\Phi_{k+1} - B\Phi_{k+1} < A\Phi_k|\Psi >_{\mathbf{H}} = -Q_k A\Phi_k, \tag{1.20}$$

where $Q_k x := B\Phi_k < x|\Psi >_{\mathbf{H}}$, $x \in H^m$.

Consider the case of a simple eigenvalue, that is $m = 1$, $\Lambda = \{\lambda\}$, $\Phi = [\varphi]$, $\Psi = [\psi]$. The finite element method builds an approximation of the solution of (1.20) which belongs to a finite-dimensional subspace $H_n := \mathrm{Span}\{e_{n,j} : j \in [\![1,n]\!]\}$ of H. This means that $\varphi_k := \sum_{j=1}^{n} \mathbf{x}_k(j)e_{n,j}$. These elements of H_n satisfy a discretized formulation of (1.20) obtained by performing the inner product of each member with $e_{n,i}$ for $i \in [\![1,n]\!]$. If ψ is chosen in H_n: $\psi := \sum_{j=1}^{n} \mathbf{c}(j)e_{n,j}$, we get a system of linear equations in the unknown $\mathbf{x}_{k+1} \in \mathbb{C}^{n\times 1}$:

$$[(\mathbf{I} - \mathbf{B}\mathbf{x}_k\mathbf{c}^*)\mathbf{A} - \lambda_k\mathbf{B}]\,\mathbf{x}_{k+1} = -\lambda_k\mathbf{B}\mathbf{x}_k, \tag{1.21}$$

where $\mathbf{A}$, $\mathbf{B} \in \mathbb{C}^{n\times n}$ and $\lambda_k \in \mathbb{C}$ are defined by $\mathbf{A}(i,j) := a(e_{n,j}, e_{n,i})$, $\mathbf{B}(i,j) :=< e_{n,j}, e_{n,i} >$, $\lambda_k := \mathbf{c}^*\mathbf{A}\mathbf{x}_k$. We remark that, if $\lambda_k \neq 0$, then the normalizing condition $< \varphi_k, \psi >= 1$ is hereditary and reads $\mathbf{c}^*\mathbf{B}\mathbf{x}_k = 1$. The matrix $\mathbf{B}$ is symmetric positive definite. Let $\mathbf{B}^{\frac{1}{2}}$ denote its symmetric positive definite square root, whose inverse will be denoted by $\mathbf{B}^{-\frac{1}{2}}$. Define the sequence $\mathbf{y}_k := \mathbf{B}^{\frac{1}{2}}\mathbf{x}_k$ and multiply each side of (1.21) by $\mathbf{B}^{-\frac{1}{2}}$ on the left. Setting $\widehat{\mathbf{A}} := \mathbf{B}^{-\frac{1}{2}}\mathbf{A}\mathbf{B}^{-\frac{1}{2}}$, $\mathbf{P}_k := \mathbf{y}_k\mathbf{c}^*$ we get the inverse-iteration-like system

$$\left[(\mathbf{I} - \mathbf{P}_k)\widehat{\mathbf{A}} - \lambda_k\mathbf{I}\right]\mathbf{y}_{k+1} = -\lambda_k\mathbf{y}_k. \tag{1.22}$$

In the general case of a multiple eigenvalue or a cluster of eigenvalues we find a Sylvester matrix problem: if

$$e_n := [e_{n,1}, \ldots, e_{n,n}], \quad \mathbf{X}_k \in \mathbb{C}^{n\times m}, \quad \Phi_k := e_n\mathbf{X}_k,$$
$$\mathbf{C} \in \mathbb{C}^{n\times m}, \quad \Psi := e_n\mathbf{C},$$

then putting $v = e_n$ in (1.19) and defining

$$\mathbf{A} := a(e_n|e_n), \quad \mathbf{B} := b(e_n|e_n), \quad \Theta_k := \mathbf{C}^*\mathbf{A}\mathbf{X}_k,$$

we get a generalization of (1.21):

$$(\mathbf{I} - \mathbf{B}\mathbf{X}_k\mathbf{C}^*)\mathbf{A}\mathbf{X}_{k+1} - \mathbf{B}\mathbf{X}_{k+1}\Theta_k = -\mathbf{B}\mathbf{X}_k\Theta_k.$$

If Λ is a singleton containing a multiple eigenvalue λ of L, then the sequence defined by $\lambda_k := (1/m)\operatorname{tr}\Theta_k$ converges to λ as $k \to \infty$, if the Newton iterations are convergent.

Consider for example the weak version of equation (1.15) in the case of the elementary one-dimensional model problem:

$$X := L^2([0,1]), \quad H := H_0^1([0,1]), \quad L\varphi := -\varphi'',$$
$$\mathcal{D}(L) := \{\varphi \in X : \varphi' \in X, \, \varphi(0) = 0 = \varphi(1)\},$$

when the discretization procedure is the finite element approximation with hat test and basis functions. Then **A** and **B** are defined by

$$\mathbf{A}(i,j) := \int_0^1 e'_{n,j}(s) e'_{n,i}(s)\, ds, \quad \mathbf{B}(i,j) := \int_0^1 e_{n,j}(s) e_{n,i}(s)\, ds.$$

If the hat functions are defined in correspondence with a uniform grid with $n+2$ points and mesh size $h_n := \dfrac{1}{n+1}$, then **A** and **B** are real symmetric tridiagonal matrices of order n:

$$\mathbf{A} = \frac{1}{h_n}\operatorname{tridiag}(-1, 2, -1), \quad \mathbf{B} = \frac{h_n}{6}\operatorname{tridiag}(1, 4, 1).$$

Practical convergence is shown in Table 2 through the relative residual at each iteration in the L^2-norm. The grid has 249 interior points and the approximate eigenspaces converge to the eigenspace corresponding to the simple eigenvalue $\lambda = (16\pi)^2$.

Consider now equation (1.20) for the two-dimensional model problem described by $\Omega :=]0,1[\times]0,1[$, $X := L^2(\Omega)$, $H := H_0^1(\Omega)$, $L\varphi := -\Delta\varphi$, and $\mathcal{D}(L) := \{\varphi \in X : \nabla\varphi \in X \times X, \, \varphi_{|\partial\Omega} = 0\}$, when the discretization procedure is the finite element approximation with hat test and basis functions over a uniform triangulation with $n := m^2$ interior points and mesh size $h_m := 1/(m+1)$, both in x and y. Then **A** and **B** are given by

$$\mathbf{A}(i,j) := \int_0^1 \int_0^1 \nabla e_{n,j}(x,y) \cdot \nabla e_{n,i}(x,y)\, dx\, dy,$$
$$\mathbf{B}(i,j) := \int_0^1 \int_0^1 e_{n,j}(x,y) e_{n,i}(x,y)\, dx\, dy.$$

It follows that

$$\mathbf{A} = \operatorname{tridiag}(-\mathbf{I}_m, \mathbf{A}_m, -\mathbf{I}_m), \quad \mathbf{A}_m := \operatorname{tridiag}(-1, 4, -1),$$
$$\mathbf{B} := \frac{h_n^2}{12}\operatorname{tridiag}(\mathbf{G}_m^\top, \mathbf{B}_m, \mathbf{G}_m),$$
$$\mathbf{B}_m := \operatorname{tridiag}(1, 6, 1) \quad \text{and} \quad \mathbf{G}_m := \operatorname{tridiag}(0, 1, 1).$$

Table 2.

Iteration	$\frac{\| \text{Residual} \|}{\| \text{Iterate} \|}$
0	$1.82E-00$
7	$8.31E-09$
8	$8.03E-14$

Table 3.

Iteration	$\frac{\| \text{Residual} \|}{\| \text{Iterate} \|}$
0	$1.64E+01$
2	$5.10E-05$
3	$4.54E-12$

Computations have been carried out with a bidimensional uniform grid with $19 \times 19 = 361$ interior points and we approximate an eigenfunction corresponding to the exact eigenvalue $\lambda = 8\pi^2$ of L. Practical convergence is shown in Table 3 through the relative residual in the L^2-norm at each iteration.

1.4 Newton Method for the Matrix Eigenvalue Problem

When computing a Schur form of a square complex matrix, the Newton-type methods constitute an alternative to the commonly used QR algorithm or can be implemented in a combined strategy. In order to approach a Schur form of a complex square matrix by a Newton-type method let us introduce the real matrix operators:

$$\mathcal{U}(\mathbf{M})(i,j) := \begin{cases} \mathbf{M}(i,j) & \text{if } i \leq j, \\ 0 & \text{otherwise,} \end{cases}$$

$$\mathcal{D}(\mathbf{M})(i,j) := \begin{cases} \mathbf{M}(i,j) & \text{if } i = j, \\ 0 & \text{otherwise.} \end{cases}$$

Let $\mathbf{A}$, $\mathbf{B}$ be matrices in $\mathbb{R}^{n\times n}$ such that $\mathbf{Z} := \mathbf{A} + \mathrm{i}\mathbf{B}$. Schur's theorem states that there exist $\mathbf{U}_\infty$, $\mathbf{V}_\infty$ in $\mathbb{R}^{n\times n}$ and $\mathbf{X}_\infty$, $\mathbf{Y}_\infty$ in $\mathrm{Im}(\mathcal{U})$, the space of upper triangular real matrices of order n, such that

$$\mathbf{Q}_\infty := \mathbf{U}_\infty + \mathrm{i}\mathbf{V}_\infty, \quad \mathbf{T}_\infty := \mathbf{X}_\infty + \mathrm{i}\mathbf{Y}_\infty$$

satisfy

$$\mathbf{Z}\mathbf{Q}_\infty = \mathbf{Q}_\infty\mathbf{T}_\infty, \quad \mathbf{Q}_\infty^*\mathbf{Q}_\infty = \mathbf{I}.$$

It can be easily checked that, if $\mathbf{D}$ is a diagonal unitary matrix, then $\mathbf{D}^*\mathbf{T}_\infty\mathbf{D}$ is still upper triangular. In other words, $\mathbf{Q}_\infty\mathbf{D}$ is a unitary matrix which triangularizes $\mathbf{Z}$ as well. Hence, $\mathbf{D}$ can be chosen so that $\mathbf{Q}_\infty$ has real diagonal entries: $\mathcal{D}(\mathbf{V}_\infty) = \mathbf{O}$. These conditions correspond to the system of matrix equations

$$\mathbf{A}\mathbf{U}_\infty - \mathbf{U}_\infty\mathbf{X}_\infty - \mathbf{B}\mathbf{V}_\infty + \mathbf{V}_\infty\mathbf{Y}_\infty := \mathbf{O},$$

$$\mathbf{BU}_\infty - \mathbf{U}_\infty \mathbf{Y}_\infty + \mathbf{AV}_\infty - \mathbf{V}_\infty \mathbf{X}_\infty := \mathbf{O},$$
$$\mathcal{U}(\mathbf{U}_\infty^\top \mathbf{U}_\infty + \mathbf{V}_\infty^\top \mathbf{V}_\infty - \mathbf{I}) := \mathbf{O},$$
$$\mathcal{D}(\mathbf{V}_\infty) + \mathcal{U}(\mathbf{U}_\infty^\top \mathbf{V}_\infty - \mathbf{V}_\infty^\top \mathbf{U}_\infty) := \mathbf{O}.$$

The real linear product space $\mathcal{B} := \mathbb{R}^{n\times n} \times \mathbb{R}^{n\times n} \times \mathrm{Im}(\mathcal{U}) \times \mathrm{Im}(\mathcal{U})$ appears to be the domain for the nonlinear operator $F : \mathcal{B} \to \mathcal{B}$,

$$[\mathbf{U}, \mathbf{V}, \mathbf{X}, \mathbf{Y}] \mapsto [\mathbf{AU} - \mathbf{BV} - \mathbf{UX} + \mathbf{VY}, \mathbf{BU} - \mathbf{UY} + \mathbf{AV} - \mathbf{VX},$$
$$\mathcal{U}(\mathbf{U}^\top \mathbf{U} + \mathbf{V}^\top \mathbf{V} - \mathbf{I}), \mathcal{D}(\mathbf{V}) + \mathcal{U}(\mathbf{U}^\top \mathbf{V} - \mathbf{V}^\top \mathbf{U})].$$

For example, if

$$\mathbf{A} := \begin{bmatrix} 0 & 0 & 1 \\ 0 & 1 & 0 \\ -1 & 0 & 0 \end{bmatrix},$$

with spectrum $\{1, \mathrm{i}, -\mathrm{i}\}$, and if

$$\mathbf{X}_0 = \mathbf{Y}_0 := \mathbf{O}, \quad \mathbf{U}_0 := \frac{1}{\sqrt{2}} \begin{bmatrix} 1 & 0 & 0 \\ 0 & 1 & 0 \\ 0 & 0 & 1 \end{bmatrix}, \quad \mathbf{V}_0 := \frac{1}{\sqrt{2}} \begin{bmatrix} 0 & 0 & 1 \\ 0 & 1 & 0 \\ 1 & 0 & 0 \end{bmatrix},$$

then Newton's method converges in 6 iterations and the QR algorithm in 15 iterations.

Both theoretical and practical aspects of this approach are being studied by the authors but some a priori convergence results are already available:

Theorem 6. (A priori convergence of Newton's method for a Schur form) *Suppose that* $\mathbf{Z}$ *has distinct eigenvalues and that there exists a unitary matrix* $\mathbf{Q}_\infty$ *triangularizing* $\mathbf{Z}$ *such that* $\mathbf{Q}_\infty(j,j) \in \mathbb{R}$ *for all* j, *and*

$$\prod_{j=1}^{n} \mathbf{Q}_\infty(j,j) \neq 0.$$

Then the hypotheses of Theorem 1 are satisfied.

The proof of this assertion can be found in [3].

References

1. M. Ahues, A note on perturbed fixed-slope iterations, *Appl. Math. Lett.* 2004 (to appear).
2. M. Ahues, A. Largillier, and B.V. Limaye, *Spectral Computations with Bounded Operators*, Chapman & Hall/CRC, Boca Raton, FL, 2001.
3. M. Ahues and A. Largillier, Newton-type methods for a Schur form (submitted for publication).

4. M. Ahues, Newton methods with Hölder-continuous derivative (submitted for publication).

5. M. Ahues and M. Telias, Refinement methods of Newton-type for approximate eigenelements of integral operators, *SIAM J. Numer. Anal.* **23** (1986), 144–159.

6. R.F. Curtain and A.J. Pritchard, *Functional Analysis in Modern Applied Mathematics*, Academic Press, London, 1977.

7. R. Kress, *Numerical Analysis*, Springer-Verlag, New York, 1998.

2 Nodal and Laplace Transform Methods for Solving 2D Heat Conduction

Ivanilda B. Aseka, Marco T. Vilhena, and Haroldo F. Campos Velho

2.1 Introduction

The reduction of the computational effort is a permanent search for engineering and science problems. This feature can be reached during the modeling process (simplified models) concerning a particular application. For example, when designing a chamber, the focus is on the minimization of temperature changes (insulated room). One does not need to know the temperature field for the entire physical domain (3D), there is only interest in the temperature level inside the room driven only by the heat flux. This problem is built up as a multi-layer heat transfer problem.

The nodal method is particularly appropriate when we are more interested in flux quantities than numerical values of a specific variable for the entire domain. The nodal method has appeared in the transport theory context, but the same methodology can be used in other applications, such as heat transfer and diffusion processes. The nodal method consists in the integration (averaging) with respect to one or more space variables. The resulting equations (integrated, or averaged) are the nodal equations, where boundary conditions are embedded in the equation system. The properties on the boundaries, expressed in the nodal equations, can be approximated by lumped analysis [1]. Finally, a time integration could be applied to the lumped system to obtain the solution. The Laplace transformation is employed here to perform the time integration. This procedure allows a semi-analytic result for the time integration. The combination of the nodal and Laplace transformation methods has been employed in the transport of neutral particles (see [2]–[4]).

The nodal method technique associated with the lumped procedure and the Laplace transformation is applied to a two-dimensional (2D) multi-layer heat transfer problem. The lumped solution is compared to the finite volume method used in [5]. For the present work, the Hermite numerical integration scheme is used for representing the boundary conditions in

The authors are grateful to the CNPq (Conselho Nacional de Desenvolvimento Científico e Tecnológico) and to FAPERGS (Rio Grande do Sul Foundation for Research Support) for their partial financial support of this work.

the nodal equations. The Hermite schemes are also compared with the temperature on the boundary, in which it is expressed as the temperature average.

2.2 Nodal Method in Multi-layer Heat Conduction

The heat equation is considered in the parallel multi-layer problem under the assumption of a perfect thermal contact along the interfaces $x = x_i$, $i = 2, 3, \ldots, n$. Each layer i is considered homogenous, isotropic, and as having constant thermal properties (ρ_i, c_{p_i}, and k_i). Initially, each layer has temperature $T_i(x, y, 0) = F_i(x, y)$ at $x_i < x < x_{i+1}$ and $y_1 < y < y_2$ for $i = 1, 2, 3, \ldots, n$ and $t = 0$. For a given time $t > 0$, the heat is transferred by convection at the four boundaries. The heat transfer coefficients at the boundaries $x = x_1$, $x = x_{n+1}$, $y = y_1$ (bottom) and $y = y_2$ (top) are, respectively, h_1, h_{n+1}, h_c, and h_d. Heat sources and sinks will not be considered here.

The mathematical formulation of this heat conduction problem for each layer is expressed as

$$\frac{\partial^2 T_i(x,y,t)}{\partial x^2} + \frac{\partial^2 T_i(x,y,t)}{\partial y^2} = \frac{1}{\alpha_i}\frac{\partial T_i(x,y,t)}{\partial t}, \tag{2.1}$$

where $x \in (x_1, x_{n+1})$, $y \in (y_1, y_2)$, and $t > 0$; here $\alpha_i = k_i/(\rho_i C_{p_i})$ is the thermal diffusivity, ρ_i the density, C_{p_i} the specific heat, k_i the thermal conductivity, and $i = 1, 2, \ldots, n$ are the indexed layers. Equation (2.1) is associated with the following boundary, interface, and initial conditions.

- Boundary conditions in the x-direction, with $y \in (y_1, y_2)$ and $t > 0$:

$$-k_1 \frac{\partial T_1(x,y,t)}{\partial x} = h_1[f_1(t) - T_1(x,y,t)] \quad \text{at} \quad x = x_1, \tag{2.1a}$$

$$k_n \frac{\partial T_n(x,y,t)}{\partial x} = h_{n+1}[f_{n+1}(t) - T_n(x,y,t)] \quad \text{at} \quad x = x_{n+1}. \tag{2.1b}$$

- Boundary conditions in the y-direction, with $x \in (x_1, x_{n+1})$ and $t > 0$:

$$-k_i \frac{\partial T_i(x,y,t)}{\partial y} = h_c[f_c(t) - T_i(x,y,t)] \quad \text{at} \quad y = y_1, \tag{2.1c}$$

$$k_i \frac{\partial T_i(x,y,t)}{\partial y} = h_d[f_d(t) - T_i(x,y,t)] \quad \text{at} \quad y = y_2. \tag{2.1d}$$

- Interface conditions ($i = 1, 2, \ldots, (n-1)$), with $y \in (y_1, y_2)$ and $t > 0$:

$$T_i(x,y,t) = T_{i+1}(x,y,t) \quad \text{at} \quad x = x_{i+1}, \tag{2.1e}$$

$$k_i \frac{\partial T_i(x,y,t)}{\partial x} = k_{i+1} \frac{\partial T_{i+1}(x,y,t)}{\partial x} \quad \text{at} \quad x = x_{i+1}. \tag{2.1f}$$

• Initial conditions, with $y \in (y_1, y_2)$ and $x \in (x_1, x_{n+1})$:

$$T_i(x, y, 0) = F_i(x, y), \quad i = 1, 2, \ldots, n. \tag{2.1g}$$

Integrating (2.1) in the y-direction between y_1 and y_2 and multiplying by $1/(\Delta y)$ results in the partial differential equation (for the variables x and t)

$$\frac{\partial^2 \tau_i(x,t)}{\partial x^2} + \frac{1}{y_2 - y_1}\left[\left.\frac{\partial T_i}{\partial y}\right|_{y=y_2} - \left.\frac{\partial T_i}{\partial y}\right|_{y=y_1}\right] = \frac{1}{\alpha_i}\frac{\partial \tau_i(x,t)}{\partial t},$$

where $\tau_i(x,t) \equiv (\Delta y)^{-1}\int_{y_1}^{y_2} T_i(x,y,t)\,dy$ and $\Delta y = y_2 - y_1$. This is called the *nodal approach* for this problem. Applying the boundary conditions, we obtain the new equation

$$\frac{\partial^2 \tau_i}{\partial x^2} - \frac{h_c}{k_i}T_i(x, y_2, t) - \frac{h_d}{k_i}T_i(x, y_1, t) = \frac{1}{\alpha_i}\frac{\partial \tau_i}{\partial t} - \frac{1}{k_i}[h_c T_c + h_d T_d(t)]. \tag{2.2}$$

Boundary conditions emerge in the nodal formulation. Some approximations need to be described to represent these terms. Table 1 lists a number of characteristics of the multi-layer region.

Table 1. Some characteristics of the multi-layer region.

Layer	Thickness [mm]	k [W m^{-1} C^{-1}]	α [m^2 s^{-1}]
Layer – 1	25	0.692	4.434×10^{-7}
Layer – 2	100	1.731	9.187×10^{-7}
Layer – 3	25	0.043	1.600×10^{-6}
Layer – 4	20	0.727	5.400×10^{-7}

2.2.1 Lumped Analysis: Standard Approach

Here, the *standard* (or *classical*) approach is to assume that the temperature at the boundaries $y = y_1$ and $y = y_2$ is equal to the average temperature; that is,

$$T_i(x, y_1, t) \approx \tau_i(x,t)\ , \tag{2.3}$$

$$T_i(x, y_2, t) \approx \tau_i(x,t)\ . \tag{2.4}$$

Substituting (2.3) and (2.4) in (2.2), we obtain for the average temperature τ_i the simplified formulation

$$\left[\frac{\partial^2}{\partial x^2} - \frac{(h_c + h_d)}{\Delta y k_i}\right]\tau_i(x,t) = \frac{1}{\alpha_i}\frac{\partial \tau_i(x,t)}{\partial t} - \frac{1}{\Delta y\, k_i}\left(h_c T_c + h_d T_d(t)\right). \tag{2.5}$$

In the next section an improvement will be addressed. The boundary conditions will be approximated by a function of the average temperature, that is, $T(x, y_1, t) \approx f[T_{av}(t)]$ and $T(x, y_2, t) \approx f[T_{av}(t)]$, where the function $f[T_{av}(t)]$ is determined from the Hermite approximation for the integrals.

2.2.2 Lumped Analysis: Improved Approach

The $H_{0,0}/H_{0,0}$ approximation. Considering the integrals defining the average temperature and the heat flux ($i = 1, \ldots, n$)

$$\frac{1}{\Delta y}\int_{y_1}^{y_2} T_i(x,y,t)\,dy = \tau_i(x,t),$$
$$\int_{y_1}^{y_2} \frac{\partial T_i(x,y,t)}{\partial y}\,dy = T_i(x,y_2,t) - T_i(x,y_1,t)$$

and using trapezoidal quadrature for the integrals, we derive the expressions ($i = 1, \ldots, n$)

$$\tau_i(x,t) \approx \frac{1}{2\Delta y}\left[T_i(x,y_1,t) + T_i(x,y_2,t)\right] \,,$$
$$T_i(x,y_2,t) - T_i(x,y_1,t) \approx \frac{1}{2}\left[\left.\frac{\partial T_i}{\partial y}\right|_{y=y_1} + \left.\frac{\partial T_i}{\partial y}\right|_{y=y_1}\right].$$

After some manipulation, and employing the boundary conditions (2.1c) and (2.1d), we obtain the boundary conditions in the form ($i = 1, .., n$)

$$T_i(x,y_1,t) = 2\Delta y\left[\frac{2k_i + h_c}{4k_i + h_c + h_d}\right]\tau_i(x,t) + \left[\frac{h_d T_d(t) - h_c T_c}{4k_i + h_c + h_d}\right], \tag{2.6}$$
$$T_i(x,y_2,t) = 2\Delta y\left[\frac{2k_i + h_d}{4k_i + h_d + h_c}\right]\tau_i(x,t) - \left[\frac{h_d T_d(t) - h_c T_c}{4k_i + h_d + h_c}\right]. \tag{2.7}$$

Clearly, (2.6) and (2.7) are *improved* expressions. Finally, substituting these expressions in (2.2), we arrive at the $H_{0,0}/H_{0,0}$ approach

$$\frac{\partial^2 \tau_i(x,t)}{\partial x^2} - \frac{4(k_i(h_c+h_d) + h_c h_d)}{k_i(4k_i + h_d + h_c)}\tau_i(x,t)$$
$$= \frac{1}{\alpha_i}\frac{\partial \tau_i}{\partial t} - \frac{1}{\Delta y\, k_i}\left[(h_c T_c + h_d T_d(t)) - \frac{h_c - h_d}{4k_i + h_d + h_c}(h_d T_d(t) - h_c T_c)\right]. \tag{2.8}$$

The $H_{1,1}/H_{0,0}$ approximation. As above, the trapezoidal quadrature is used to obtain the approximation $H_{1,1}$ for the average temperature, while the approximation $H_{0,0}$ is employed to estimate the heat flux. From this consideration, the relation between the average temperature and the temperature at the boundary is

$$\tau_i(x,t) \cong \frac{1}{2\Delta y}\left(T_i\big|_{y=y_1} + T_i\big|_{y=y_2}\right) + \frac{1}{12}\left[\frac{\partial T_i}{\partial y}\bigg|_{y=y_1} - \frac{\partial T_i}{\partial y}\bigg|_{y=y_2}\right].$$

On applying the boundary conditions (2.1c) and (2.1d) to the above equation and using the $H_{0,0}$ approximation, we find that the temperatures are expressed as ($i = 1, \ldots, n$)

$$T_i(x,y_1,t) = \frac{U}{V}\tau_i(x,t) + \frac{W}{V} + \frac{U}{12k_iV}\left[h_dT_d(t) + h_cT_c\right], \tag{2.9}$$

where

$$\begin{aligned} U &= 12k_i\Delta y(2k_i + h_c), \\ V &= (6k_i + \Delta y h_d)(2k_i + h_c) + (6k_i + \Delta y h_c)(2k_i + h_d), \\ W &= (6k_i + \Delta y h_c)(h_dT_d(t) - h_cT_c), \end{aligned}$$

and

$$T_i(x,y_2,t) = \frac{Z}{V}\tau_i(x,t) - \frac{X}{V} + \frac{Z}{12k_iV}\left[h_dT_d(t) + h_cT_c\right], \tag{2.10}$$

where

$$\begin{aligned} Z &= 12k_i\Delta y(2k_i + h_d), \\ X &= (6k_i + \Delta y h_d)[h_dT_d(t) - h_cT_c]. \end{aligned}$$

Expressions (2.9) and (2.10) represent the average temperature and the temperature at the surface for the $H_{1,1}/H_{0,0}$ approximation; therefore, the $H_{1,1}/H_{0,0}$ formulation is written as

$$\begin{aligned} &\frac{\partial^2\tau_i(x,t)}{\partial x^2} - \frac{24(k_ih_c + k_ih_d + h_ch_d)}{B}\tau(x,t) \\ &= \frac{1}{\alpha_i}\frac{\partial\tau_i(x,t)}{\partial t} - \frac{1}{\Delta y k_i}(h_cT_c + h_dT_d(t)) \\ &\quad + (h_dT_d(t) - h_cT_c)\left[\left(\frac{h_c}{D}(6k_i + \Delta y\, h_d) - \frac{h_d}{B}(6k_i + \Delta y\, h_c)\right)\right] \\ &\quad - \frac{1}{\Delta y k_i}\frac{h_dT_d(t) + h_cT_c}{12k_i}\left(h_c\frac{C}{D} + h_d\frac{A}{B}\right). \end{aligned} \tag{2.11}$$

We now have the set of 1D differential equations (2.5), (2.8), and (2.11) to solve, subject to the boundary conditions

$$-k_1\frac{\partial\tau_1(x,t)}{\partial x} = h_a[T_a(t) - \tau_1(x,t)] \quad \text{at } x = x_1, \quad t > 0,$$

$$k_n \frac{\partial \tau_n(x,t)}{\partial x} = h_b[T_b - \tau_n(x,t)] \quad \text{at} \quad x = x_{n+1}, \quad t > 0,$$

the interface conditions ($i = 1, 2, \ldots, n-1$)

$$\tau_i(x,t) = \tau_{i+1}(x,t) \quad \text{at} \quad x = x_{i+1}, \quad t > 0,$$
$$k_i \frac{\partial \tau_i(x,t)}{\partial x} = k_{i+1} \frac{\partial \tau_{i+1}(x,t)}{\partial x} \quad \text{at} \quad x = x_{i+1}, \quad t > 0,$$

and initial conditions ($i = 1, 2, \ldots, n$)

$$\tau_i(x,0) = G_{0,i}(x), \quad x_i < x < x_{i+1},$$

where

$$G_{0,i}(x) = \frac{1}{\Delta y} \int_{y_1}^{y_2} F_i(x,y)\, dy.$$

2.2.3. Time Integration: Laplace Transformation Method

The Laplace transform of a function $\tau_i(x,t)$ is defined by

$$\bar{\tau}_i(x,s) \equiv \mathcal{L}\{\tau_i(x,t)\} = \int_0^{\infty} \tau_i(x,t)\, \mathrm{e}^{-st}\, dt.$$

Applying the Laplace transformation to the nodal equation results in

$$\frac{d^2 \bar{\tau}_i(x,s)}{dx^2} - \left(\beta_i + \frac{s}{\alpha_i}\right) \bar{\tau}_i(x,s) = -\frac{1}{\alpha_i} G_{0,i}(x) - \left[\frac{1}{\Delta y\, k_i}\right] \gamma_i\,, \qquad (2.12)$$

which satisfies the transformed boundary conditions and the interface conditions ($i = 1, 2, 3$). The parameters β_i and γ_i are expressed as

$$\beta_i = \begin{cases} \dfrac{h_c + h_d}{\Delta y\, k_i} & \text{for the standard approach,} \\ \dfrac{4[k_i\, \Delta y\, (h_c + h_d) + h_c\, h_d]}{k_i\, (4k_i + h_d + h_c)} & \text{for the } H_{0,0}/H_{0,0} \text{ approach,} \\ \dfrac{h_d U_{\text{down}} + h_c U_{\text{upper}}}{k_i\, \Delta y\, V} & \text{for the } H_{1,1}/H_{0,0} \text{ approach,} \end{cases}$$

$$\gamma_i = \begin{cases} h_d \bar{T}_d(s) + h_c \dfrac{T_c}{s} & \text{for the standard approach,} \\ \left(h_d \bar{T}_d(s) + h_c \dfrac{T_c}{s}\right) - \dfrac{h_d - h_c}{4k_i + h_d + h_c} \left(h_c \dfrac{T_c}{s} - h_d \bar{T}_d(s)\right) & \\ & \text{for the } H_{0,0}/H_{0,0} \text{ approach,} \\ \left[h_d \bar{T}_d(s) + h_c \dfrac{T_c}{s}\right] + \dfrac{h_d \bar{Z}(s) - h_c \bar{w}(s)}{k_i\, \Delta y\, V} & \\ & \text{for the } H_{1,1}/H_{0,0} \text{ approach.} \end{cases}$$

The solution of (2.12) can be written as

$$\begin{aligned}\bar{\tau}_i(x,s) &= A_i e^{-R_i x} + B_i e^{R_i x} \\ &\quad - \frac{e^{-R_i x}}{2R_i}\int_{x_i}^{x} e^{R_i \xi}\left[-\frac{G_{0,i}(\xi)}{\alpha_i} - \left(\frac{1}{\Delta y\, k_i}\right)\gamma_i(s)\right] d\xi \\ &\quad + \frac{e^{R_i x}}{2R_i}\int_{x_i}^{x} e^{-R_i \xi}\left[-\frac{G_{0,i}(\xi)}{\alpha_i} - \left(\frac{1}{\Delta y\, k_i}\right)\gamma_i(s)\right] d\xi,\end{aligned}$$

where $R_i = \sqrt{\beta_i + (s/\alpha_i)}$ and $\mathrm{e}^{-R_i x}$ and $e^{R_i x}$ are linearly independent functions. Another representation for this solution is

$$\begin{aligned}\bar{\tau}_i(x,s) &= A_i e^{-R_i x} + B_i e^{R_i x} + I_i(x) \\ &\quad - \frac{\gamma_i(s)}{2\, R_i\, \Delta y\, k_i}\left[2 - e^{-R_i(x-x_i)} - e^{-R_i(x_i-x)}\right],\end{aligned}$$

where

$$\begin{aligned}I_i(x) = \frac{1}{2R_i\alpha_i}\Bigg[\int_{x_i}^{x} e^{-R_i(x-\xi)}\, G_{0,i}(\xi)\, d\xi \\ - \int_{x_i}^{x} e^{R_i(x-\xi)}\, G_{0,i}(\xi)\, d\xi\Bigg].\end{aligned}$$

The coefficients A_i and B_i $(i = 1, 2, 3, 4)$ are determined from the boundary and interface conditions, which yield a linear system.

Finally, the solution $\tau_i(x,t)$ is obtained by applying the inverse Laplace transformation, that is,

$$\tau_i(x,t) = \frac{1}{2\pi j}\int_{c-j\infty}^{c+j\infty} e^{st}\, \bar{\tau}_i(x,s)\, ds.$$

Changing the variable $s = p/t$ and using a Gaussian quadrature formula [6] for the Bromwich integral yields

$$\tau_i(x,t) = \frac{1}{2\pi j}\int_{c'-j\infty}^{c'+j\infty} e^{p}\, \frac{\bar{\tau}_i(x,p/t)}{t}\, dp \approx \sum_{k=1}^{n} w_k\, \frac{p_k}{t}\, \bar{\tau}_i(x,p_k/t),$$

where $c' = c/t$. Numerical values for the weights w_k and nodes p_k can be found in the literature [7]. Substituting the variable s by p_k/t allows us to compute the integration constants A_i and B_i by solving $4 \times n$ linear systems (n is the quadrature order for numerical inversion of the Laplace transform) to obtain $\tau_i(x,t)$.

2.3 Numerical Results

Our numerical example consists of a four-layer region. The whole domain is 17 cm (width) × 100 cm (height). The features for each layer are displayed in Table 1. The external and internal convection coefficients are, respectively, $h_c = h_d = 16.95\ \mathrm{W\,m^{-2}\,C^{-1}}$ and $h_b = h_c = 8.26\ \mathrm{W\,m^{-2}\,C^{-1}}$.

The temperature inside the chamber is assumed to be constant at $T_b = 24°\mathrm{C}$, while the external temperature is changing with time—a diurnal cycle, according to the *sun-air temperature*. This is a fictitious temperature, where the solar radiation is taken into account [8]. The data for the sun-air temperature (T_{SA}) are those for 40°N latitude on July 21 and $\alpha/h_0 = 0.026$ (see Tables 2 and 3). In order to have a continuous function for T_{SA}, a piecewise polynomial interpolation is used with 5 intervals in a day: [00:00–05:00], [05:00–12:00], [12:00–16:00], [16:00–19:00], and [19:00–24:00].

Table 2. The sun-air temperature for a vertical surface.

time $[h]$	temperature $[^0\mathrm{C}]$	time $[h]$	temperature $[^0\mathrm{C}]$
1	25.430	13	40.446
2	24.880	14	46.682
3	24.440	15	50.860
4	24.110	16	52.350
5	24.000	17	50.618
6	25.104	18	43.948
7	26.382	19	31.416
8	27.918	20	29.830
9	29.764	21	28.620
10	31.700	22	27.520
11	33.752	23	26.640
12	35.850	24	25.980

Table 3. The sun-air temperature for a horizontal surface.

time $[h]$	temperature $[^0\mathrm{C}]$	time $[h]$	temperature $[^0\mathrm{C}]$
1	25.430	13	54.642
2	24.880	14	53.624
3	24.440	15	50.886
4	24.110	16	46.604
5	24.000	17	41.128
6	25.104	18	35.290
7	26.382	19	31.286
8	27.918	20	29.830
9	29.764	21	28.620
10	31.700	22	27.520
11	33.752	23	26.640
12	53.946	24	25.980

For the initial condition, the function $G_{0,i}(x)$ is represented by a second-degree polynomial, which is in good agreement with the true initial temperature field. Here, the same periods of time are used to split the day. Each period represents a new evolution problem, where the end of a period is the initial condition for the next period, and so on. The least square estimation was used to compute the polynomial coefficients for the interpolation.

Table 4. The heat flux (W) for the first, second, third, and fourth days, from the $H_{1,1}/H_{0,0}$ approach.

hour	Day 1	Day 2	Day 3	Day 4
1	0.1269	10.4372	10.5311	10.5319
2	0.3013	8.9478	9.0250	9.0256
3	0.4548	7.6100	7.6733	7.6738
4	0.5204	6.4112	6.4632	6.4637
5	0.5074	5.3486	5.3913	5.3917
6	0.4897	4.4660	4.5011	4.5014
7	0.5472	3.8123	3.8411	3.8414
8	0.7888	3.4698	3.4934	3.4936
9	1.2567	3.4580	3.4774	3.4775
10	1.9635	3.7709	3.7868	3.7870
11	2.9075	4.3915	4.4046	4.4047
12	4.6784	5.8969	5.9076	5.9077
13	6.5352	7.5356	7.5444	7.5445
14	8.3510	9.1724	9.1796	9.1797
15	10.5553	11.2298	11.2357	11.2358
16	13.1129	13.6667	13.6715	13.6716
17	15.6895	16.1442	16.1482	16.1482
18	17.8605	18.2338	18.2371	18.2372
19	19.1369	19.4434	19.4461	19.4461
20	19.0072	19.2589	19.2611	19.2611
21	17.5949	17.8016	17.8034	17.8034
22	15.7441	15.9138	15.9153	15.9153
23	13.8571	13.9964	13.9977	13.9977
24	12.0699	12.1842	12.1853	12.1853

The average heat flux for the first four simulation days is shown in Table 4. After the fourth day, the transient processes disappear. Only the results obtained employing the $H_{1,1}/H_{0,0}$ approach, which produces better results, are shown. For comparison, Table 5 shows the heat flux for the fourth day computed using the finite volume method (FVM), the Gauss–Laplace transformation scheme (G-L), and the method developed here with the $H_{1,1}/H_{0,0}$ formulation and 8 quadrature points for numerical inversion of the Laplace transform. Both the FVM and G-L approaches use the finite volume method for domain decomposition, but the FVM uses the forward Euler method for time integration, whereas the G-L uses the Laplace transformation with numerical inversion.

This method also allows us to calculate the temperature inside the wall.

Table 5. The heat flux (W) for the fourth day, from FVM, G-L, and $H_{1,1}/H_{0,0}$.

hour	FVM	G-L	$Q_{H_{11}/H_{00}}$
1	10.188	12.652	10.531
2	8.686	11.223	9.025
3	7.339	9.887	7.673
4	6.140	8.643	6.463
5	5.093	7.492	5.391
6	4.295	6.422	4.501
7	3.802	5.497	3.841
8	3.644	4.813	3.493
9	3.833	4.417	3.477
10	4.357	4.318	3.787
11	5.195	4.513	4.404
12	8.200	4.394	5.907
13	10.602	4.593	7.544
14	12.323	5.490	9.179
15	14.255	7.166	11.235
16	16.194	9.889	13.671
17	18.010	13.038	16.148
18	19.318	16.193	18.237
19	19.818	18.581	19.446
20	19.250	19.339	19.261
21	17.629	18.587	17.803
22	15.655	17.214	15.915
23	13.689	15.672	13.997
24	11.854	14.132	12.185

2.4. Final Remarks

Starting from a lumped analysis, where *simplified* models are designed, this paper introduces a semi-analytic approach, combining the Laplace transformation and lumped approach. In addition, three different schemes for lumped analysis are considered, and the results are compared against a full numerical formulation based on the finite volume method.

The method presented in this paper is effective for computing the heat flux through the wall, and it reduces the computational effort. One remarkable feature of this new method is that domain discretization is not necessary. The Gaussian approach for numerical inversion of the Laplace transform performed well.

Ours is a new procedure for this problem, which has until now been solved only by using fully standard numerical schemes. The application of this method to a 3D domain and other orthogonal coordinate systems is straightforward.

References

1. R.M. Cotta and M.D. Mikhailov, *Heat Conduction,* Wiley, New York, 1997.

2. E. Hauser, Study and solution of the transport equation by LTS_N method for higher angular quadrature order, D.Sc. Dissertation, Graduate Program in Mechanical Engineering, Federal University of Rio Grande do Sul, Porto Alegre (RS), Brazil, 1997 (Portuguese).

3. R. Pazos, M.T. Vilhena, and E. Hauser, Solution and study of the two-dimensional nodal neutron transport equation, in *Proc. Internat. Conf. on Nuclear Energy,* 2002.

4. M.T. Vilhena, L.B. Barichello, J. Zabadal, C. Segatto, A. Cardona, and R. Pazos, Solution to the multidimensional linear transport equation by spectral methods, *Progr. Nuclear Energy* **35** (1999), 275–291.

5. P.O. Beyer, Transient heat conduction for multi-layer walls, Ph.D. Dissertation, Graduate Program in Mechanical Engineering, Federal University of Rio Grande do Sul, Porto Alegre (RS), Brazil, 1998 (Portuguese).

6. M. Heydarian, N. Mullineux, and J.R. Reed, Solution of parabolic partial differential equations, *Appl. Math. Modelling* **5** (1981), 448–449.

7. A.H. Stroud and D. Secrest, *Gaussian Quadrature Formulas,* Prentice Hall, Englewood Cliffs, NJ, 1966.

8. *Handbook of Fundamentals*, 1993, Amer. Soc. of Heating, Refrigeration, and Air-Conditioning Engineers (ASHRAE), Atlanta, 1993.

3 The Cauchy Problem in the Bending of Thermoelastic Plates

Igor Chudinovich and Christian Constanda

3.1 Introduction

The main advantage of plate theories is the reduction of the original three-dimensional problem to a two-dimensional mathematical model. This not only makes the essential effects of the phenomenon of bending more prominent, but also simplifies the analytic arguments and the numerical computation algorithms. Starting with Kirchhoff's, many such theories have been proposed, each more refined than the preceding one in terms of mathematical sophistication and range of results. In this paper, we study the initial value problem for a very large (therefore, mathematically infinite) elastic plate with transverse shear deformation, as described in [1] and later generalized in [2] to account for thermal effects. We use the linearity of the governing equations to split the full investigation into two cases, namely that of homogeneous initial conditions and that of a homogeneous system. We show that the solution of the model can be represented in terms of some "initial" potentials with densities related to the prescribed data.

The corresponding problem for adiabatic plate deformation was discussed in [3].

3.2 Prerequisites

Suppose that the plate occupies a region $\mathbb{R}^2 \times [-h_0/2, h_0/2]$, $h_0 = \text{const}$, in $\mathbb{R}^3$. The displacement field at a point x' and $t \geq 0$ is $v(x',t) = (v_1(x',t), v_2(x',t), v_3(x',t))^{\mathrm{T}}$, where the superscript T denotes matrix transposition, and the temperature in the plate is $\tau(x',t)$. If $x' = (x, x_3)$, $x = (x_1, x_2) \in \mathbb{R}^2$, then for a plate with transverse shear deformation we assume [1] that

$$v(x',t) = (x_3 u_1(x,t), x_3 u_2(x,t), u_3(x,t))^{\mathrm{T}}$$

and use the temperature "averaged" across the thickness by means of the formula [2]

$$u_4(x,t) = \frac{1}{h^2 h_0} \int_{-h_0/2}^{h_0/2} x_3 \tau(x, x_3, t)\, dx_3, \quad h^2 = \frac{h_0^2}{12}.$$

Then $U(x,t) = (u(x,t)^{\mathrm{T}}, u_4(x,t))^{\mathrm{T}}$, $u(x,t) = (u_1(x,t), u_2(x,t), u_3(x,t))^{\mathrm{T}}$, satisfies

$$\mathcal{B}_0 \partial_t^2 U(x,t) + \mathcal{B}_1 \partial_t U(x,t) + \mathcal{A} U(x,t) = \mathcal{Q}(x,t), \quad (x,t) \in G, \tag{3.1}$$

where $G = \mathbb{R}^2 \times (0,\infty)$, $\mathcal{B}_0 = \mathrm{diag}\{\rho h^2, \rho h^2, \rho, 0\}$, $\partial_t = \partial/\partial t$, $\rho = \mathrm{const} > 0$ is the density of the material,

$$\mathcal{B}_1 = \begin{pmatrix} 0 & 0 & 0 & 0 \\ 0 & 0 & 0 & 0 \\ 0 & 0 & 0 & 0 \\ \eta\partial_1 & \eta\partial_2 & 0 & \varkappa^{-1} \end{pmatrix}, \quad \mathcal{A} = \begin{pmatrix} & & & h^2\gamma\partial_1 \\ & A & & h^2\gamma\partial_2 \\ & & & 0 \\ 0 & 0 & 0 & -\Delta \end{pmatrix},$$

$$A = \begin{pmatrix} -h^2\mu\Delta - h^2(\lambda+\mu)\partial_1^2 + \mu & -h^2(\lambda+\mu)\partial_1\partial_2 & \mu\partial_1 \\ -h^2(\lambda+\mu)\partial_1\partial_2 & -h^2\mu\Delta - h^2(\lambda+\mu)\partial_2^2 + \mu & \mu\partial_2 \\ -\mu\partial_1 & -\mu\partial_2 & -\mu\Delta \end{pmatrix},$$

$\partial_\alpha = \partial/\partial_\alpha$, $\alpha = 1,2$, $\eta, \varkappa$, and γ are positive physical constants, λ and μ are the Lamé coefficients satisfying $\lambda + \mu > 0$, $\mu > 0$, and $\mathcal{Q}(x,t) = (q(x,t)^{\mathrm{T}}, q_4(x,t))^{\mathrm{T}}$, where $q(x,t) = (q_1(x,t), q_2(x,t), q_3(x,t))^{\mathrm{T}}$ is a combination of the forces and moments acting on the plate and its faces and $q_4(x,t)$ is a combination of the averaged heat source density and the temperature and heat flux on the faces.

In terms of smooth functions, the Cauchy problem for (3.1) consists in finding $U(x,t) \in \mathrm{C}^2(G)$, $u \in \mathrm{C}^1(\bar{G})$, $u_4 \in \mathrm{C}(\bar{G})$, such that

$$\begin{aligned} \mathcal{B}_0 \partial_t^2 U(x,t) + \mathcal{B}_1 \partial_t U(x,t) + \mathcal{A} U(x,t) &= \mathcal{Q}(x,t), \quad (x,t) \in G, \\ U(x,0) = U_0(x), \quad \partial_t u(x,0) &= \psi(x), \quad x \in \mathbb{R}^2, \end{aligned} \tag{3.2}$$

where $U_0(x) = (\varphi(x)^{\mathrm{T}}, \theta(x))^{\mathrm{T}}$, $\varphi(x) = (\varphi_1(x), \varphi_2(x), \varphi_3(x))^{\mathrm{T}}$, and the initial "velocity" $\psi(x) = (\psi_1(x), \psi_2(x), \psi_3(x))^{\mathrm{T}}$ are given.

For $\kappa > 0$, we denote by $\mathbb{H}_{1,\kappa}(G)$ the space of all four-component distributions $U(x,t)$ with norm

$$\|U\|_{1,\kappa;G}^2 = \int_G e^{-2\kappa t} \Big\{ |U(x,t)|^2 + |\partial_t U(x,t)|^2 + \sum_{i=1}^4 |\nabla u_i(x,t)|^2 \Big\}\, dx\, dt,$$

or, equivalently,

$$\Big\{ \int_G e^{-2\kappa t} \big[(1+|\xi|)^2 |\tilde{U}(\xi,t)|^2 + |\partial_t \tilde{U}(\xi,t)|^2 \big]\, d\xi\, dt \Big\}^{1/2},$$

where $\tilde{U}(\xi,t) = (\tilde{u}(\xi,t)^{\mathrm{T}}, \tilde{u}_4(\xi,t))^{\mathrm{T}}$, $\tilde{u}(\xi,t) = (\tilde{u}_1(\xi,t), \tilde{u}_2(\xi,t), \tilde{u}_3(\xi,t))^{\mathrm{T}}$, is the Fourier transform of $U(x,t)$ with respect to x.

Let $W(x,t)=(w(x,t)^{\mathrm{T}},w_4(x,t))^{\mathrm{T}}\in \mathrm{C}_0^\infty(\bar{G})$. Multiplying the ith equation in (3.2) by $\bar{w}_i$ and the conjugate form of the fourth equation by $h^2\gamma\eta^{-1}w_4$, integrating over G, and adding the results together yields

$$\begin{aligned}\int\limits_G \big[&(B_0\partial_t^2 u,w)+(Au,w)+h^2\gamma\eta^{-1}\varkappa^{-1}(w_4,\partial_t u_4)\\ &-h^2\gamma\eta^{-1}(w_4,\Delta u_4)+h^2\gamma(w_4,\partial_t \operatorname{div} u)+h^2\gamma(\nabla u_4,w)\big]\,dx\,dt\\ &=\int\limits_G \big[(q,w)+h^2\gamma\eta^{-1}(w_4,q_4)\big]\,dx\,dt,\end{aligned}\tag{3.3}$$

where $B_0=\operatorname{diag}\{\rho h^2,\rho h^2,\rho\}$, $(\cdot\,,\cdot)$ is the inner product in the vector space $\mathbb{C}^m$, and $(\cdot\,,\cdot)$ is the inner product in $\big[L^2(\mathbb{R}^2)\big]^m$ for any $m\in\mathbb{N}$. Using integration by parts in (3.3) and the initial conditions from (3.2), we find that

$$\begin{aligned}\int\limits_0^\infty \big[&a(u,w)-(B_0^{1/2}\partial_t u,B_0^{1/2}\partial_t w)_0+h^2\gamma\eta^{-1}\varkappa^{-1}(w_4,\partial_t u_4)_0\\ &+h^2\gamma\eta^{-1}(\nabla w_4,\nabla u_4)_0-h^2\gamma(\nabla w_4,\partial_t u)_0+h^2\gamma(\nabla u_4,w)_0\big]\,dt\\ &=(B_0\psi,\gamma_0 w)_0+\int\limits_0^\infty \big[(q,w)_0+h^2\gamma\eta^{-1}(w_4,q_4)_0\big]\,dt,\end{aligned}\tag{3.4}$$

where γ_0 is the continuous trace operator from the Sobolev space of index $m\in\mathbb{N}$ and with weight $\exp(-2\kappa t)$, $t>0$, of functions on G to the corresponding standard Sobolev space of index $m-1/2$ of functions (vector functions) defined in $\mathbb{R}^2$, and $a(u,w)=2\int\limits_{\mathbb{R}^2}E(u,w)\,dx$, where

$$\begin{aligned}2E(u,w)&=h^2E_0(u,w)+h^2\mu(\partial_2 u_1+\partial_1 u_2)(\partial_2\bar{w}_1+\partial_1\bar{w}_2)\\ &\quad+\mu[(u_1+\partial_1 u_3)(\bar{w}_1+\partial_1\bar{w}_3)+(u_2+\partial_2 u_3)(\bar{w}_2+\partial_2\bar{w}_3)],\\ E_0(u,w)&=(\lambda+2\mu)\big[(\partial_1 u_1)(\partial_1\bar{w}_1)+(\partial_2 u_2)(\partial_2\bar{w}_2)\big]\\ &\quad+\lambda\big[(\partial_1 u_1)(\partial_2\bar{w}_2)+(\partial_2 u_2)(\partial_1\bar{w}_1)\big].\end{aligned}$$

It is easy to see that if $f\in \mathrm{C}^2(\mathbb{R}^2)$ and $g\in \mathrm{C}_0^\infty(\mathbb{R}^2)$, then $(Af,g)_0=a(f,g)$. Guided by (3.4), we say that $U(x,t)$ is a weak solution of (3.2) if $U\in \mathbb{H}_{1,\kappa}(G)$ for some $\kappa>0$, U satisfies (3.4) for any $W\in \mathrm{C}_0^\infty(\bar{G})$, and $\gamma_0 U=U_0(x)$.

Theorem 1. *The Cauchy problem* (3.2) *has at most one weak solution* $U\in\mathbb{H}_{1,\kappa}(G)$.

Owing to linearity, the solution of (3.2) can be written as the sum of the solutions of two simpler problems: the Cauchy problem for the homogeneous system (3.1) with the prescribed initial data and the Cauchy problem for the nonhomogeneous system (3.1) with zero initial data.

3.3 Homogeneous System

Suppose that $\mathcal{Q}(x,t)=0$. Then (3.4) becomes

$$\begin{aligned}\int_0^\infty \big[a(u,w) &- (B_0^{1/2}\partial_t u, B_0^{1/2}\partial_t w)_0 + h^2\gamma\eta^{-1}\varkappa^{-1}(w_4,\partial_t u_4)_0\\ &+ h^2\gamma\eta^{-1}(\nabla w_4,\nabla u_4)_0 - h^2\gamma(\nabla w_4,\partial_t u)_0\\ &+ h^2\gamma(\nabla u_4, w)_0\big]\,dt = (B_0\psi,\gamma_0 w)_0. \qquad (3.5)\end{aligned}$$

We want to find $U\in \mathbb{H}_{1,\kappa}(G)$ satisfying (3.5) for all $W\in C_0^\infty(\bar G)$ and $\gamma_0 U = U_0(x) = (\varphi(x)^{\mathrm{T}},\theta(x))^{\mathrm{T}}$.

We consider a matrix $D(x,t)$ of fundamental solutions for (3.1), that is, a matrix such that

$$\begin{aligned}\mathcal{B}_0\partial_t^2 D(x,t) + \mathcal{B}_1\partial_t D(x,t) + \mathcal{A}D(x,t) &= \delta(x,t)I, \quad (x,t)\in\mathbb{R}^3,\\ D(x,t) &= 0, \quad t<0,\end{aligned}$$

where I is the identity (4×4)-matrix and δ is the Dirac delta.

We write $D(x,t)=\chi(t)\Phi(x,t)$, $\tilde D(\xi,t)=\chi(t)\tilde\Phi(\xi,t)$, where $\chi(t)$ is the characteristic function of the positive semi-axis.

The initial potential of the first kind of density $F(x)=(f(x)^{\mathrm{T}},f_4(x))^{\mathrm{T}}$, $f(x)=(f_1(x),f_2(x),f_3(x))^{\mathrm{T}}$, is defined by

$$\mathcal{J}(x,t) = (\mathcal{J}F)(x,t) = \int_{\mathbb{R}^2} D(x-y,t)F(y)\,dy = \int_{\mathbb{R}^2}\Phi(x-y,t)F(y)\,dy, \quad t>0.$$

The initial potential of the second kind of density $G(x)=(g(x)^{\mathrm{T}},g_4(x))^{\mathrm{T}}$, $g(x)=(g_1(x),g_2(x),g_3(x))^{\mathrm{T}}$, is defined by

$$\begin{aligned}\mathcal{E}(x,t) = (\mathcal{E}G)(x,t) &= \int_{\mathbb{R}^2}\partial_t D(x-y,t)G(y)\,dy\\ &= \int_{\mathbb{R}^2}\partial_t\Phi(x-y,t)G(y)\,dy = \partial_t(\mathcal{J}G)(x,t), \quad t>0.\end{aligned}$$

Theorem 2. *If* $\varphi\in H_3(\mathbb{R}^2)$, $\theta\in H_2(\mathbb{R}^2)$, *and* $\psi\in H_1(\mathbb{R}^2)$, *and if, in addition,* $f=B_0\psi$, $f_4=\varkappa^{-1}\theta+\eta\,\mathrm{div}\,\varphi$, $g=B_0\psi$, *and* $g_4=0$, *then*

$$\mathcal{L}(x,t) = (\mathcal{J}F)(x,t) + (\mathcal{E}G)(x,t)$$

is the solution of (3.5) *in* $\mathbb{H}_{1,\kappa}(G)$ *for any* $\kappa>0$ *and satisfies the initial condition* $\gamma_0\mathcal{L}=(\varphi^{\mathrm{T}},\theta)^{\mathrm{T}}$ *and the estimate*

$$\|\mathcal{L}\|_{1,\kappa;G} \le c\big\{\|\varphi\|_3 + \|\theta\|_2 + \|\psi\|_1\big\},$$

where $c=\mathrm{const}>0$ *is independent of the data functions.*

We are interested in the the relationship between the smoothness of the solution of the Cauchy problem and that of the given initial data. We choose to examine this, for example, in the space $\mathbb{H}_{m,\kappa}(G)$ consisting of elements $U = (u^{\mathrm{T}}, u_4)^{\mathrm{T}}$ with norm

$$\|U\|'_{m,\kappa;G} = \Big\{ \int_G e^{-2\kappa t} \Big[(1+|\xi|)^{2m} |\tilde{U}(\xi,t)|^2 + \sum_{k=1}^m (1+|\xi|)^{2(m-k)} \times \big(|\partial_t^k \tilde{u}(\xi,t)|^2 + |\partial_t^{k-1} \tilde{u}_4(\xi,t)|^2 \big) \Big] \, d\xi \, dt \Big\}^{1/2}, \quad m \in \mathbb{N}.$$

Theorem 3. *If*

$$\varphi \in H_{m+1}(\mathbb{R}^2), \quad \theta \in H_m(\mathbb{R}^2), \quad \psi \in H_m(\mathbb{R}^2), \quad m = 1, 2,$$

$$\varphi \in H_{2m-1}(\mathbb{R}^2), \quad \theta \in H_{2m-2}(\mathbb{R}^2), \quad \psi \in H_{2m-3}(\mathbb{R}^2), \quad m \geq 3,$$

and $f = B_0\psi$, $f_4 = \varkappa^{-1}\theta + \eta \operatorname{div} \varphi$, $g = B_0\varphi$, *and* $g_4 = 0$, *then*

$$\mathcal{L}(x,t) = (\mathcal{J}F)(x,t) + (\mathcal{E}G)(x,t)$$

is the solution of (3.5) *in* $\mathbb{H}'_{m,\kappa}(G)$ *for any* $\varkappa > 0$, *and*

$$\|\mathcal{L}\|'_{m,\kappa;G} \leq c(\|\varphi\|_{m+1} + \|\theta\|_m + \|\psi\|_m), \quad m = 1, 2,$$

$$\|\mathcal{L}\|'_{m,\kappa;G} \leq c(\|\varphi\|_{2m-1} + \|\theta\|_{2m-2} + \|\psi\|_{2m-3}), \quad m \geq 3,$$

where $c = \text{const} > 0$ *is independent of the data functions.*

3.4 Homogeneous Initial Data

Let $\varphi(x) = \theta(x) = \psi(x) \equiv 0$. Then the variational equation (3.4) takes the form

$$\int_0^\infty \big[a(u,w) - (B_0^{1/2}\partial_t u, B_0^{1/2}\partial_t w)_0 + h^2\gamma\eta^{-1}\varkappa^{-1}(w_4, \partial_t u_4)_0 + h^2\gamma\eta^{-1}(\nabla w_4, \nabla u_4)_0 - h^2\gamma(\nabla w_4, \partial_t u)_0 + h^2\gamma(\nabla u_4, w)_0 \big] \, dt$$
$$= \int_0^\infty \big[(q,w)_0 + h^2\gamma\eta^{-1}(w_4, q_4)_0 \big] \, dt \quad \forall W \in C_0^\infty(\bar{G}) \tag{3.6}$$

and we seek for it a solution $U \in \mathbb{H}_{1,\kappa}(G)$ such that $\gamma_0 U = 0$.

The Laplace transform with respect to t of a function $u(x,t)$ is

$$\hat{u}(x,p) = \int_0^\infty e^{-pt} u(x,t) \, dt.$$

We consider the area potential $\mathcal{U}(x,t)$ of density $Q(x,t) = (q(x,t)^{\mathrm{T}}, q_4(x,t))^{\mathrm{T}}$, $q(x,t) = (q_1(x,t), q_2(x,t), q_3(x,t))^{\mathrm{T}}$, of class $C_0^\infty(G)$, defined by

$$\mathcal{U}(x,t) = (\mathcal{U}Q)(x,t) = \int_G D(x-y,t-\tau)Q(y,\tau)\,dy\,d\tau, \quad (x,t) \in G.$$

Let $\mathbb{C}_\kappa = \{p = \sigma + i\tau : \sigma > \kappa\}$, $\kappa \in \mathbb{R}$, and let k, l, $m \in \mathbb{R}$. We introduce a number of necessary spaces of (vector or scalar) distributions, as follows.

(i) $H_m(\mathbb{R}^2)$ is the standard Sobolev space, with norm

$$\|u\|_m = \Big\{ \int_{\mathbb{R}^2} (1+|\xi|^2)^m |\tilde{u}(\xi)|^2\,d\xi \Big\}^{1/2}.$$

(ii) $H_{m,p}(\mathbb{R}^2)$, $p \in \mathbb{C}$, is the space that coincides with $H_m(\mathbb{R}^2)$ as a set but is endowed with the norm

$$\|u\|_{m,p} = \Big\{ \int_{\mathbb{R}^2} (1+|\xi|^2+|p|^2)^m |\tilde{u}(\xi)|^2\,d\xi \Big\}^{1/2}.$$

(iii) $\mathcal{H}^{\mathcal{L}}_{m,k,\kappa}(\mathbb{R}^2)$ and $H^{\mathcal{L}}_{m,k,\kappa}(\mathbb{R}^2)$ are the spaces of functions $\hat{u}(x,p)$ that, regarded as mappings from $\mathbb{C}_\kappa$ to $H_m(\mathbb{R}^2)$, are holomorphic; they are equipped, respectively, with the norms

$$[\hat{u}]^2_{m,k,\kappa} = \sup_{\sigma>\kappa} \int_{-\infty}^{\infty} (1+|p|^2)^k \|\breve{u}(\xi,p)\|^2_m\,d\tau < \infty,$$

$$\|\hat{u}\|^2_{m,k,\kappa} = \sup_{\sigma>\kappa} \int_{-\infty}^{\infty} (1+|p|^2)^k \|\breve{u}(\xi,p)\|^2_{m,p}\,d\tau < \infty,$$

where $\breve{u}(\xi,p)$ is the Fourier transform of $\hat{u}(x,p)$ with respect to x.

(iv) $\mathcal{H}^{\mathcal{L}^{-1}}_{m,k,\kappa}(G)$ and $H^{\mathcal{L}^{-1}}_{m,k,\kappa}(G)$ are the spaces of the inverse Laplace transforms $u(x,t)$ of $\hat{u}(x,p) \in \mathcal{H}^{\mathcal{L}}_{m,k,\kappa}(\mathbb{R}^2)$ and $\hat{u}(x,p) \in H^{\mathcal{L}}_{m,k,\kappa}(\mathbb{R}^2)$, endowed with the norms

$$[u]_{m;k,\kappa;G} = [\hat{u}]_{m,k,\kappa}, \quad \|u\|_{m;k,\kappa;G} = \|\hat{u}\|_{m,k,\kappa}.$$

(v) $\mathcal{H}^{\mathcal{L}^{-1}}_{m;k,l;\kappa}(G) = \mathcal{H}^{\mathcal{L}^{-1}}_{m,k,\kappa}(G) \times \mathcal{H}^{\mathcal{L}^{-1}}_{m,l;\kappa}(G)$ is the space of all $U(x,t) = (u(x,t)^{\mathrm{T}}, u_4(x,t))^{\mathrm{T}}$, $u(x,t) = (u_1(x,t), u_2(x,t), u_3(x,t))^{\mathrm{T}}$, with norm

$$[U]_{m;k,l,\kappa;G} = [u]_{m;k,\kappa;G} + [u_4]_{m;l,\kappa;G}.$$

(vi) $H^{\mathcal{L}^{-1}}_{m;k,l;\kappa}(G) = H^{\mathcal{L}^{-1}}_{m,k,\kappa}(G) \times H^{\mathcal{L}^{-1}}_{m,l;\kappa}(G)$ is the space of functions U of the same form as in (v) but is equipped with the norm

$$\|U\|_{m;k,l,\kappa;G} = \|u\|_{m;k,\kappa;G} + \|u_4\|_{m;\,l,\kappa;G}.$$

(vii) $\mathbb{H}^{\mathcal{L}^{-1}}_{1,\kappa}(G) = H^{\mathcal{L}^{-1}}_{1;0,0;\kappa}(G)$, with $\|U\|_{1,\kappa;G} = \|U\|_{1;0,0,\kappa;G}$; it is obvious that this is the subspace of $\mathbb{H}_{1,\kappa}(G)$ consisting of all $U = (u^{\mathrm{T}}, u_4)^{\mathrm{T}}$ such that $\gamma_0 U = 0$.

Theorem 4. *For any $\mathcal{Q} = (q^{\mathrm{T}}, q_4)^{\mathrm{T}} \in \mathcal{H}^{\mathcal{L}^{-1}}_{-1;1,1;\kappa}(G)$, $\kappa > 0$, equation* (3.6) *has a unique solution $U = \mathcal{U}\mathcal{Q} \in \mathbb{H}^{\mathcal{L}^{-1}}_{1,\kappa}(G)$. If $\mathcal{Q} \in \mathcal{H}^{\mathcal{L}^{-1}}_{-1;k,k;\kappa}(G)$, then $U \in H^{\mathcal{L}^{-1}}_{1;k-1,k-1;\kappa}(G)$ and*

$$\|U\|_{1;k-1,k-1,\kappa;G} \le c[\mathcal{Q}]_{-1;k,k,\kappa;G},$$

where $c = \mathrm{const} > 0$ is independent of $\mathcal{Q}$ and k but may depend on κ.

Full details of the proofs of these assertions will appear in a future publication.

References

1. C. Constanda, *A Mathematical Analysis of Bending of Plates with Transverse Shear Deformation,* Longman/Wiley, Harlow-New York, 1990.
2. P. Schiavone and R.J. Tait, Thermal effects in Mindlin-type plates, *Quart. J. Mech. Appl. Math.* **46** (1993), 27–39.
3. I. Chudinovich and C. Constanda, The Cauchy problem in the theory of plates with transverse shear deformation, *Math. Models Methods Appl. Sci.* **10** (2000), 463–477.

4 Mixed Initial-boundary Value Problems for Thermoelastic Plates

Igor Chudinovich and Christian Constanda

4.1 Introduction

A study is made of the time-dependent bending of an elastic plate with transverse shear deformation [1] under external forces and moments, internal heat sources, homogeneous initial conditions, and nonhomogeneous mixed boundary conditions, when thermal effects are taken into account (see [2]). The problems, solved by means of a variational method coupled with the Laplace transformation technique, are shown to have unique stable solutions in appropriate spaces of distributions. The corresponding results in the absence of heat sources can be found in [3]–[7].

4.2 Prerequisites

Consider a homogeneous and isotropic elastic material occupying a region $\bar{S} \times [-h_0/2, h_0/2] \subset \mathbb{R}^3$, $S \subset \mathbb{R}^2$. The displacement of a point x' at time $t \geq 0$ is characterized by a vector $v(x', t) = (v_1(x', t), v_2(x', t), v_3(x', t))^{\mathrm{T}}$, where the superscript T denotes matrix transposition. The temperature in the plate is denoted by $\theta(x', t)$. Let $x' = (x, x_3)$, $x = (x_1, x_2) \in \bar{S}$. The process of bending is described by a field of the form [1]

$$v(x', t) = (x_3 u_1(x, t), x_3 u_2(x, t), u_3(x, t))^{\mathrm{T}},$$

coupled with the "averaged" temperature [2]

$$u_4(x, t) = \frac{1}{h^2 h_0} \int_{-h_0/2}^{h_0/2} x_3 \theta(x, x_3, t)\, dx_3, \quad h^2 = \frac{h_0^2}{12}.$$

Then the vector

$$U(x, t) = (u(x, t)^{\mathrm{T}}, u_4(x, t))^{\mathrm{T}}, \quad u(x, t) = (u_1(x, t), u_2(x, t), u_3(x, t))^{\mathrm{T}},$$

is a solution of

$$\mathcal{B}_0 \partial_t^2 U(x, t) + \mathcal{B}_1 \partial_t U(x, t) + \mathcal{A} U(x, t) = \mathcal{Q}(x, t),$$
$$(x, t) \in G = S \times (0, \infty), \tag{4.1}$$

where $\mathcal{B}_0 = \operatorname{diag}\{\rho h^2, \rho h^2, \rho, 0\}$, $\partial_t = \partial/\partial t$, $\rho > 0$ is the constant density of the material,

$$\mathcal{B}_1 = \begin{pmatrix} 0 & 0 & 0 & 0 \\ 0 & 0 & 0 & 0 \\ 0 & 0 & 0 & 0 \\ \eta\partial_1 & \eta\partial_2 & 0 & \varkappa^{-1} \end{pmatrix}, \quad \mathcal{A} = \begin{pmatrix} & & & h^2\gamma\partial_1 \\ & A & & h^2\gamma\partial_2 \\ & & & 0 \\ 0 & 0 & 0 & -\Delta \end{pmatrix},$$

$$A = \begin{pmatrix} -h^2\mu\Delta - h^2(\lambda+\mu)\partial_1^2 + \mu & -h^2(\lambda+\mu)\partial_1\partial_2 & \mu\partial_1 \\ -h^2(\lambda+\mu)\partial_1\partial_2 & -h^2\mu\Delta - h^2(\lambda+\mu)\partial_2^2 + \mu & \mu\partial_2 \\ -\mu\partial_1 & -\mu\partial_2 & -\mu\Delta \end{pmatrix},$$

$\partial_\alpha = \partial/\partial x_\alpha$, $\alpha = 1, 2$, η, $\varkappa$, and γ are positive physical constants, λ and μ are the Lamé coefficients of the material satisfying $\lambda + \mu > 0$, $\mu > 0$, and $\mathcal{Q}(x,t) = (q(x,t)^{\mathrm{T}}, q_4(x,t))^{\mathrm{T}}$, where $q(x,t) = (q_1(x,t), q_2(x,t), q_3(x,t))^{\mathrm{T}}$ is a combination of the forces and moments acting on the plate and its faces and $q_4(x,t)$ is a combination of the averaged heat source density and the temperature and heat flux on the faces.

Without loss of generality [8], we assume that

$$U(x,0) = 0, \quad \partial_t u(x,0) = 0, \quad x \in S. \tag{4.2}$$

Suppose that the boundary ∂S of S is a simple, closed, piecewise smooth contour described in terms of four open arcs counted counterclockwise as ∂S_i, $i = 1, \dots, 4$, such that

$$\partial S = \cup_{i=1}^4 \overline{\partial S_i}, \quad \partial S_i \cap \partial S_j = \emptyset, \quad i \neq j, \quad i, j = 1, \dots, 4.$$

We make the notation $(i, j = 1, 2, 3, 4)$

$$\Gamma = \partial S \times (0, \infty), \quad \Gamma_i = \partial S_i \times (0, \infty),$$
$$\partial S_{ij} = \partial S_i \cup \partial S_j \cup (\overline{\partial S_i} \cap \overline{\partial S_j}), \quad \Gamma_{ij} = \partial S_{ij} \times (0, \infty).$$

We consider the boundary conditions

$$u(x,t) = f(x,t), \quad u_4(x,t) = f_4(x,t), \quad (x,t) \in \Gamma_1, \tag{4.3}$$
$$u(x,t) = f(x,t), \quad \partial_n u_4(x,t) = g_4(x,t), \quad (x,t) \in \Gamma_2, \tag{4.4}$$

where $n = n(x) = (n_1(x), n_2(x), n_3(x))^{\mathrm{T}}$ is the outward unit normal to ∂S and $\partial_n = \partial/\partial n$.

We also need the expression of the moment-force boundary operator [1], which is

$$T = \begin{pmatrix} h^2\left[(\lambda+2\mu)n_1\partial_1 + \mu n_2\partial_2\right] & h^2(\lambda n_1\partial_2 + \mu n_2\partial_1) & 0 \\ h^2(\mu n_1\partial_2 + \lambda n_2\partial_1) & h^2\left[(\lambda+2\mu)n_2\partial_2 + \mu n_1\partial_1\right] & 0 \\ \mu n_1 & \mu n_2 & \mu\partial_n \end{pmatrix},$$

and consider the additional boundary conditions

$$Tu(x,t) - h^2\gamma n(x)u_4(x,t) = g(x,t), \quad \partial_n u_4(x,t) = g_4(x,t), \quad (x,t) \in \Gamma_3, \tag{4.5}$$

$$Tu(x,t) - h^2\gamma n(x)u_4(x,t) = g(x,t), \quad u_4(x,t) = f_4(x,t), \quad (x,t) \in \Gamma_4. \tag{4.6}$$

The functions $f(x,t)$, $f_4(x,t)$, $g(x,t)$, and $g_4(x,t)$ in (4.3)–(4.6) are prescribed.

Let S^+ and S^- be the interior and exterior domains bounded by ∂S, respectively, and let $G^\pm = S^\pm \times (0,\infty)$. The interior and exterior initial-boundary value problems (TM$^\pm$) require us to find $U \in \mathrm{C}^2(G^\pm) \cap \mathrm{C}^1(\bar{G}^\pm)$ satisfying (4.1) in $G^\pm$, (4.2) in $S^\pm$, and (4.3)–(4.6).

4.3 The Parameter-dependent Problems

We denote by $\hat{s}(x,p)$ the Laplace transform of a function $s(x,t)$. The Laplace-transformed problems (TM$^\pm$) are elliptic boundary value problems (TM$_p^\pm$) that depend on the complex parameter p. They consist in finding $\hat{U} \in \mathrm{C}^2(S^\pm) \cap \mathrm{C}^1(\bar{S}^\pm)$ such that

$$p^2\mathcal{B}_0\hat{U}(x,p) + p\,\mathcal{B}_1\hat{U}(x,p) + \mathcal{A}\hat{U}(x,p) = \hat{\mathcal{Q}}(x,p), \quad x \in S^\pm, \tag{4.7}$$

and

$$\begin{aligned}
\hat{u}(x,p) = \hat{f}(x,p), \quad \hat{u}_4(x,p) = \hat{f}_4(x,p), \quad x \in \partial S_1,\\
\hat{u}(x,p) = \hat{f}(x,p), \quad \partial_n\hat{u}_4(x,p) = \hat{g}_4(x,p), \quad x \in \partial S_2,\\
T\hat{u}(x,p) - h^2\gamma n(x)\hat{u}_4(x,p) = \hat{g}(x,p), \quad \partial_n\hat{u}_4(x,p) = \hat{g}_4(x,p), \quad x \in \partial S_3,\\
T\hat{u}(x,p) - h^2\gamma n(x)\hat{u}_4(x,p) = \hat{g}(x,p), \quad \hat{u}_4(x,p) = \hat{f}_4(x,p), \quad x \in \partial S_4.
\end{aligned}$$

A number of spaces of distributions are introduced for $m \in \mathbb{R}$ and $p \in \mathbb{C}$. Thus,

$H_m(\mathbb{R}^2)$ is the Sobolev space of all $\hat{v}_4(x)$ with norm

$$\|\hat{v}_4\|_m = \left\{ \int_{\mathbb{R}^2} (1+|\xi|^2)^m |\tilde{v}_4(\xi)|^2\, d\xi \right\}^{1/2},$$

where $\tilde{v}_4(\xi)$ is the Fourier transform of $\hat{v}_4(x)$;

$\mathbf{H}_{m,p}(\mathbb{R}^2)$: the space of all $\hat{v}(x)$ which coincides with $[H_m(\mathbb{R}^2)]^3$ as a set but is endowed with the norm

$$\|\hat{v}\|_{m,p} = \left\{ \int_{\mathbb{R}^2} (1+|\xi|^2+|p|^2)^m |\tilde{v}(\xi)|^2\, d\xi \right\}^{1/2};$$

$H_m(S^\pm)$ and $\mathbf{H}_{m,p}(S^\pm)$ are the spaces of the restrictions to $S^\pm$ of all $\hat{v}_4 \in H_m(\mathbb{R}^2)$ and $\hat{v} \in \mathbf{H}_{m,p}(\mathbb{R}^2)$, respectively, with norms

$$\|\hat{u}_4\|_{m;S^\pm} = \inf_{\hat{v}_4 \in H_m(\mathbb{R}^2):\, \hat{v}_4|_{S^\pm} = \hat{u}_4} \|\hat{v}_4\|_m,$$

$$\|\hat{u}\|_{m,p;S^\pm} = \inf_{\hat{v} \in \mathbf{H}_{m,p}(\mathbb{R}^2):\, \hat{v}|_{S^\pm} = \hat{u}} \|\hat{v}\|_{m,p};$$

$H_{-m}(S^\pm)$ and $\mathbf{H}_{-m,p}(S^\pm)$ are the duals of $\mathring{H}_m(S^\pm)$ and $\mathring{\mathbf{H}}_{m,p}(S^\pm)$ with respect to the duality generated by the inner products in $L^2(S^\pm)$ and $\left[L^2(S^\pm)\right]^3$;

$H_{1/2}(\partial S)$ and $\mathbf{H}_{1/2,p}(\partial S)$ are the spaces of the traces on ∂S of all $\hat{u}_4 \in H_1(S^+)$ and $\hat{u} \in \mathbf{H}_{1,p}(S^+)$, with norms

$$\|\hat{f}_4\|_{1/2;\partial S} = \inf_{\hat{u}_4 \in H_1(S^+):\, \hat{u}_4|_{\partial S} = \hat{f}_4} \|\hat{u}_4\|_{1;S^+},$$

$$\|\hat{f}\|_{1/2,p;\partial S} = \inf_{\hat{u} \in \mathbf{H}_{1,p}(S^+):\, \hat{u}|_{\partial S} = \hat{f}} \|\hat{u}\|_{1,p;S^+};$$

$H_{-1/2}(\partial S)$ and $\mathbf{H}_{-1/2,p}(\partial S)$ are the duals of $H_{1/2}(\partial S)$ and $\mathbf{H}_{1/2,p}(\partial S)$ with respect to the duality generated by the inner products in $L^2(\partial S)$ and $[L^2(\partial S)]^3$, with norms $\|\hat{g}_4\|_{-1/2,\partial S}$ and $\|\hat{g}\|_{-1/2,p;\partial S}$.

We denote by the same symbol $\gamma^\pm$ the continuous (uniformly with respect to $p \in \mathbb{C}$) trace operators from $H_1(S^\pm)$ to $H_{1/2}(\partial S)$ and from $\mathbf{H}_{1,p}(S^\pm)$ to $\mathbf{H}_{1/2,p}(\partial S)$, by $\partial\tilde{S} \subset \partial S$ any open part of ∂S with $\operatorname{mes} \partial\tilde{S} > 0$, and by $\tilde{\pi}$ the operator of restriction from ∂S to $\partial\tilde{S}$. The following spaces are also needed.

$H_{\pm 1/2}(\partial\tilde{S})$ and $\mathbf{H}_{\pm 1/2,p}(\partial\tilde{S})$ are the spaces of the restrictions to $\partial\tilde{S}$ of all the elements of $H_{\pm 1/2}(\partial S)$ and $\mathbf{H}_{\pm 1/2,p}(\partial S)$, respectively, with norms

$$\|\hat{e}_4\|_{\pm 1/2;\partial\tilde{S}} = \inf_{\hat{r}_4 \in H_{\pm 1/2}(\partial S):\, \tilde{\pi}\hat{r}_4 = \hat{e}_4} \|\hat{r}_4\|_{\pm 1/2;\partial S},$$

$$\|\hat{e}\|_{\pm 1/2,p;\partial\tilde{S}} = \inf_{\hat{r} \in \mathbf{H}_{\pm 1/2,p}(\partial S):\, \tilde{\pi}\hat{r} = \hat{e}} \|\hat{r}\|_{\pm 1/2,p;\partial S};$$

$\mathring{H}_{\pm 1/2}(\partial\tilde{S})$ and $\mathring{\mathbf{H}}_{\pm 1/2,p}(\partial\tilde{S})$ are, respectively, the subspaces of $H_{\pm 1/2}(\partial S)$ and $\mathbf{H}_{\pm 1/2,p}(\partial S)$ consisting of all the elements with support in $\overline{\partial\tilde{S}}$; we may denote the norms of $\hat{e}_4 \in \mathring{H}_{\pm 1/2}(\partial\tilde{S})$ and $\hat{e} \in \mathring{H}_{\pm 1/2}(\partial\tilde{S})$ by $\|\hat{e}_4\|_{\pm 1/2,\partial S}$ and $\|\hat{e}\|_{\pm 1/2,p;\partial S}$, and remark that $H_{\pm 1/2}(\partial\tilde{S})$ are the duals of $\mathring{H}_{\mp 1/2}(\partial\tilde{S})$ and that $\mathbf{H}_{\pm 1/2,p}(\partial\tilde{S})$ are the duals of $\mathring{\mathbf{H}}_{\mp 1/2,p}(\partial\tilde{S})$ with respect to the duality generated by the inner products in $L^2(\partial\tilde{S})$ and $[L^2(\partial\tilde{S})]^3$.

Consider operators π_i and π_{ij}, $i, j = 1, \dots, 4$, of restriction from ∂S to ∂S_i and from ∂S to ∂S_{ij}. One further batch of spaces is now introduced.

$H_1(S^\pm, \partial S_{23})$ and $\mathbf{H}_{1,p}(S^\pm, \partial S_{34})$ are, respectively, the subspaces of $H_1(S^\pm)$ and $\mathbf{H}_{1,p}(S^\pm)$ consisting of all $\hat{u}_4 \in H_1(S^\pm)$ and $\hat{u} \in \mathbf{H}_{1,p}(S^\pm)$ such that $\pi_{41}\gamma^\pm \hat{u}_4 = 0$ and $\pi_{12}\gamma^\pm \hat{u} = 0$;

$H_{-1}(S^\pm, \partial S_{23})$ and $\mathbf{H}_{-1,p}(S^\pm, \partial S_{34})$ are the duals of $H_1(S^\pm, \partial S_{23})$ and $\mathbf{H}_{1,p}(S^\pm, \partial S_{34})$ with respect to the original dualities, with the norms of $\hat{q}_4 \in H_{-1}(S^\pm, \partial S_{23})$ and $\hat{q} \in \mathbf{H}_{-1,p}(S^\pm, \partial S_{34})$ denoted by $[\hat{q}_4]_{-1;S^\pm,\partial S_{23}}$ and $[\hat{q}]_{-1,p;S^\pm,\partial S_{34}}$;

$\mathcal{H}_{1,p}(S^\pm) = \mathbf{H}_{1,p}(S^\pm) \times H_1(S^\pm)$, with the norm of its elements $\hat{U} = (\hat{u}^{\mathrm{T}}, \hat{u}_4)^{\mathrm{T}}$ defined by $|||\hat{U}|||_{1,p;S^\pm} = \|\hat{u}\|_{1,p;S^\pm} + \|\hat{u}_4\|_{1;S^\pm}$;

$\mathcal{H}_{1,p}(S^\pm; \partial S_{34}, \partial S_{23}) = \mathbf{H}_{1,p}(S^\pm, \partial S_{34}) \times H_1(S^\pm, \partial S_{23})$, which is a subspace of $\mathcal{H}_{1,p}(S^\pm)$.

Let $\kappa > 0$, and let $\mathbb{C}_\kappa = \{p = \sigma + i\tau \in \mathbb{C} : \sigma > \kappa\}$. Throughout what follows, c is a generic positive constant occurring in estimates, which is independent of the functions in those estimates and of $p \in \mathbb{C}_\kappa$, but may depend on κ. Also, let $(\cdot\,,\cdot)_{0;S^\pm}$, $(\cdot\,,\cdot)_{0;\partial S}$, and $(\cdot\,,\cdot)_{0;\partial\tilde{S}}$ be the inner products in $[L^2(S^\pm)]^m$, $[L^2(\partial S)]^m$, and $[L^2(\partial\tilde{S})]^m$ for all $m \in \mathbb{N}$, and let $\|\cdot\|_{0;S^\pm}$, $\|\cdot\|_{0;\partial S}$, and $\|\cdot\|_{0;\partial\tilde{S}}$ be the norms on the same spaces.

Consider a classical solution $\hat{U}(x,p) = (\hat{u}(x,p)^{\mathrm{T}}, \hat{u}_4(x,p))^{\mathrm{T}}$ of either of the problems ($\mathrm{TM}_p^\pm$), of class $\mathrm{C}^2(S^\pm) \cap \mathrm{C}^1(\bar{S}^\pm)$. We choose any function (with compact support in the case of S^-) $\hat{W}(x,p) = (\hat{w}(x,p)^{\mathrm{T}}, \hat{w}_4(x,p))^{\mathrm{T}}$, $\hat{W} \in \mathrm{C}_0^\infty(\bar{S}^\pm)$, such that $\hat{w}(x,p) = 0$ for $x \in \partial S_{12}$ and $\hat{w}_4(x,p) = 0$ for $x \in \partial S_{41}$, and multiply (4.7) by $\hat{W} \in [L^2(S^\pm)]^4$ to find that

$$\Upsilon_{\pm,p}(\hat{U}, \hat{W}) = (\hat{\mathcal{Q}}, \hat{W})_{0;S^\pm} \pm L(\hat{W}), \tag{4.8}$$

where

$$\begin{aligned}
\Upsilon_{\pm,p}(\hat{U}, \hat{W}) &= a_\pm(\hat{u}, \hat{w}) + (\nabla\hat{u}_4, \nabla\hat{w}_4)_{0;S^\pm} + p^2(B_0^{1/2}\hat{u}, B_0^{1/2}\hat{w})_{0;S^\pm}\\
&\quad + \varkappa^{-1}p(\hat{u}_4, \hat{w}_4)_{0;S^\pm} - h^2\gamma(\hat{u}_4, \operatorname{div}\hat{w})_{0;S^\pm} + \eta p(\operatorname{div}\hat{u}, \hat{w}_4)_{0;S^\pm},
\end{aligned}$$

$$a_\pm(\hat{u}, \hat{w}) = 2\int\limits_{S^\pm} E(\hat{u}, \hat{w})\,dx,$$

$$\begin{aligned}
2E(\hat{u}, \hat{w}) &= h^2E_0(\hat{u}, \hat{w}) + h^2\mu(\partial_2\hat{u}_1 + \partial_1\hat{u}_2)(\partial_2\bar{\hat{w}}_1 + \partial_1\bar{\hat{w}}_2)\\
&\quad + \mu\big[(\hat{u}_1 + \partial_1\hat{u}_3)(\bar{\hat{w}}_1 + \partial_1\bar{\hat{w}}_3) + (\hat{u}_2 + \partial_2\hat{u}_3)(\bar{\hat{w}}_2 + \partial_2\bar{\hat{w}}_3)\big],\\
E_0(\hat{u}, \hat{w}) &= (\lambda + 2\mu)\big[(\partial_1\hat{u}_1)(\partial_1\bar{\hat{w}}_1) + (\partial_2\hat{u}_2)(\partial_2\bar{\hat{w}}_2)\big]\\
&\quad + \lambda\big[(\partial_1\hat{u}_1)(\partial_2\bar{\hat{w}}_2) + (\partial_2\hat{u}_2)(\partial_1\bar{\hat{w}}_1)\big],\\
B_0 &= \operatorname{diag}\{\rho h^2, \rho h^2, \rho\}, \quad L(\hat{W}) = (\hat{g}_4, \hat{w}_4)_{0;\partial S_{23}} + (\hat{g}, \hat{w})_{0;\partial S_{34}}.
\end{aligned}$$

Hence, the variational problems ($\mathrm{TM}_p^\pm$) consist in finding $\hat{U} \in \mathcal{H}_{1,p}(S^\pm)$ that satisfies (4.8) for any $\hat{W} \in \mathcal{H}_{1,p}(S^\pm; \partial S_{34}, \partial S_{23})$ and

$$\pi_{12}\gamma^\pm\hat{u} = \hat{f}, \quad \pi_{41}\gamma^\pm\hat{u}_4 = \hat{f}_4.$$

Theorem 1. *For any* $\hat{q} \in \mathbf{H}_{-1,p}(S^\pm, \partial S_{34})$, $\hat{q}_4 \in H_{-1}(S^\pm, \partial S_{23})$, $\hat{f} \in \mathbf{H}_{1/2,p}(\partial S_{12})$, $\hat{f}_4 \in H_{1/2}(\partial S_{41})$, $\hat{g} \in \mathbf{H}_{-1/2,p}(\partial S_{34})$, *and* $\hat{g}_4 \in H_{-1/2}(\partial S_{23})$, $p \in \mathbb{C}_\kappa$, $\kappa > 0$, *problems* $(\mathrm{TM}_p^\pm)$ *have unique solutions* $\hat{U}(x,p) \in \mathcal{H}_{1,p}(S^\pm)$, *and these solutions satisfy*

$$\begin{aligned}|||\hat{U}|||_{1,p;S^\pm} \le c\big\{ &|p|[\hat{q}]_{-1,p;S^\pm,\partial S_{34}} + [\hat{q}_4]_{-1;S^\pm,\partial S_{23}} \\ &+ |p|\big(\|\hat{f}\|_{1/2,p;\partial S_{12}} + \|\hat{f}_4\|_{1/2,\partial S_{41}}\big) \\ &+ |p|\|\hat{g}\|_{-1/2,p;\partial S_{34}} + \|\hat{g}_4\|_{-1/2,\partial S_{23}}\big\}.\end{aligned}$$

One last set of spaces has to be introduced for $\partial\tilde{S} \subset \partial S$, $\kappa > 0$, and $k \in \mathbb{R}$.

$\mathbf{H}_{\pm 1/2}(\partial\tilde{S})$, $\mathbf{H}_1(S^\pm)$, and $\mathbf{H}_{-1}(S^\pm, \partial S_{34})$ are $\mathbf{H}_{\pm 1/2,p}(\partial\tilde{S})$, $\mathbf{H}_{1,p}(S^\pm)$, and $\mathbf{H}_{-1,p}(S^\pm, \partial S_{34})$ with $p = 0$, with norms denoted by $\|\cdot\|_{\pm 1/2;\partial\tilde{S}}$, $\|\cdot\|_{1;S^\pm}$, and $[\cdot]_{-1;S^\pm,\partial S_{34}}$;

$\mathbf{H}^{\mathcal{L}}_{\pm 1/2,k,\kappa}(\partial\tilde{S})$, $\mathbf{H}^{\mathcal{L}}_{1,k,\kappa}(S^\pm)$, and $\mathbf{H}^{\mathcal{L}}_{-1,k,\kappa}(S^\pm, \partial S_{34})$ consist of all $\hat{e}(x,p)$, $\hat{u}(x,p)$, and $\hat{q}(x,p)$ that define holomorphic mappings $\hat{e}(x,p) : \mathbb{C}_\kappa \to \mathbf{H}_{\pm 1/2}(\partial\tilde{S})$, $\hat{u}(x,p) : \mathbb{C}_\kappa \to \mathbf{H}_1(S^\pm)$, and $\hat{q}(x,p) : \mathbb{C}_\kappa \to \mathbf{H}_{-1}(S^\pm, \partial S_{34})$, and for which

$$\|\hat{e}\|^2_{\pm 1/2,k,\kappa;\partial\tilde{S}} = \sup_{\sigma>\kappa} \int_{-\infty}^{\infty} (1+|p|^2)^k \|\hat{e}(x,p)\|^2_{\pm 1/2,p;\partial\tilde{S}}\, d\tau < \infty,$$

$$\|\hat{u}\|^2_{1,k,\kappa;S^\pm} = \sup_{\sigma>\kappa} \int_{-\infty}^{\infty} (1+|p|^2)^k \|\hat{u}(x,p)\|^2_{1,p;S^\pm}\, d\tau < \infty,$$

$$[\hat{q}]^2_{-1,k,\kappa;S^\pm,\partial S_{34}} = \sup_{\sigma>\kappa} \int_{-\infty}^{\infty} (1+|p|^2)^k [\hat{q}(x,p)]^2_{-1,p;S^\pm,\partial S_{34}}\, d\tau < \infty;$$

$H^{\mathcal{L}}_{\pm 1/2,k,\kappa}(\partial\tilde{S})$, $H^{\mathcal{L}}_{1,k,\kappa}(S^\pm)$, and $H^{\mathcal{L}}_{-1,k,\kappa}(S^\pm, \partial S_{23})$ consist of all $\hat{e}_4(x,p)$, $\hat{u}_4(x,p)$, and $\hat{q}_4(x,p)$ that define holomorphic mappings $\hat{e}_4(x,p) : \mathbb{C}_\kappa \to H_{\pm 1/2}(\partial\tilde{S})$, $\hat{u}_4(x,p) : \mathbb{C}_\kappa \to H_1(S^\pm)$, and $\hat{q}_4(x,p) : \mathbb{C}_\kappa \to H_{-1}(S^\pm, \partial S_{23})$, and for which

$$\|\hat{e}_4\|^2_{\pm 1/2,k,\kappa;\partial\tilde{S}} = \sup_{\sigma>\kappa} \int_{-\infty}^{\infty} (1+|p|^2)^k \|\hat{e}_4(x,p)\|^2_{\pm 1/2,\partial\tilde{S}}\, d\tau < \infty,$$

$$\|\hat{u}_4\|^2_{1,k,\kappa;S^\pm} = \sup_{\sigma>\kappa} \int_{-\infty}^{\infty} (1+|p|^2)^k \|\hat{u}_4(x,p)\|^2_{1;S^\pm}\, d\tau < \infty,$$

$$[\hat{q}_4]^2_{-1,k,\kappa;S^\pm,\partial S_{23}} = \sup_{\sigma>\kappa} \int_{-\infty}^{\infty} (1+|p|^2)^k [\hat{q}_4(x,p)]^2_{-1;S^\pm,\partial S_{23}}\, d\tau < \infty;$$

$\mathcal{H}_1(S^\pm) = \mathbf{H}_1(S^\pm) \times H_1(S^\pm)$, with norms $|||\hat{U}|||_{1;S^\pm} = \|\hat{u}\|_{1;S^\pm} + \|\hat{u}_4\|_{1;S^\pm}$; $\mathcal{H}^{\mathcal{L}}_{1,k,l,\kappa}(S^\pm) = \mathbf{H}^{\mathcal{L}}_{1,k,\kappa}(S^\pm) \times H^{\mathcal{L}}_{1,l,\kappa}(S^\pm)$, with the norms $|||\hat{U}|||_{1,k,l,\kappa;S^\pm} = \|\hat{u}\|_{1,k,\kappa;S^\pm} + \|\hat{u}_4\|_{1,l,\kappa;S^\pm}$.

Theorem 2. *Let $\kappa > 0$ and $l \in \mathbb{R}$. If*

$$\hat{q}(x,p) \in \mathbf{H}^{\mathcal{L}}_{-1,l+1,\kappa}(S^\pm, \partial S_{34}), \quad \hat{q}_4(x,p) \in H^{\mathcal{L}}_{-1,l,\kappa}(S^\pm, \partial S_{23}),$$
$$\hat{f}(x,p) \in \mathbf{H}^{\mathcal{L}}_{1/2,l+1,\kappa}(\partial S_{12}), \quad \hat{f}_4(x,p) \in H^{\mathcal{L}}_{1/2,l+1,\kappa}(\partial S_{41}),$$
$$\hat{g}(x,p) \in \mathbf{H}^{\mathcal{L}}_{-1/2,l+1,\kappa}(\partial S_{34}), \quad \hat{g}_4(x,p) \in H^{\mathcal{L}}_{-1/2,l,\kappa}(\partial S_{23}),$$

then the (weak) solutions $\hat{U}(x,p) = (\hat{u}(x,p)^{\mathrm{T}}, \hat{u}_4(x,p))^{\mathrm{T}}$ of the problems $(\mathrm{TM}_p^\pm)$ belong to $\mathcal{H}^{\mathcal{L}}_{1,l,l,\kappa}(S^\pm)$ and

$$\begin{aligned} |||\hat{U}|||_{1,l,l,\kappa;S^\pm} \le c\big\{ & [\hat{q}]_{-1,l+1,\kappa;S^\pm,\partial S_{34}} + [\hat{q}_4]_{-1,l,\kappa;S^\pm,\partial S_{23}} \\ & + \|\hat{f}\|_{1/2,l+1,\kappa;\partial S_{12}} + \|\hat{f}_4\|_{1/2,l+1,\kappa;\partial S_{41}} \\ & + \|\hat{g}\|_{-1/2,l+1,\kappa;\partial S_{34}} + \|\hat{g}_4\|_{-1/2,l,\kappa;\partial S_{23}}\big\}. \end{aligned}$$

4.4 The Main Results

For $\kappa > 0$ and $k, l \in \mathbb{R}$, we define

$$\mathbf{H}^{\mathcal{L}^{-1}}_{1,k,\kappa}(G^\pm), \quad H^{\mathcal{L}^{-1}}_{1,l,\kappa}(G^\pm), \quad \mathcal{H}^{\mathcal{L}^{-1}}_{1,k,l,\kappa}(G^\pm) = \mathbf{H}^{\mathcal{L}^{-1}}_{1,k,\kappa}(G^\pm) \times H^{\mathcal{L}^{-1}}_{1,l,\kappa}(G^\pm),$$
$$\mathbf{H}^{\mathcal{L}^{-1}}_{-1,l,\kappa}(G^\pm, \Gamma_{34}), \quad H^{\mathcal{L}^{-1}}_{-1,l,\kappa}(G^\pm, \Gamma_{23}), \quad \mathbf{H}^{\mathcal{L}^{-1}}_{1/2,l,\kappa}(\Gamma_{12}),$$
$$H^{\mathcal{L}^{-1}}_{1/2,l,\kappa}(\Gamma_{41}), \quad \mathbf{H}^{\mathcal{L}^{-1}}_{-1/2,l,\kappa}(\Gamma_{34}), \quad H^{\mathcal{L}^{-1}}_{-1/2,l,\kappa}(\Gamma_{23})$$

to be the spaces consisting of the inverse Laplace transforms of the elements of

$$\mathbf{H}^{\mathcal{L}}_{1,k,\kappa}(S^\pm), \quad H^{\mathcal{L}}_{1,l,\kappa}(S^\pm), \quad \mathcal{H}^{\mathcal{L}}_{1,k,l,\kappa}(S^\pm) = \mathbf{H}^{\mathcal{L}}_{1,k,\kappa}(S^\pm) \times H^{\mathcal{L}}_{1,l,\kappa}(S^\pm),$$
$$\mathbf{H}^{\mathcal{L}}_{-1,l,\kappa}(S^\pm, \partial S_{34}), \quad H^{\mathcal{L}}_{-1,l,\kappa}(S^\pm, \partial S_{23}), \quad \mathbf{H}^{\mathcal{L}}_{1/2,l,\kappa}(\partial S_{12}),$$
$$H^{\mathcal{L}}_{1/2,l,\kappa}(\partial S_{41}), \quad \mathbf{H}^{\mathcal{L}}_{-1/2,l,\kappa}(\partial S_{34}), \quad H^{\mathcal{L}}_{-1/2,l,\kappa}(\partial S_{23}),$$

respectively, with norms

$$\|u\|_{1,k,\kappa;G^\pm} = \|\hat{u}\|_{1,k,\kappa;S^\pm}, \quad \|u_4\|_{1,l,\kappa;G^\pm} = \|\hat{u}_4\|_{1,l,\kappa;S^\pm},$$
$$|||U|||_{1,k,l,\kappa;G^\pm} = |||\hat{U}|||_{1,k,l,\kappa;S^\pm},$$
$$[q]_{-1,l,\kappa;G^\pm,\Gamma_{34}} = [\hat{q}]_{1,l,\kappa;S^\pm,\partial S_{34}}, \quad [q_4]_{-1,l,\kappa;G^\pm,\Gamma_{23}} = [\hat{q}_4]_{1,l,\kappa;S^\pm,\partial S_{23}},$$
$$\|f\|_{1/2,l,\kappa;\Gamma_{12}} = \|\hat{f}\|_{1/2,l,\kappa;\partial S_{12}}, \quad \|f_4\|_{1/2,l,\kappa;\Gamma_{41}} = \|\hat{f}_4\|_{1/2,l,\kappa;\partial S_{41}},$$
$$\|g\|_{-1/2,l,\kappa;\Gamma_{34}} = \|\hat{g}\|_{-1/2,l,\kappa;\partial S_{34}}, \quad \|g_4\|_{-1/2,l,\kappa;\Gamma_{23}} = \|\hat{g}_4\|_{-1/2,l,\kappa;\partial S_{23}}.$$

We extend the notation $\gamma^{\pm}$ to the trace operators from $G^{\pm}$ to Γ, and π_{ij} to the operators of restriction from Γ to its parts Γ_{ij}, $i, j = 1, 2, 3, 4$.

$U \in \mathcal{H}^{\mathcal{L}^{-1}}_{1,0,0,\kappa}(G^{\pm})$, $U(x,t) = (u(x,t)^{\mathrm{T}}, u_4(x,t))^{\mathrm{T}}$, is called a weak solution of ($\mathrm{TM}^{\pm}$) if

(i) $\gamma_0 u = 0$, where γ_0 is the trace operator on $S^{\pm} \times \{t = 0\}$;
(ii) $\pi_{12}\gamma^{\pm} u = f(x,t)$ and $\pi_{41}\gamma^{\pm} u_4 = f_4(x,t)$;
(iii) U satisfies

$$\Upsilon_{\pm}(U, W) = \int_0^{\infty} (\mathcal{Q}, W)_{0;S^{\pm}}\, dt \pm L(W),$$

where

$$\begin{aligned}\Upsilon_{\pm}(U, W) = \int_0^{\infty} \big\{ & a_{\pm}(u, w) + (\nabla u_4, \nabla w_4)_{0;S^{\pm}} \\ & - (B_0^{1/2}\partial_t u, B_0^{1/2}\partial_t w)_{0;S^{\pm}} - \varkappa^{-1}(u_4, \partial_t w_4)_{0;S^{\pm}} \\ & - h^2\gamma(u_4, \operatorname{div} w)_{0;S^{\pm}} - \eta(\operatorname{div} u, \partial_t w_4)_{0;S^{\pm}} \big\}\, dt,\end{aligned}$$

$$L(W) = \int_0^{\infty} \big\{ (g, w)_{0;\partial S_{34}} + (g_4, w_4)_{0;\partial S_{23}} \big\}\, dt,$$

for all $W \in \mathrm{C}_0^{\infty}(\bar{G}^{\pm})$, $W(x,t) = (w(x,t)^{\mathrm{T}}, w_4(x,t))^{\mathrm{T}}$, such that $w(x,t) = 0$ for $(x,t) \in \Gamma_{12}$ and $w_4(x,t) = 0$ for $(x,t) \in \Gamma_{41}$.

Theorem 3. *Let $U(x,t) = \mathcal{L}^{-1}\hat{U}(x,p)$ be the inverse Laplace transform of the weak solution $\hat{U}(x,p)$ of either of the problems ($\mathrm{TM}_p^{\pm}$). If*

$$\begin{aligned} q(x,t) &\in \mathbf{H}^{\mathcal{L}^{-1}}_{-1,l+1,\kappa}(G^{\pm}, \Gamma_{34}), \quad q_4(x,t) \in H^{\mathcal{L}^{-1}}_{-1,l,\kappa}(G^{\pm}, \Gamma_{23}), \\ f(x,t) &\in \mathbf{H}^{\mathcal{L}^{-1}}_{1/2,l+1,\kappa}(\Gamma_{12}), \quad f_4(x,t) \in H^{\mathcal{L}^{-1}}_{1/2,l+1,\kappa}(\Gamma_{41}), \\ g(x,t) &\in \mathbf{H}^{\mathcal{L}^{-1}}_{-1/2,l+1,\kappa}(\Gamma_{34}), \quad g_4(x,t) \in H^{\mathcal{L}^{-1}}_{-1/2,l,\kappa}(\Gamma_{23}), \end{aligned}$$

where $\kappa > 0$ and $l \in \mathbb{R}$, then $U \in \mathcal{H}^{\mathcal{L}^{-1}}_{1,l,l,\kappa}(G^{\pm})$ and

$$\begin{aligned} |||U|||_{1,l,l,\kappa;G^{\pm}} \le c\big\{ & [q]_{-1,l+1,\kappa;G^{\pm},\Gamma_{34}} + [q_4]_{-1,l,\kappa;G^{\pm},\Gamma_{23}} \\ & + \|f\|_{1/2,l+1,\kappa;\Gamma_{12}} + \|f_4\|_{1/2,l+1,\kappa;\Gamma_{41}} \\ & + \|g\|_{-1/2,l+1,\kappa;\Gamma_{34}} + \|g_4\|_{-1/2,l,\kappa;\Gamma_{23}} \big\}. \end{aligned}$$

If, in addition, $l \ge 0$, then U is a weak solution of the corresponding problem ($\mathrm{TM}^{\pm}$).

Theorem 4. *Each of the problems* ($\mathrm{TM}^{\pm}$) *has at most one weak solution.*

Full details of the proofs of these assertions will appear in a future publication.

References

1. C. Constanda, *A Mathematical Analysis of Bending of Plates with Transverse Shear Deformation,* Longman/Wiley, Harlow-New York, 1990.

2. P. Schiavone and R.J. Tait, Thermal effects in Mindlin-type plates, *Quart. J. Mech. Appl. Math.* **46** (1993), 27–39.

3. I. Chudinovich and C. Constanda, The Cauchy problem in the theory of plates with transverse shear deformation, *Math. Models Methods Appl. Sci.* **10** (2000), 463–477.

4. I. Chudinovich and C. Constanda, Nonstationary integral equations for elastic plates, *C.R. Acad. Sci. Paris Sér. I* **329** (1999), 1115–1120.

5. I. Chudinovich and C. Constanda, Boundary integral equations in dynamic problems for elastic plates, *J. Elasticity* **68** (2002), 73–94.

6. I. Chudinovich and C. Constanda, Time-dependent boundary integral equations for multiply connected plates, *IMA J. Appl. Math.* **68** (2003), 507–522.

7. I. Chudinovich and C. Constanda, *Variational and Potential Methods for a Class of Linear Hyperbolic Evolutionary Processes,* Springer-Verlag, London, 2005.

8. I. Chudinovich, C. Constanda, and J. Colín Venegas, The Cauchy problem in the theory of thermoelastic plates with transverse shear deformation, *J. Integral Equations Appl.* **16** (2004), 321–342.

5 On the Structure of the Eigenfunctions of a Vibrating Plate with a Concentrated Mass and Very Small Thickness

Delfina Gómez, Miguel Lobo, and Eugenia Pérez

5.1 Introduction and Statement of the Problem

Let Ω and B be two bounded domains of $\mathbb{R}^2$ with smooth boundaries which we denote by $\partial\Omega$ and Γ, respectively. For simplicity, we consider that both Ω and B contain the origin. Let ε be a positive parameter that tends to zero. Let us consider εB and $\varepsilon\Gamma$, the homothetics of B and Γ, respectively, with ratio ε. We assume that $\varepsilon\bar{B}$ is contained in Ω.

We consider the vibrations of a homogeneous and isotropic plate occupying the domain $\bar{\Omega} \times [-h_0/2, h_0/2]$ of $\mathbb{R}^3$ that contains a small region of high density, the so-called *concentrated mass*. The size of this small region, $\varepsilon\bar{B} \times [-h_0/2, h_0/2]$, depends on the parameter ε and the density is of order $O(\varepsilon^{-m})$ in this part and $O(1)$ outside; m is a positive parameter. Let h_0 be the plate thickness, $0 < h_0 \ll \operatorname{diam}\Omega$.

We consider the associated spectral problem in the framework of the Reissner–Mindlin plate model

$$\begin{gathered}
M^\varepsilon_{\alpha\beta,\beta} - M^\varepsilon_{3\alpha} = h^2\zeta^\varepsilon u^\varepsilon_\alpha, \quad M^\varepsilon_{3\beta,\beta} = \zeta^\varepsilon u^\varepsilon_3 \quad \text{in } \Omega - \varepsilon\bar{B}, \ \alpha = 1,2, \\
M^\varepsilon_{\alpha\beta,\beta} - M^\varepsilon_{3\alpha} = h^2\zeta^\varepsilon\varepsilon^{-m} u^\varepsilon_\alpha, \quad M^\varepsilon_{3\beta,\beta} = \zeta^\varepsilon\varepsilon^{-m} u^\varepsilon_3 \quad \text{in } \varepsilon B, \ \alpha = 1,2, \\
[u^\varepsilon_i] = \left[M^\varepsilon_{i\beta} n_\beta\right] = 0 \quad \text{on } \varepsilon\Gamma, \ i = 1,2,3, \\
u^\varepsilon_i = 0 \quad \text{on } \partial\Omega, \ i = 1,2,3.
\end{gathered} \tag{5.1}$$

Here, $u^\varepsilon = (u^\varepsilon_1, u^\varepsilon_2, u^\varepsilon_3)^t$, where $(x_3 u^\varepsilon_1(x), x_3 u^\varepsilon_2(x), u^\varepsilon_3(x))^t$ is the *displacement vector*, $x = (x_1, x_2) \in \Omega$, u^ε_1 and u^ε_2 are the component angles, M^ε is the moments matrix of elements

$$M^\varepsilon_{\alpha\beta} = h^2(\lambda u^\varepsilon_{\gamma,\gamma}\delta_{\alpha\beta} + \mu(u^\varepsilon_{\alpha,\beta} + u^\varepsilon_{\beta,\alpha})),$$

M^ε_3 is the shear force vector of components

$$M^\varepsilon_{3\alpha} = \mu(u^\varepsilon_\alpha + u^\varepsilon_{3,\alpha}),$$

This work has been partially supported by DGES, BFM2001–1266.

λ and μ are the Lamé constants of the material, $n = (n_1(x), n_2(x))$, $x \in \varepsilon\Gamma$, is the unit outward normal to $\varepsilon\Gamma$, $\delta_{\alpha\beta}$ is the Kronecker symbol, $v_{i,\alpha} = \partial v_i / \partial x_\alpha$, $h^2 = h_0^2/12$, and the brackets denote the jump across $\varepsilon\Gamma$ of the enclosed quantities. Here and in what follows, Greek and Latin subscripts take the values 1, 2, and 1, 2, 3, respectively, and the convention of summation over repeated indices is understood. See, for example, [1]–[3] for more details on the Reissner–Mindlin plate model.

For each fixed $\varepsilon > 0$, problem (5.1) is a standard eigenvalue problem for a positive, self-adjoint anticompact operator. Let us consider

$$0 < \zeta_1^\varepsilon \leq \zeta_2^\varepsilon \leq \cdots \leq \zeta_n^\varepsilon \leq \cdots \xrightarrow{n\to\infty} \infty,$$

the sequence of eigenvalues, with the classical convention of repeated eigenvalues. Let $\{u^{\varepsilon,n}\}_{n=1}^{\infty}$ be the corresponding eigenfunctions, which are assumed to be an orthonormal basis in $(H_0^1(\Omega))^3$.

In this paper, we address the asymptotic behavior of the eigenvalues and eigenfunctions of the spectral problem (5.1) when the parameters ε and h tend to zero. Let us refer to [2] and [4] for a vibrating plate, without concentrated masses, whose thickness h tends to zero, and to [5] and [6] for a vibrating plate with a concentrated mass. Also, see [7] for the study of the vibrations of a plate with a concentrated mass in the framework of the Kirchhoff–Love plate model. We mention [8]–[10] for general references on vibrating systems with concentrated masses. Assuming that h depends on ε, out of all the possible relations between ε and h, here we consider the more realistic relation where $h = \varepsilon^r$ with $r \geq 1$.

Depending on the values of the parameters m (related to the density) and r (related to the thickness), there is a different asymptotic behavior of the eigenelements $(\zeta_n^\varepsilon, u^{\varepsilon,n})$ when $\varepsilon \to 0$. Here, we consider the case where $m > 4$ and $r \geq 1$.

By using the minimax principle, we obtain (see [6] for the proof):

$$C\varepsilon^{m+2r-2} |\log \varepsilon|^{-1} \leq \zeta_n^\varepsilon \leq C_n \varepsilon^{m-4+2r}, \quad \text{for each fixed } n = 1, 2, \ldots,$$

where C and C_n are constants independent of ε, and $C_n \to \infty$ as $n \to \infty$. The eigenvalues ζ_n^ε of order $O(\varepsilon^{m-4+2r})$ are the so-called *low frequencies*, and, in [6] we prove that, depending on whether r is $r = 1$ or $r > 1$, there are different limit problems characterizing the limit values of $\zeta_n^\varepsilon / \varepsilon^{m-4+2r}$ and the corresponding eigenfunctions.

As a matter of fact, in the case where $r = 1$, the values $\zeta_n^\varepsilon / \varepsilon^{m-2}$ are approached by the eigenvalues of a spectral problem for the Reissner–Mindlin plate model in an unbounded domain (see (5.8)). Instead, when $r > 1$, a Kirchhoff–Love plate model in an unbounded domain describes the asymptotic behavior of $\zeta_n^\varepsilon / \varepsilon^{m-4+2r}$ (see (5.38)). In order to make clear this behavior, and for the sake of completeness, in Theorem 1 of Section 5.2 and Theorem 2 of Section 5.3 we summarize the results obtained in [6], for $r = 1$ and $r > 1$, respectively, on the convergence of the eigenvalues $\zeta_n^\varepsilon / \varepsilon^{m-4+2r}$ and the corresponding eigenfunctions.

The aim of this paper is to provide information on the structure of the associated eigenfunctions, which is not clear from the above-mentioned results. We use asymptotic expansions and matching principles to obtain the composite expansion of the eigenfunctions (see [10] and [11] for the technique). In particular, we show that in both cases, $r=1$ and $r>1$, the first two components of the associated vibrations for the low frequencies always have a local character, that is, the first two components of the displacement vector, $x_3 u_1^{\varepsilon,n}(x)$ and $x_3 u_2^{\varepsilon,n}(x)$, are only significant in a region near the concentrated mass, i.e., for $|x| = O(\varepsilon)$, while they are very small at the distance $O(1)$ of the concentrated mass (see formulas (5.28) and (5.32), (5.33) for $r=1$ and (5.46) for $r>1$). However, this local character does not always hold for the last component of the displacement $u_3^{\varepsilon,n}(x)$ (see formulas (5.30) and (5.34), (5.35) for $r=1$ and (5.47) for $r>1$). We address the case $r=1$ in Section 5.2 and the case $r>1$ in Section 5.3. Below we introduce some notation, which will be used throughout the paper.

First, we observe that problem (5.1) can be written in the form:

$$\begin{gathered} A_{\alpha j} u_j^\varepsilon = \zeta^\varepsilon h^2 u_\alpha^\varepsilon, \quad A_{3j} u_j^\varepsilon = \zeta^\varepsilon u_3^\varepsilon \quad \text{in } \Omega - \varepsilon \bar{B}, \ \alpha = 1,2, \\ A_{\alpha j} u_j^\varepsilon = \zeta^\varepsilon \varepsilon^{-m} h^2 u_\alpha^\varepsilon, \quad A_{3j} u_j^\varepsilon = \zeta^\varepsilon \varepsilon^{-m} u_3^\varepsilon \quad \text{in } \varepsilon B, \ \alpha = 1,2, \\ [u_i^\varepsilon] = \left[N_{ij} u_j^\varepsilon\right] = 0 \quad \text{on } \varepsilon\Gamma, \ i = 1,2,3, \\ u_i^\varepsilon = 0 \quad \text{on } \partial\Omega, \ i = 1,2,3, \end{gathered} \tag{5.2}$$

where $A = (A_{ij})_{i,j=1,2,3}$ is the differential operator on Ω defined by

$$A = \begin{pmatrix} -h^2\mu\Delta - h^2(\lambda+\mu)\partial_1^2 + \mu & -h^2(\lambda+\mu)\partial_1\partial_2 & \mu\partial_1 \\ -h^2(\lambda+\mu)\partial_1\partial_2 & -h^2\mu\Delta - h^2(\lambda+\mu)\partial_2^2 + \mu & \mu\partial_2 \\ -\mu\partial_1 & -\mu\partial_2 & -\mu\Delta \end{pmatrix} \tag{5.3}$$

and $N = (N_{ij})_{i,j=1,2,3}$ is the boundary operator on $\varepsilon\Gamma$ defined by

$$N = \begin{pmatrix} h^2(\lambda+2\mu)n_1\partial_1 + h^2\mu n_2\partial_2 & h^2\mu n_2\partial_1 + h^2\lambda n_1\partial_2 & 0 \\ h^2\lambda n_2\partial_1 + h^2\mu n_1\partial_2 & h^2\mu n_1\partial_1 + h^2(\lambda+2\mu)n_2\partial_2 & 0 \\ \mu n_1 & \mu n_2 & \mu(n_1\partial_1 + n_2\partial_2) \end{pmatrix}. \tag{5.4}$$

Then, considering the spectral parameter $\eta = \zeta\varepsilon^{4-2r-m}$ and the *local variable* $y = x/\varepsilon$, we define the new functions $(V_1^\varepsilon, V_2^\varepsilon, V_3^\varepsilon) = (U_1^\varepsilon, U_2^\varepsilon, U_3^\varepsilon/\varepsilon)$ where $U^\varepsilon(y)$ are the eigenfunctions of (5.2) written in the local variable y. Thus, problem (5.2) for $h = \varepsilon^r$ reads

$$\begin{gathered} \hat{A}_{\alpha j}^\varepsilon V_j^\varepsilon = \eta^\varepsilon \varepsilon^{m+2r-2} V_\alpha^\varepsilon, \quad \hat{A}_{3j}^\varepsilon V_j^\varepsilon = \eta^\varepsilon \varepsilon^m V_3^\varepsilon \quad \text{in } \varepsilon^{-1}\Omega - \bar{B}, \ \alpha = 1,2, \\ \hat{A}_{\alpha j}^\varepsilon V_j^\varepsilon = \eta^\varepsilon \varepsilon^{2r-2} V_\alpha^\varepsilon, \quad \hat{A}_{3j}^\varepsilon V_j^\varepsilon = \eta^\varepsilon V_3^\varepsilon \quad \text{in } B, \ \alpha = 1,2, \\ [V_i^\varepsilon] = \left[\hat{N}_{ij}^\varepsilon V_j^\varepsilon\right] = 0 \quad \text{on } \Gamma, \ i = 1,2,3, \\ V_i^\varepsilon = 0 \quad \text{on } \partial(\varepsilon^{-1}\Omega), \ i = 1,2,3, \end{gathered} \tag{5.5}$$

where $\hat{A}^\varepsilon = (\hat{A}^\varepsilon_{ij})_{i,j=1,2,3}$ is the operator

$$\hat{A}^\varepsilon = \begin{pmatrix} -\mu\Delta_y - (\lambda+\mu)\partial^2_{y1} + \varepsilon^{2-2r}\mu & -(\lambda+\mu)\partial_{y1}\partial_{y2} & \varepsilon^{2-2r}\mu\partial_{y1} \\ -(\lambda+\mu)\partial_{y1}\partial_{y2} & -\mu\Delta_y - (\lambda+\mu)\partial^2_{y2} + \varepsilon^{2-2r}\mu & \varepsilon^{2-2r}\mu\partial_{y2} \\ -\varepsilon^{2-2r}\mu\partial_{y1} & -\varepsilon^{2-2r}\mu\partial_{y2} & -\varepsilon^{2-2r}\mu\Delta_y \end{pmatrix} \tag{5.6}$$

and $\hat{N}^\varepsilon = (\hat{N}^\varepsilon_{ij})_{i,j=1,2,3}$ is the operator

$$\hat{N}^\varepsilon = \begin{pmatrix} (\lambda+2\mu)n_1\partial_{y1} + \mu n_2\partial_{y2} & \mu n_2\partial_{y1} + \lambda n_1\partial_{y2} & 0 \\ \lambda n_2\partial_{y1} + \mu n_1\partial_{y2} & \mu n_1\partial_{y1} + (\lambda+2\mu)n_2\partial_{y2} & 0 \\ \varepsilon^{2-2r}\mu n_1 & \varepsilon^{2-2r}\mu n_2 & \varepsilon^{2-2r}\mu(n_1\partial_{y1} + n_2\partial_{y2}) \end{pmatrix}; \tag{5.7}$$

$n = (n_1(y), n_2(y))$ $(y \in \Gamma)$ is the unit outward normal to Γ, and $\partial_{y\alpha}$ denotes $\partial/\partial y_\alpha$.

5.2 Asymptotics in the Case $r = 1$

If we formally pass to the limit in (5.5) when $\varepsilon \to 0$, we obtain the following spectral problem:

$$\begin{aligned} \hat{A}_{ij}V_j &= 0 \quad \text{in } \mathbb{R}^2 - \bar{B},\ \ i = 1,2,3, \\ \hat{A}_{ij}V_j &= \eta V_i \quad \text{in } B,\ \ i = 1,2,3, \\ [V_i] &= \left[\hat{N}_{ij}V_j\right] = 0 \quad \text{on } \Gamma,\ \ i = 1,2,3, \\ V_\alpha &\to c_\alpha \quad \text{as } |y| \to \infty,\ \ \alpha = 1,2, \\ V_3 &= -c_\alpha y_\alpha + c_3 + O(\log|y|) \quad \text{as } |y| \to \infty; \end{aligned} \tag{5.8}$$

where $\hat{A} = (\hat{A}_{ij})_{i,j=1,2,3}$ and $\hat{N} = (\hat{N}_{ij})_{i,j=1,2,3}$ are the operators defined by (5.6) and (5.7), respectively, for $r = 1$, and c_i, $i = 1,2,3$, are some unknown but well-determined constants.

We observe that, once the value of the parameter h has been set at 1 (i.e., $h \equiv 1$ in (5.3) and (5.4)), $\hat{A}$ and $\hat{N}$ are the operators defined by (5.3) and (5.4), respectively, associated with a Reissner–Mindlin plate model in the whole space $\mathbb{R}^2$ for $h = 1$ where the condition at the infinity that we consider ensures that (5.8) is a well-posed problem and has a discrete spectrum (see [1] and [6]). Let us denote by

$$0 = \eta_1 \le \eta_2 \le \eta_3 \le \cdots \le \eta_n \le \cdots \xrightarrow{n\to\infty} \infty$$

the sequence of eigenvalues of (5.8) with the classical convention of repeated eigenvalues.

The following theorem states the convergence of the eigenvalues of (5.2) and their corresponding eigenfunctions written in the local variable $y = x/\varepsilon$ to the limit problem (5.8).

Theorem 1. *Let ζ_n^ε be the eigenvalues of problem (5.2) for $h = \varepsilon$. If $m > 4$, then for each n, the values $\zeta_n^\varepsilon/\varepsilon^{m-2}$ converge as $\varepsilon \to 0$ to the eigenvalue η_n of (5.8). Furthermore, for any eigenfunction V of (5.8) associated with η_k, verifying $\|V\|_{(L^2(B))^3} = 1$, there exists U^ε, U^ε being a linear combination of eigenfunctions of (5.2) associated with the eigenvalues converging towards η_k, such that U_α^ε and $U_3^\varepsilon/\varepsilon$ converge in $L^2(B)$ towards V_α and V_3, respectively.*

In addition, if U^ε is an eigenfunction of (5.2) associated with eigenvalues ζ^ε such that $\zeta^\varepsilon/\varepsilon^{m-2}$ converge as $\varepsilon \to 0$ to η^*, satisfying

$$\varepsilon^m \|U_\alpha^\varepsilon\|^2_{L^2(\varepsilon^{-1}\Omega-\bar{B})} + \varepsilon^{m-2}\|U_3^\varepsilon\|^2_{L^2(\varepsilon^{-1}\Omega-\bar{B})} + \|U_\alpha^\varepsilon\|^2_{L^2(B)} + \varepsilon^{-2}\|U_3^\varepsilon\|^2_{L^2(B)} = 1,$$

and U_α^ε and $U_3^\varepsilon/\varepsilon$ converge weakly in $L^2(B)$ to V_α^* and V_3^*, respectively, with $V^* \neq 0$, then U_α^ε and $U_3^\varepsilon/\varepsilon$ converge in $L^2(B)$ to V_α^* and V_3^*, respectively, where V^* is an eigenfunction of (5.8) associated with η^*.

We refer to [6] for the proof of Theorem 1, which is based on certain known results of the Spectral Perturbation Theory (see [9] and [10]).

Next, using techniques of asymptotic expansions and matching principles, we study the structure of the eigenfunctions of problem (5.2) for $h = \varepsilon$.

We postulate an asymptotic expansion for the eigenvalues ζ^ε,

$$\zeta^\varepsilon = \eta\varepsilon^{m-2} + o(\varepsilon^{m-2}), \tag{5.9}$$

and, for the corresponding eigenfunctions u^ε, we postulate an outer expansion in $\Omega - \{0\}$

$$u^\varepsilon(x) = \alpha_0(\varepsilon)\left[u^0(x) + \varepsilon u^1(x) + \varepsilon^2 u^2(x) + o(\varepsilon^2)\right] \tag{5.10}$$

and a local expansion in a neighborhood of $x = 0$

$$U_\alpha^\varepsilon(y) = V_\alpha(y) + o(1) \quad \alpha = 1, 2, \tag{5.11}$$

$$U_3^\varepsilon(y) = \varepsilon V_3(y) + o(\varepsilon), \tag{5.12}$$

where y is the local variable $y = x/\varepsilon$.

Since $m > 4$ and $r = 1$, replacing expansions (5.9), (5.11), and (5.12) in the spectral problem (5.2) written in the local variable $y = x/\varepsilon$, that is, problem (5.5) where $\eta^\varepsilon = \zeta^\varepsilon\varepsilon^{2-m}$ and $(V_1^\varepsilon, V_2^\varepsilon, V_3^\varepsilon) = (U_1^\varepsilon, U_2^\varepsilon, U_3^\varepsilon/\varepsilon)$, we obtain that (η, V) satisfies equations $(5.8)_1$-$(5.8)_3$ and, consequently, we set that it is an eigenelement of the so-called *local problem* (5.8).

On the other hand, replacing expansions (5.9) and (5.10) in problem (5.2) for $h = \varepsilon$, we collect coefficients of the same powers of ε and gather equations satisfied by u^j. At a first step, we have that u^0 verifies

$$u_\alpha^0 + u_{3,\alpha}^0 = 0 \quad \text{in } \Omega - \{0\}, \quad \alpha = 1, 2, \tag{5.13}$$

$$u_{\alpha,\alpha}^0 + \Delta u_3^0 = 0 \quad \text{in } \Omega - \{0\}, \tag{5.14}$$

$$u_i^0 = 0 \quad \text{on } \partial\Omega, \quad i = 1, 2, 3. \tag{5.15}$$

Let us note that for any u_3^0 smooth function in $\Omega - \{0\}$ such that $u_3^0 = \partial u_3^0/\partial n = 0$ on $\partial\Omega$, the function defined by $u^0 = (-u_{3,1}^0, -u_{3,2}^0, u_3^0)^t$ satisfies (5.13)–(5.15) and we need to compute higher-order terms of the asymptotic expansion (5.10) in order to obtain further restrictions for u_3^0.

In a second step, we obtain the same equations (5.13)–(5.15) for u^1. Following the process, we have

$$\tilde{A}_{\alpha\beta}u_\beta^0 + \mu u_\alpha^2 + \mu u_{3,\alpha}^2 = 0 \quad \text{in } \Omega - \{0\}, \quad \alpha = 1,2, \tag{5.16}$$

$$\mu u_{\alpha,\alpha}^2 + \mu\Delta u_3^2 = 0 \quad \text{in } \Omega - \{0\}, \tag{5.17}$$

$$u_i^2 = 0 \quad \text{on } \partial\Omega, \quad i = 1,2,3,$$

where $\tilde{A} = (\tilde{A}_{\alpha\beta})_{\alpha,\beta=1,2}$ is the two-dimensional elasticity operator

$$\tilde{A} = \begin{pmatrix} -\mu\Delta - (\lambda+\mu)\partial_1^2 & -(\lambda+\mu)\partial_1\partial_2 \\ -(\lambda+\mu)\partial_1\partial_2 & -\mu\Delta - (\lambda+\mu)\partial_2^2 \end{pmatrix}. \tag{5.18}$$

Then, combining (5.16) and (5.17) yields

$$\partial_\alpha(\tilde{A}_{\alpha\beta}u_\beta^0) = 0 \quad \text{in } \Omega - \{0\}, \tag{5.19}$$

and, on account of (5.13) and (5.15), we deduce that u_3^0 is a solution of

$$\begin{aligned} &\Delta^2 u_3^0 = 0 \quad \text{in } \Omega - \{0\}, \\ &u_3^0 = \frac{\partial u_3^0}{\partial n} = 0 \quad \text{on } \partial\Omega. \end{aligned} \tag{5.20}$$

Thus, once we determine a function u_3^0 verifying (5.20), and the order function $\alpha_0(\varepsilon)$, we obtain u_α^0 for $\alpha = 1,2$ by (5.13) and the leading term in the outer expansion (5.10) is determined. Let us determine u_3^0 and $\alpha_0(\varepsilon)$ such that the matching for the expansions (5.10)–(5.12) holds.

First, we observe that u_3^0 satisfies equation (5.20) in Ω except at the origin, and we can think of solutions of the biharmonic operator which are singular at the origin. That is, we can look for the type of singularity of u_3^0 at the origin among the fundamental solution of the biharmonic operator and its derivatives. In this way, we look for $u_3^0(x)$ in the form

$$u_3^0(x) = a_1\frac{\partial}{\partial x_1}(|x|^2\log|x|) + a_2\frac{\partial}{\partial x_2}(|x|^2\log|x|) + F(x), \tag{5.21}$$

where a_1, a_2 are some constants to be determined by matching, and $F(x)$ is a regular function in Ω; namely, F is the solution of the problem

$$\begin{aligned} &\Delta^2 F = 0 \quad \text{in } \Omega, \\ &F = u^* \quad \text{on } \partial\Omega, \\ &\frac{\partial F}{\partial n} = \frac{\partial u^*}{\partial n} \quad \text{on } \partial\Omega, \end{aligned} \tag{5.22}$$

where the function u^* is defined in a neighborhood of the boundary $\partial\Omega$ by

$$\begin{aligned} u^*(x_1, x_2) &= -a_1 \frac{\partial}{\partial x_1}(|x|^2 \log |x|) - a_2 \frac{\partial}{\partial x_2}(|x|^2 \log |x|) \\ &= -(a_1 x_1 + a_2 x_2)(2 \log \sqrt{x_1^2 + x_2^2} + 1). \end{aligned}$$

Second, (5.21) and (5.13) lead us to the formulas

$$u_3^0(x_1, x_2) = (a_1 x_1 + a_2 x_2)(2 \log \sqrt{x_1^2 + x_2^2} + 1) + F(x_1, x_2), \quad (5.23)$$

$$\begin{aligned} u_1^0(x_1, x_2) &= -u_{3,1}^0(x_1, x_2) \\ &= -a_1(2 \log \sqrt{x_1^2 + x_2^2} + 1) - \frac{2(a_1 x_1 + a_2 x_2) x_1}{x_1^2 + x_2^2} \\ &\qquad - \frac{\partial F}{\partial x_1}(x_1, x_2), \end{aligned} \quad (5.24)$$

and

$$\begin{aligned} u_2^0(x_1, x_2) &= -u_{3,2}^0(x_1, x_2) \\ &= -a_2(2 \log \sqrt{x_1^2 + x_2^2} + 1) - \frac{2(a_1 x_1 + a_2 x_2) x_2}{x_1^2 + x_2^2} \\ &\qquad - \frac{\partial F}{\partial x_2}(x_1, x_2). \end{aligned} \quad (5.25)$$

Therefore, the constants a_1 and a_2 determine u_i^0, for $i = 1, 2, 3$.

Finally, the matching principle for the first two components of u^ε

$$\lim_{\substack{\varepsilon \to 0 \\ \text{fixed } y}} \alpha_0(\varepsilon) u_\alpha^0(\varepsilon y) = \lim_{\substack{\varepsilon \to 0 \\ \text{fixed } x}} V_\alpha\left(\frac{x}{\varepsilon}\right), \quad \alpha = 1, 2, \quad (5.26)$$

gives us α_0, a_1, and a_2. Indeed, taking into account formulas (5.24) and (5.25) for u_α^0, and the behavior at the infinity of V_α where V is an eigenfunction of (5.8) associated with η, we deduce that (5.26) is satisfied for

$$\alpha_0(\varepsilon) = \frac{1}{|\log \varepsilon|}, \quad c_1 = 2a_1, \quad \text{and} \quad c_2 = 2a_2. \quad (5.27)$$

Then u_3^0 and $\alpha_0(\varepsilon)$ in (5.23) and (5.27), respectively, are well-determined functions and, consequently, the leading terms in the expansions (5.10)–(5.12) are also determined.

By the construction of u_3^0, the matching condition (5.26) for the first two components of the asymptotic expansions of u^ε holds and, hence, the composite expansion in Ω for these components is

$$u_\alpha^\varepsilon(x) \sim \frac{1}{|\log \varepsilon|} u_\alpha^0(x) + V_\alpha\left(\frac{x}{\varepsilon}\right) - c_\alpha, \quad \alpha = 1, 2, \quad (5.28)$$

where (V_1, V_2, V_3) is an eigenfunction of (5.8) associated with η, and u_α^0 for $\alpha = 1, 2$ are defined by (5.24) and (5.25), respectively, with $a_1 = c_1/2$, $a_2 = c_2/2$, and F the solution of (5.22). Obviously, the convention that $\log|x|$ is replaced by $\log\varepsilon$ when $|x| \leq \varepsilon$ should be understood in (5.28).

In order to obtain the composite expansion for the third component of the displacements u^ε, we recall the matching principle in the intermediate variable, namely,

$$\lim_{\substack{\varepsilon\to 0\\ \text{fixed } \xi\neq 0}} \alpha_0(\varepsilon)u_3^0(\xi\varepsilon|\log\varepsilon|) = \lim_{\substack{\varepsilon\to 0\\ \text{fixed } \xi\neq 0}} \varepsilon V_3(\xi|\log\varepsilon|), \tag{5.29}$$

where

$$\xi = \frac{x}{\varepsilon|\log\varepsilon|}$$

is an intermediate variable between x and y, with $y = x/\varepsilon$ (see Section 3.3 in [11]). Taking into account (5.27), (5.23), and the behavior at infinity of V_3, where V_3 is the third component of an eigenfunction V of (5.8) associated with η, condition (5.29) is also satisfied, and the composite expansion in Ω for u_3^ε is:

$$u_3^\varepsilon(x) \sim \varepsilon V_3\left(\frac{x}{\varepsilon}\right) + \chi\left(\frac{x}{\varepsilon|\log\varepsilon|}\right)\left[\frac{1}{|\log\varepsilon|}u_3^0(x) - \varepsilon V_3\left(\frac{x}{\varepsilon}\right)\right], \tag{5.30}$$

where $\chi(\xi)$ is a smooth function satisfying

$$\chi(\xi) = \begin{cases} 0 & \text{for } |\xi| \leq a, \\ 1 & \text{for } |\xi| \geq b, \end{cases} \tag{5.31}$$

with $0 < a < b$ any fixed constants.

Note that formula (5.28) shows that the vibrations corresponding to the first two components $u_1^\varepsilon, u_2^\varepsilon$ always have a local character, that is, the displacements $x_3u_1^\varepsilon(x)$ and $x_3u_2^\varepsilon(x)$ are only significant in a region near the concentrated mass (for $|x| = O(\varepsilon)$) while they are very small at the distance $O(1)$ of the concentrated mass. Indeed, the displacements $x_3u_1^\varepsilon(x)$ and $x_3u_2^\varepsilon(x)$ are of order $O(\varepsilon)$ for $|x| = O(\varepsilon)$,

$$x_3u_\alpha^\varepsilon(x) \sim x_3\left[\frac{1}{|\log\varepsilon|}u_\alpha^0(x) + V_\alpha\left(\frac{x}{\varepsilon}\right) - c_\alpha\right] = O(\varepsilon) \quad \text{for } |x| = O(\varepsilon), \tag{5.32}$$

and they are of order $O(\varepsilon/|\log\varepsilon|)$ for $|x| = O(1)$,

$$x_3u_\alpha^\varepsilon(x) \sim x_3\frac{1}{|\log\varepsilon|}u_\alpha^0(x) = O\left(\frac{\varepsilon}{|\log\varepsilon|}\right) \quad \text{for } |x| = O(1). \tag{5.33}$$

Nevertheless, for the third component u_3^ε, the order of magnitude of the displacements can be larger outside the concentrated mass if the constants c_1 and c_2 are different from zero. In this case, $c_1 \neq 0$ or $c_2 \neq 0$, formula

(5.30) allows us to assert that the third component of the displacement $u_3^\varepsilon(x)$ is of order $O(\varepsilon)$ for $|x| = O(\varepsilon)$,

$$u_3^\varepsilon(x) \sim \varepsilon V_3\Big(\frac{x}{\varepsilon}\Big) = O(\varepsilon) \quad \text{for } |x| = O(\varepsilon), \tag{5.34}$$

and it is of order $O(1/|\log\varepsilon|)$ for $|x| = O(1)$,

$$u_3^\varepsilon(x) \sim \frac{1}{|\log\varepsilon|} u_3^0(x) = O\left(\frac{1}{|\log\varepsilon|}\right) \quad \text{for } |x| = O(1). \tag{5.35}$$

Remark 1. We observe that in the case where $c_1 = c_2 = 0$, that is, the constants c_α appearing in the condition at infinity for the local problem (5.8) simultaneously take the value zero, formulas (5.21)–(5.27) allow us to choose functions $u_i^0 = 0$, for $i = 1, 2, 3$. In this case, since (5.28) and (5.30) read

$$u_\alpha^\varepsilon(x) \sim x_3 V_\alpha\Big(\frac{x}{\varepsilon}\Big), \quad \alpha = 1, 2,$$

$$u_3^\varepsilon(x) \sim \varepsilon V_3\Big(\frac{x}{\varepsilon}\Big),$$

the local character of the three components of the displacement is maintained. We note that now we do not need to use the intermediate variable, and that the result complements that in [12].

Remark 2. Let us note that the local character of the vibrations corresponding with the eigenelements $(\zeta_n^\varepsilon/\varepsilon^{m-2}, u_n^\varepsilon)$, deduced by means of asymptotic expansions and matching principles in this section, is in good agreement with Theorem 1, where we need to use the local variable $y = x/\varepsilon$ to prove the convergence. Indeed, we observe that the functions given by (5.28) and (5.30), approaching the eigenfunctions u^ε associated with $\zeta^\varepsilon/\varepsilon^{m-2}$, satisfy

$$\left\| V_\alpha(y) - \left(\frac{1}{|\log\varepsilon|} u_\alpha^0(\varepsilon y) + V_\alpha(y) - c_\alpha\right)\right\|_{L^2(B(0,R))} \to 0 \tag{5.36}$$

and

$$\left\| V_3(y) - \left(V_3(y) + \chi\Big(\frac{y}{|\log\varepsilon|}\Big)\left[\frac{1}{\varepsilon|\log\varepsilon|} u_3^0(\varepsilon y) - V_3(y)\right]\right)\right\|_{L^2(B(0,R))} \to 0\,, \tag{5.37}$$

as $\varepsilon \to 0$, for any fixed constant R, $R > 0$. Then we can assert that Theorem 1 justifies, in some way, the asymptotic expansions obtained in this section.

5.3 Asymptotics in the Case $r > 1$

We consider the following asymptotic expansions for the eigenvalues η^ε of (5.5), and for their corresponding eigenfunctions V^ε:

$$\eta^\varepsilon = \eta + o(1), \quad V^\varepsilon = V + o(1).$$

Replacing both expressions in (5.5), for $r > 1$ and $m > 4$, we obtain that $V_\alpha = -V_{3,\alpha}$. Moreover, taking limits in the variational formulation of (5.5) for the test functions $W \in (\mathcal{D}(\mathbb{R}^2))^3$ such that $W_\alpha = -W_{3,\alpha}$, we obtain that V_3 verifies the spectral problem:

$$\begin{aligned}
(\lambda + 2\mu)\Delta^2 V_3 = 0 \quad &\text{in } \mathbb{R}^2 - \bar{B}, \\
(\lambda + 2\mu)\Delta^2 V_3 = \eta V_3 \quad &\text{in } B, \\
[V_3] = \left[\frac{\partial V_3}{\partial n}\right] = [\Delta V_3] = \left[\frac{\partial(\Delta V_3)}{\partial n}\right] = 0 \quad &\text{on } \Gamma, \\
V_3 = c_\alpha y_\alpha + c_3 + O(\log|y|) \quad &\text{as } |y| \to \infty,
\end{aligned} \tag{5.38}$$

where c_i, for $i = 1, 2, 3$, are some unknown but well-determined constants. As in (5.8), this condition at infinity is a consequence of general results for the solutions of elliptic systems with a finite energy integral (see [6] and [13] for details).

We observe that problem (5.38) is an eigenvalue problem for a Kirchhoff–Love plate model in an unbounded domain and has a discrete, nonnegative spectrum:

$$0 = \eta_1 = \eta_2 = \eta_3 \leq \eta_4 \leq \eta_5 \leq \cdots \leq \eta_n \leq \cdots \xrightarrow{n \to \infty} \infty,$$

denote the eigenvalues of (5.38) with the classical convention of repeated eigenvalues.

The following result states the convergence of the eigenvalues of (5.2) for $h = \varepsilon^r$ and $r > 1$ (see [6] for its proof):

Theorem 2. *Let ζ_n^ε be the eigenvalues of problem (5.2) for $h = \varepsilon^r$. If $m > 4$ and $r > 1$, for each n, the values $\zeta_n^\varepsilon / \varepsilon^{m-4+2r}$ converge as $\varepsilon \to 0$ to the eigenvalue η_n of (5.38). Moreover, for any eigenfunction V_3 of (5.38) associated with η_k and satisfying $\|V_3\|_{L^2(B)} = 1$, there exists a linear combination U^ε of eigenfunctions of (5.2) associated with the eigenvalues converging towards η_k, such that $U_3^\varepsilon/\varepsilon$ converges to V_3 in $L^2(B)$.*

In addition, if U^ε is an eigenfunction of (5.2) associated with eigenvalues ζ^ε such that $\zeta^\varepsilon/\varepsilon^{m-4+2r}$ converges to η^* as $\varepsilon \to 0$, satisfying

$$\varepsilon^{m+2r-2}\|U_\alpha^\varepsilon\|^2_{L^2(\varepsilon^{-1}\Omega - \bar{B})} + \varepsilon^{m-2}\|U_3^\varepsilon\|^2_{L^2(\varepsilon^{-1}\Omega - \bar{B})} + \varepsilon^{2r-2}\|U_\alpha^\varepsilon\|^2_{L^2(B)} + \varepsilon^{-2}\|U_3^\varepsilon\|^2_{L^2(B)} = 1,$$

and $U_3^\varepsilon/\varepsilon$ converges to $V_3^* \neq 0$ weakly in $L^2(B)$, then U_α^ε and $U_3^\varepsilon/\varepsilon$ converge in $L^2(B)$ to V_α^* and V_3^*, respectively, where $V_\alpha^* = -V_{3,\alpha}^*$ and V_3^* is an eigenfunction of (5.38) associated with η^*.

In a similar way to the case where $r = 1$, we study the structure of the eigenfunctions of problem (5.2) for $h = \varepsilon^r$ with $r > 1$. We briefly outline here the main steps of the proof.

We postulate an asymptotic expansion for the eigenvalues ζ^ε,

$$\zeta^\varepsilon = \eta\varepsilon^{m-4+2r} + o(\varepsilon^{m-4+2r}), \tag{5.39}$$

an outer expansion for the corresponding eigenfunctions u^ε in $\Omega - \{0\}$,

$$u^\varepsilon(x) = \alpha_0(\varepsilon)\left[u^0(x) + \varepsilon^r u^r(x) + \varepsilon^{2r}u^{2r}(x) + o(\varepsilon^{2r})\right], \tag{5.40}$$

and a local expansion in a neighborhood of the concentrated mass:

$$U_\alpha^\varepsilon(y) = V_\alpha(y) + o(1) \quad \alpha = 1, 2, \tag{5.41}$$

$$U_3^\varepsilon(y) = \varepsilon V_3(y) + o(\varepsilon). \tag{5.42}$$

As outlined at the beginning of the section, replacing asymptotic expansions (5.39) and (5.41), (5.42) in (5.5) for $h = \varepsilon^r$ with $r > 1$ and $m > 4$, $\eta^\varepsilon = \zeta^\varepsilon\varepsilon^{4-2r-m}$ and $(V_1^\varepsilon, V_2^\varepsilon, V_3^\varepsilon) = (U_1^\varepsilon, U_2^\varepsilon, U_3^\varepsilon/\varepsilon)$ leads us to $V_\alpha = -V_{3,\alpha}$, where V_3 is an eigenfunction of (5.38) associated with η.

On the other hand, replacing expansions (5.39) and (5.40) in (5.2) for $h = \varepsilon^r$ with $r > 1$ and collecting coefficients of the same power of ε, we have that u^0 and u^r satisfy equations (5.13)–(5.15) while u^{2r} verifies

$$\tilde{A}_{\alpha\beta}u_\beta^0 + \mu u_\alpha^{2r} + \mu u_{3,\alpha}^{2r} = 0 \quad \text{in } \Omega - \{0\}, \quad \alpha = 1, 2, \tag{5.43}$$

$$\mu u_{\alpha,\alpha}^{2r} + \mu\Delta u_3^{2r} = 0 \quad \text{in } \Omega - \{0\}, \tag{5.44}$$

$$u_i^{2r} = 0 \quad \text{on } \partial\Omega, \quad i = 1, 2, 3,$$

where $\tilde{A} = (\tilde{A}_{\alpha\beta})_{\alpha,\beta=1,2}$ is the two-dimensional elasticity operator defined by (5.18). Now, by virtue of (5.43) and (5.44) we get (5.19) and combining this equation with (5.13) and (5.15), we conclude that $u_\alpha^0 = -u_{3,\alpha}^0$ where u_3^0 verifies problem (5.20).

As in Section 5.2, we look for u_3^0 in the form (5.21) where a_1, a_2 are certain constants to be determined by matching outer and local expansions, and $F(x)$ is the solution of (5.22). Thus, taking into account formulas (5.24) and (5.25), the asymptotic behavior at infinity of V_3, V_3 being an eigenfunction of (5.38), and the fact that $V_\alpha = -V_{3,\alpha}$, we deduce that (5.26) is satisfied if we take the order function $\alpha_0(\varepsilon)$ and the constants a_1, a_2 to be

$$\alpha_0(\varepsilon) = \frac{1}{|\log\varepsilon|}, \quad c_1 = -2a_1, \qquad c_2 = -2a_2, \tag{5.45}$$

and the leading terms in expansions (5.39)–(5.42) are determined.

Introducing the intermediate variable $\xi = x/(\varepsilon|\log\varepsilon|)$, we note that the matching condition (5.29) for the local and outer expansions of u_3^ε also holds, so the composite expansions for the components u_i^ε in Ω are now

$$u_\alpha^\varepsilon(x) \sim \frac{1}{|\log\varepsilon|} u_\alpha^0(x) + V_\alpha\Big(\frac{x}{\varepsilon}\Big) + c_\alpha, \quad \alpha = 1,2, \tag{5.46}$$

and

$$u_3^\varepsilon(x) \sim \varepsilon V_3\Big(\frac{x}{\varepsilon}\Big) + \chi\Big(\frac{x}{\varepsilon|\log\varepsilon|}\Big)\Big[\frac{1}{|\log\varepsilon|}u_3^0(x) - \varepsilon V_3\Big(\frac{x}{\varepsilon}\Big)\Big], \tag{5.47}$$

where $V_\alpha = -V_{3,\alpha}$, V_3 is an eigenfunction of (5.38) associated with η, u_3^0 is defined by (5.23), and u_α^0, $\alpha = 1,2$, are defined by (5.24) and (5.25), respectively, with $a_1 = -c_1/2$, $a_2 = -c_2/2$, and F the solution of (5.22).

Formula (5.46) allows us to assert that, for $r > 1$, the displacements $x_3 u_1^\varepsilon$ and $x_3 u_2^\varepsilon$ are of order $O(\varepsilon^r)$ for $|x| = O(\varepsilon)$ and of order $O(\varepsilon^r/|\log\varepsilon|)$ for $|x| = O(1)$ (see (5.32) and (5.33) to compare when $r = 1$). Thus, also, for $r > 1$, the first two components of the vibrations associated with the low frequencies are localized near the concentrated masses. The same can be said for the third component of the displacement, $u_3^\varepsilon(x)$, in the case where $c_1 = c_2 = 0$, while the order of magnitude can be larger outside the concentrated mass if one of the constants c_1 or c_2 is different from zero. In this last case, on account of (5.47), $u_3^\varepsilon(x)$ is of order $O(\varepsilon)$ for $|x| = O(\varepsilon)$ and of order $O(1/|\log\varepsilon|)$ for $|x| = O(1)$ (see (5.34) and (5.35) and Remark 1 to compare when $r = 1$).

Remark 3. As in Section 5.2 and Remark 2, we can assert that Theorem 2 justifies the asymptotic expansions in this section. Indeed, the functions, given by (5.46) and (5.47), approaching the eigenfunctions u^ε associated with $\zeta^\varepsilon/\varepsilon^{m-4+2r}$, satisfy (5.36) and (5.37) as $\varepsilon \to 0$, for any fixed constant $R > 0$, where now $V_\alpha = -V_{3,\alpha}$ and V_3 is an eigenfunction of (5.38) associated with η.

References

1. C. Constanda, *A Mathematical Analysis of Bending of Plates with Transverse Shear Deformation,* Longman-Wiley, Harlow-New York, 1990.
2. Ph. Destuynder and M. Salaun, *Mathematical Analysis of Thin Plates Models,* Springer-Verlag, Heidelberg, 1996.
3. H. Reismann, *Elastic Plates. Theory and Application,* Wiley, New York, 1988.
4. I.S. Zorin and S.A. Nazarov, Edge effect in the bending of a thin three-dimensional plate, *J. Appl. Math. Mech.* **53** (1989), 500–507.
5. D. Gómez, M. Lobo, and E. Pérez, On a vibrating plate with a concentrated mass, *C.R. Acad. Sci. Paris Sér. IIb* **328** (2000), 495–500.

6. D. Gómez, M. Lobo, and E. Pérez, On the vibrations of a plate with concentrated mass and very small thickness, *Math. Methods Appl. Sci.* **26** (2003), 27–65.

7. Yu.D. Golovaty, Spectral properties of oscillatory systems with adjoined masses, *Trans. Moscow Math. Soc.* **54** (1993), 23–59.

8. M. Lobo and E. Pérez, Local problems for vibrating systems with concentrated masses: a review, *C.R. Mécanique* **331** (2003), 303–317.

9. O.A. Oleinik, A.S. Shamaev, and G.A. Yosifian, *Mathematical Problems in Elasticity and Homogenization,* North-Holland, London, 1992.

10. J. Sanchez-Hubert and E. Sanchez-Palencia, *Vibration and Coupling of Continuous Systems. Asymptotic Methods,* Springer-Verlag, Heidelberg, 1989.

11. W. Eckhaus, *Asymptotic Analysis of Singular Perturbations,* North-Holland, Amsterdam, 1979.

12. D. Gómez, M. Lobo, and E. Pérez, Estudio asintótico de las vibraciones de placas muy delgadas con masas concentradas, in *Proceedings XVIII CEDYA/VIII CMA,* Dept. Enginyeria Informàtica i Matemàtiques, Universitat Rovira i Virgili, Tarragona, 2003.

13. V.A. Kondratiev and O.A. Oleinik, On the behaviour at infinity of solutions of elliptic systems with a finite energy integral, *Arch. Rational Mech. Anal.* **99** (1987), 75–89.

6 A Finite-dimensional Stabilized Variational Method for Unbounded Operators

Charles W. Groetsch

6.1 Introduction

Stabilization problems inevitably arise in the solution of inverse problems that are phrased in infinite-dimensional function spaces. The direct versions of these problems typically involve highly smoothing operators and consequently the inversion process is usually highly ill posed. This ill-posed problem can be viewed on a theoretical level (and often on a quite practical level) as evaluating an unbounded operator on some data space (an extension of the range of the direct operator). A typical case involves the solution of a linear inverse problem phrased as a Fredholm integral equation of the first kind

$$x(s) = \int_{\Omega} k(s,t)y(t)\,dt$$

or, in operator form,

$$Ky = x,$$

where K is a compact linear operator acting between Hilbert spaces. In this case the conventional solution is $y = Lx$, where $L = K^{\dagger}$, the Moore–Penrose generalized inverse of K, is a closed densely defined unbounded linear operator.

In the integral equation example just mentioned the solution of the inverse problem is given implicitly as the minimal norm least squares solution of the Fredholm integral equation of the first kind. We now give a couple of concrete examples of model problems from inverse heat flow theory of solutions of inverse problems which are given explicitly as values of some unbounded operator.

Suppose a uniform bar, identified with the interval $[0,\pi]$, is heated to an initial temperature distribution $g(x)$ for $x \in [0,\pi]$ while the endpoints of the bar are kept at temperature zero. For suitable choices of constants the evolution of the space-time temperature distribution of the bar, $u(x,t)$, is governed by the one-dimensional heat equation

$$\frac{\partial u}{\partial t} = \frac{\partial^2 u}{\partial x^2}, \quad 0 < x < \pi, \quad t > 0$$

This work was supported in part by the Charles Phelps Taft Foundation.

and satisfies the boundary conditions $u(0,t) = u(\pi,t) = 0$. Suppose we observe the temperature distribution $f(x)$ of the bar at some later time, say $t = 1$, that is, the function $f(x) = u(x,1)$ is observed, and we wish to reconstruct the initial distribution $g(x)$. Separation of variables leads to a solution of the form

$$u(x,t) = \sum_{n=1}^{\infty} c_n e^{-n^2 t} \sin nx$$

and therefore

$$f(x) = \sum_{n=1}^{\infty} c_n e^{-n^2} \sin nx,$$

where

$$c_m = \frac{2}{\pi} e^{m^2} \int_0^{\pi} f(y) \sin my \, dy.$$

We then see that

$$u(x,t) = \frac{2}{\pi} \sum_{n=1}^{\infty} e^{n^2} e^{-n^2 t} \int_0^{\pi} f(y) \sin ny \sin nx \, dy$$

and hence

$$g(x) = u(x,0) = \frac{2}{\pi} \sum_{n=1}^{\infty} e^{n^2} \int_0^{\pi} f(y) \sin ny \sin nx \, dy.$$

That is, $g = Lf$, where

$$Lf(x) = \frac{2}{\pi} \sum_{n=1}^{\infty} e^{n^2} \sin nx \int_0^{\infty} f(y) \sin ny \, dy.$$

In other words, the solution g of the inverse problem is obtained from the data f via the *unbounded* operator L defined on functions f in the subspace

$$\mathcal{D}(L) = \left\{ f \in L^2[0,\pi] : \sum_{m=1}^{\infty} e^{2m^2} a_m^2 < \infty, \quad a_m = \int_0^{\pi} f(y) \sin my \, dy \right\}.$$

Here the instability is apparent: small (in L^2-norm) perturbations of the data f can, because of the factors e^{n^2} in the kernel, be expressed as very large changes in the solution g.

The problem of determining a spatially distributed source term from the temperature distribution at a specific time provides another example of the solution of an inverse problem given as the value of an unbounded operator. If in the model

$$\frac{\partial u}{\partial t} = \frac{\partial^2 u}{\partial x^2} + g(x), \quad 0 < x < \pi, \quad 0 < t,$$

where $u(x,t)$ is subject to the boundary and initial conditions

$$u(0,t) = u(\pi,t) = 0, \quad u(x,0) = 0,$$

one wishes to reconstruct the source distribution $g(x)$ from the spacial temperature distribution at some later time, say $f(x) = u(x,1)$, one is led to the explicit representation

$$g(x) = Lf(x) = \frac{2}{\pi} \sum_{n=1}^{\infty} \frac{n^2}{1 - e^{-n^2}} \sin nx \int_0^{\pi} f(s) \sin ns \, ds.$$

That is, $g = Lf$, where L is the linear operator on $L^2[0,\pi]$ with domain

$$\mathcal{D}(L) = \left\{ f : \sum_{m=1}^{\infty} m^4 a_m^2 < \infty, \quad a_m = \frac{2}{\pi} \int_0^{\pi} f(s) \sin ms \, ds \right\}.$$

It is a simple matter to verify that the operators L which provide the solutions of the inverse problems in these examples are closed, densely defined, and unbounded.

In this paper we treat a theoretical aspect, specifically convergence theory, of an abstract finite element method for stable approximate evaluation of closed linear operators on a Hilbert space. We investigate the convergence of certain stabilized finite-dimensional approximations to the true solution. In [1] a treatment, in a more general context, of the convergence of finite-dimensional approximations to a stabilized infinite-dimensional approximation of the solution can be found.

6.2 Background

Our problem, in the abstract, is to evaluate a closed unbounded operator $L : \mathcal{D}(L) \subseteq H_1 \to H_2$ defined on a dense subspace $\mathcal{D}(L)$ of a Hilbert space H_1 at a vector $x \in \mathcal{D}(L)$. The rub is that we have only an approximation $x^\delta \in H_1$ satisfying $\|x - x^\delta\| \leq \delta$, where $\delta > 0$ is a known error bound. These approximate data typically represent some measured "rough" version of the true data vector x and generally $x^\delta \notin \mathcal{D}(L)$ and hence we may not apply the operator directly to the available data. Even in the unlikely case that $x^\delta \in \mathcal{D}(L)$ for all δ, we are not guaranteed that $Lx^\delta \to Lx$ as $\delta \to 0$ since L is unbounded. This is the classic instability problem that arises so often in solving linear inverse problems.

In [2] some stabilization techniques for approximate evaluation of unbounded operators are unified in a general scheme based on spectral theory and a theorem of von Neumann. This theorem states that if L is closed and densely defined then the operator $\widetilde{L}$ defined by

$$\widetilde{L} = (I + L^* L)^{-1}$$

is bounded and self-adjoint and the operator $L\widetilde{L}$ is bounded (see [3]). This result suggests stabilizing the evaluation of Lx by an approximation of the form

$$L\widetilde{L}S_\alpha(\widetilde{L})x,$$

where $S_\alpha(t)$ is a parameterized class of functions on (0,1] that approximates the function $1/t$ in an appropriate sense. This is the basis of the analysis in [2]. In the case when only an approximate data vector x^δ is available, the process may be viewed as a data smoothing step

$$x^\delta \mapsto \widetilde{L}S_\alpha(\widetilde{L})x^\delta$$

followed by a stabilized approximation of Lx

$$Lx \approx L\widetilde{L}S_\alpha(\widetilde{L})x^\delta,$$

where the operator $L\widetilde{L}S_\alpha(\widetilde{L})$ acting on the data vector x^δ is bounded.

The analysis in [2] is carried out in the context of an infinite-dimensional Hilbert space. In this paper we investigate a finite-dimensional realization of a particular stabilization method known as the Tikhonov–Morozov method.

6.3 The Tikhonov–Morozov Method

The family of functions

$$S_\alpha(t) = (\alpha + (1-\alpha)t)^{-1}, \quad 0 < \alpha < 1,$$

leads to the stable approximation

$$\begin{aligned} Lx &\approx L\widetilde{L}(\alpha I + (1-\alpha)\widetilde{L})^{-1}x^\delta \\ &= L(I + \alpha L^* L)^{-1}x^\delta =: Lx^\delta_\alpha, \end{aligned}$$

known as the Tikhonov–Morozov method [4], given approximate data $x^\delta \in H_1$ satisfying $\|x - x^\delta\| \leq \delta$. Under suitable conditions on the true data $x \in D(L)$ and with appropriate choice of the regularization parameter $\alpha = \alpha(\delta)$ it can be shown (see [2]) that

$$\|Lx - Lx^\delta_\alpha\| = O(\delta^{2/3}).$$

However, for the important class of operators having compact resolvent this order of approximation can not be improved except in the trivial case in which the true data is in the null space of the operator (see [5]). One of the attractive features of this method is that the "smoothed" data x_α^δ has a variational characterization, namely it is the minimizer over $\mathcal{D}(L)$ of the functional

$$\Phi_\alpha(z; x^\delta) = \|z - x^\delta\|^2 + \alpha\|Lz\|^2.$$

It is a routine matter to show that for each $x^\delta \in H_1$ the functional $\Phi_\alpha(\cdot; x^\delta)$ has a unique minimizer x_α^δ over $\mathcal{D}(L)$ and this minimizer enjoys an extra order of smoothness in that it necessarily lies in $\mathcal{D}(L^*L)$.

6.4 An Abstract Finite Element Method

The variational characterization of the Tikhonov–Morozov approximation x_α^δ as the minimizer over $\mathcal{D}(L)$ of the functional

$$\Phi_\alpha(z; x^\delta) = \|z - x^\delta\|^2 + \alpha\|Lz\|^2$$

suggests the possibility of using finite element methods to effectively compute the approximations. To this end, suppose $\{V_m\}_{m=1}^\infty$ is a sequence of finite-dimensional subspaces of H_1 satisfying

$$V_1 \subseteq V_2 \subseteq \ldots \subseteq \mathcal{D}(L) \quad \text{and} \quad \overline{\cup_{m=1}^\infty V_m} = H_1.$$

Given $x \in \mathcal{D}(L)$, the finite element approximation to Lx will be $Lx_{\alpha,m}$, where

$$x_{\alpha,m} = \text{argmin}_{z \in V_m} \|z - x\|^2 + \alpha\|Lz\|^2.$$

Since V_m is finite dimensional, such a minimizer exists and is unique. Suppose $\dim V_m = n(m)$ and that $\{\varphi_1^{(m)}, \ldots, \varphi_{n(m)}^{(m)}\}$ is a basis for V_m. Then the coefficients $\{c_j^{(m)}\}$ of the approximation

$$x_{\alpha,m} = \sum_{j=1}^{n(m)} c_j^{(m)} \varphi_j^{(m)}$$

are determined by the conditions

$$\frac{d}{dt}\Phi_\alpha(x_{\alpha,m} + t\varphi_i^{(m)}; x)|_{t=0} = 0, \quad i = 1, \ldots, n(m),$$

which are equivalent to the system of linear algebraic equations

$$\sum_{j=1}^{n(m)} [\langle \varphi_i^{(m)}, \varphi_j^{(m)} \rangle + \alpha \langle L\varphi_i^{(m)}, L\varphi_j^{(m)} \rangle] c_j^{(m)}$$
$$= \langle x, \varphi_i^{(m)} \rangle, \quad i = 1, \ldots, n(m).$$

When only an approximation $x^\delta \in H_1$ is available satisfying $\|x-x^\delta\| \le \delta$, the minimizer of the functional $\Phi_\alpha(\cdot; x^\delta)$ over V_m is denoted by $x^\delta_{\alpha,m}$:

$$x^\delta_{\alpha,m} = \operatorname{argmin}_{z\in V_m} \Phi_\alpha(z; x^\delta).$$

As a first step in our analysis, we determine a stability bound for

$$\|Lx_{\alpha,m} - Lx^\delta_{\alpha,m}\|.$$

The stability bound turns out to be the same as that found for the approximation in infinite-dimensional space. We will employ the inner product $[\cdot,\cdot]$ defined on $\mathcal{D}(L)$ by

$$[u,w] = \langle u,w\rangle + \alpha\langle Lu, Lw\rangle,$$

where α is a fixed positive number, and the associated norm $|u| = \sqrt{[u,u]}$. Note that since L is closed, $\mathcal{D}(L)$ is a Hilbert space when endowed with this inner product.

Theorem 1. *If $x \in \mathcal{D}(L) \subseteq H_1$ and $x^\delta \in H_1$ satisfies $\|x - x^\delta\| \le \delta$, then*

$$\|Lx_{\alpha,m} - Lx^\delta_{\alpha,m}\| \le \delta/\sqrt{\alpha}.$$

Proof. The necessary condition

$$\frac{d}{dt}\Phi(x_{\alpha,m} + tv; x)|_{t=0} = 0,$$

for all $v \in V_m$ gives

$$\langle x_{\alpha,m} - x, v\rangle + \alpha\langle Lx_{\alpha,m}, Lv\rangle = 0, \tag{6.1}$$

for all $v \in V_m$ and similarly

$$\langle x^\delta_{\alpha,m} - x, v\rangle + \alpha\langle Lx^\delta_{\alpha,m}, Lv\rangle = 0,$$

for all $v \in V_m$. The condition (6.1) may be expressed in terms of the inner product $[\cdot,\cdot]$ in the following way:

$$\begin{aligned}[x_{\alpha,m} - x, v] &= \langle x_{\alpha,m} - x, v\rangle + \alpha\langle L(x_{\alpha,m} - x), Lv\rangle \\ &= -\alpha\langle Lx, Lv\rangle, \end{aligned} \tag{6.2}$$

for all $v \in V_m$. On the other hand,

$$\begin{aligned}[x^\delta_{\alpha,m} - x, v] &= \langle x^\delta_{\alpha,m} - x, v\rangle + \alpha\langle Lx^\delta_{\alpha,m} - Lx, Lv\rangle \\ &= \langle x^\delta - x, v\rangle + \langle x^\delta_{\alpha,m} - x^\delta, v\rangle + \alpha\langle L(x^\delta_{\alpha,m} - x), Lv\rangle \\ &= \langle x^\delta - x, v\rangle - \alpha\langle Lx, Lv\rangle \end{aligned}$$

and therefore,

$$[x^\delta_{\alpha,m} - x_{\alpha,m}, v] = \langle x^\delta - x, v\rangle, \tag{6.3}$$

for all $v \in V_m$. In particular, setting $v = x^\delta_{\alpha,m} - x_{\alpha,m}$ in (6.3), and applying the Cauchy–Schwarz inequality, one obtains

$$\begin{aligned} \|x^\delta_{\alpha,m} - x_{\alpha,m}\|^2 + \alpha\|Lx^\delta_{\alpha,m} - Lx_{\alpha,m}\|^2 &= |x^\delta_{\alpha,m} - x_{\alpha,m}|^2 \\ &\leq \delta\|x^\delta_{\alpha,m} - x_{\alpha,m}\|. \end{aligned}$$

Therefore,

$$\|x^\delta_{\alpha,m} - x_{\alpha,m}\| \leq \delta$$

and hence

$$\alpha\|Lx^\delta_{\alpha,m} - Lx_{\alpha,m}\|^2 \leq \delta^2,$$

giving the result.

We see from this theorem that the condition $\delta/\sqrt{\alpha} \to 0$, combined with a condition that ensures that $\|Lx_{\alpha,m} - Lx\| \to 0$, will guarantee the convergence of the stabilized finite element approximations to Lx.

The remaining development requires an analysis of the difference between the finite-dimensional approximation $x_{\alpha,m}$ and the infinite-dimensional approximation x_α using exact data $x \in \mathcal{D}(L)$ which is characterized by

$$x_\alpha = \mathrm{argmin}_{z\in\mathcal{D}(L)}\|z - x\|^2 + \alpha\|Lz\|^2.$$

This is equivalent to

$$0 = \langle x_\alpha - x, v\rangle + \alpha\langle Lx_\alpha, Lv\rangle = [x_\alpha - x, v] + \alpha\langle Lx, v\rangle$$

for all $v \in \mathcal{D}(L)$. The corresponding finite element approximation $x_{\alpha,m}$ satisfies (6.2), that is,

$$0 = [x_{\alpha,m} - x, v] + \alpha\langle Lx, v\rangle$$

for all $v \in V_m$. Subtracting, we find that

$$[x_\alpha - x_{\alpha,m}, v] = 0, \tag{6.4}$$

for all $v \in V_m$. We can express this in a geometrical way by saying that $x_{\alpha,m}$ is the $[\cdot,\cdot]$-orthogonal projection of x_α onto the finite-dimensional subspace V_m. That is,

$$x_{\alpha,m} = \mathbf{P}_m x_\alpha, \tag{6.5}$$

where $\mathbf{P}_m : \mathcal{D}(L) \to V_m$ is the orthogonal projector of the Hilbert space $\mathcal{D}(L)$, equipped with the inner product $[\cdot,\cdot]$, onto the subspace $V_m \subseteq \mathcal{D}(L)$.

Let P_m be the (ordinary) orthogonal projector of H_1 onto V_m. One may bound the quantity $\|Lx_\alpha - Lx_{\alpha,m}\|$ in terms of the two quantities

$$\beta_m = \|(I - P_m)\widetilde{L}\|$$

and

$$\gamma_m = \|L(I - P_m)\widetilde{L}\|.$$

We note that since $L\widetilde{L}$, LP_m, and $\widetilde{L}$ are all bounded linear operators, both of these quantities are finite. We begin with a result that requires relatively modest hypotheses on the true data x.

Theorem 2. *If $x \in \mathcal{D}(L^*L)$, then*

$$\|Lx_\alpha - Lx_{\alpha,m}\|^2 \le (\beta_m^2/\alpha + \gamma_m^2)\|x + L^*Lx\|^2.$$

Proof. First we note that $x \in \mathcal{D}(L^*L)$ if and only if $x = \widetilde{L}w$ where $w = x + L^*Lx$. From the characterization (6.5) we have

$$\begin{aligned}\alpha\|Lx_\alpha - Lx_{\alpha,m}\|^2 &\le |x_\alpha - x_{\alpha,m}|^2 \\ &= |x_\alpha - \mathbf{P}_m x_\alpha|^2 \le |x_\alpha - P_m x_\alpha|^2 \\ &= \|(I - P_m)x_\alpha\|^2 + \alpha\|L(I - P_m)x_\alpha\|^2.\end{aligned}$$

But

$$\begin{aligned}x_\alpha &= (I + \alpha L^*L)^{-1}x = \widetilde{L}(\alpha I + (1-\alpha)\widetilde{L})^{-1}x \\ &= \widetilde{L}(\alpha I + (1-\alpha)\widetilde{L})^{-1}\widetilde{L}w.\end{aligned}$$

Also, $\|(\alpha I + (1-\alpha)\widetilde{L})^{-1}\widetilde{L}\| \le 1$. Therefore,

$$\alpha\|Lx_\alpha - Lx_{\alpha,m}\|^2 \le (\|(I - P_m)\widetilde{L}\|^2 + \alpha\|L(I - P_m)\widetilde{L}\|^2)\|w\|^2,$$

that is,

$$\|Lx_\alpha - Lx_{\alpha,m}\|^2 \le (\beta_m^2/\alpha + \gamma_m^2)\|w\|^2.$$

We now need a well-known consequence of the uniform boundedness principle.

Lemma 1. *Suppose $\{A_n\}$ is a sequence of bounded linear operators satisfying $A_n x \to 0$ as $n \to \infty$ for each $x \in H_1$. If K is a compact linear operator, then $\|A_n K\| \to 0$ as $n \to \infty$.*

Suppose L^*L has compact resolvent, i.e., $\widetilde{L}$ is compact. Applying the previous lemma with $A_n = I - P_n$, we see that

$$\beta_n \to 0 \quad \text{as} \quad n \to \infty.$$

If we assume that

$$LP_n z \to Lz \quad \text{for} \quad z \in \mathcal{D}(L^*L), \tag{6.6}$$

then we can also apply the lemma to the operator $A_n = L(I - P_n)\widetilde{L}$ to find that

$$\gamma_n \to 0 \quad \text{as} \quad n \to \infty.$$

We may now give a basic convergence and regularity result for the stabilized finite element approximations.

Theorem 3. *Suppose that $\widetilde{L}$ is compact, that condition (6.6) holds, and that $x \in \mathcal{D}(L^*L)$. If $\alpha = \alpha(\delta) \to 0$ as $\delta \to 0$ and $m = m(\alpha) \to \infty$ as $\alpha \to 0$ in such a way that $\delta/\sqrt{\alpha} \to 0$, $\gamma_m \to 0$, and $\beta_m^2/\alpha \to 0$, then*

$$Lx_{\alpha,m}^{\delta} \to Lx.$$

Proof. Note that

$$\begin{aligned}\|Lx_{\alpha,m}^{\delta} - Lx\| &\leq \|Lx_{\alpha,m}^{\delta} - Lx_{\alpha,m}\| + \|Lx_{\alpha,m} - Lx_{\alpha}\| + \|Lx_{\alpha} - Lx\| \\ &\leq \delta/\sqrt{\alpha} + O(\sqrt{\beta_m^2/\alpha + \gamma_m^2}) + \|Lx_{\alpha} - Lx\|.\end{aligned}$$

It is well known [5] that $\|Lx_{\alpha} - Lx\| \to 0$ as $\alpha \to 0$ (see, e.g., [2]), and the result follows.

Remark. If we are willing to assume more on the true data, namely that $x \in R(\widetilde{L}^{\nu})$ for some $\nu \geq 1$, then minor modifications of the argument above give the bound

$$\|Lx_{\alpha,m}^{\delta} - Lx\| \leq \delta/\sqrt{\alpha} + O\left(\sqrt{\beta_m^2(\nu)/\alpha + \gamma_m^2(\nu)}\right) + \|Lx_{\alpha} - Lx\|,$$

where

$$\beta_m(\nu) = \|(I - P_m)\widetilde{L}^{\nu}\|$$

and

$$\gamma_m(\nu) = \|L(I - P_m)\widetilde{L}^{\nu}\|.$$

It is well known that if $x \in \mathcal{D}(LL^*L)$, then $\|Lx_{\alpha} - Lx\| = O(\alpha)$. For completeness we supply the argument. If $x \in \mathcal{D}(LL^*L)$, then $Lx = \widehat{L}w$, where $w = (I + LL^*)Lx$ and $\widehat{L} := (I + LL^*)^{-1}$ is bounded and self-adjoint. Therefore,

$$\begin{aligned}\|Lx_{\alpha} - Lx\| &= \|L\widetilde{L}(\alpha I + (1-\alpha)\widetilde{L})^{-1}x - Lx\| \\ &= \|(\widehat{L}(\alpha I + (1-\alpha)\widehat{L})^{-1} - I)\widehat{L}w\| \\ &= \alpha\|(\widehat{L} - I)(\alpha I + (1-\alpha)\widehat{L})^{-1}\widehat{L}\|\|w\| \leq \alpha\|w\|.\end{aligned}$$

We therefore obtain the following assertion.

Corollary. *If condition (6.6) holds and if $x \in \mathcal{D}(LL^*L)$, $\gamma_m = O(\alpha)$, $\beta_m = O(\alpha^{3/2})$, and $\alpha \sim \delta^{2/3}$, then*

$$\|Lx_{\alpha,m}^{\delta} - Lx\| = O(\delta^{2/3}).$$

This corollary shows that in principle the finite element approximations are capable of achieving the optimal order of convergence possible for the

Tikhonov–Morozov method. As a specific but simple instance of the corollary, consider the case of the operator that maps the temperature distribution $f(x)$ at time $t = 1$ to the distributed forcing term $g(x)$. Let

$$V_m = \text{span}\{\varphi_1, \varphi_2, \ldots, \varphi_m\},$$

where $\varphi_j(x) = \sqrt{2/\pi}\sin jx$. Then

$$LP_n z = \sum_{j=1}^{n} \frac{j^2}{1 - e^{-j^2}} \langle z, \varphi_j \rangle \to Lz$$

for each $z \in \mathcal{D}(L)$ and hence condition (6.6) is satisfied. Also, elementary estimates show that $\beta_m \leq m^{-4}$ and $\gamma_m \leq m^{-2}$. Therefore, a choice of stabilization parameter of the form $\alpha \sim \delta^{2/3}$ along with a choice of subspace dimension of the form $m \sim \delta^{-1/3}$ result in the optimal order of approximation $O(\delta^{2/3})$.

References

1. V.P. Tanana, Necessary and sufficient conditions of convergence of finite-dimensional approximations for L-regularized solutions of operator equations, *J. Inverse Ill-posed Problems* **8** (2000), 449–457.
2. C.W. Groetsch, Spectral methods for linear inverse problems with unbounded operators, *J. Approx. Theory* **70** (1992), 16–28.
3. F. Riesz and B. Sz.-Nagy, *Functional Analysis*, Ungar, New York, 1955.
4. V.A. Morozov, *Methods for Solving Incorrectly Posed Problems*, Springer-Verlag, New York, 1984.
5. C.W. Groetsch and O. Scherzer, The optimal order of convergence for stable evaluation of differential operators, *Electronic J. Diff. Equations* **3** (1993), 1–12.

7 A Converse Result for the Tikhonov–Morozov Method

Charles W. Groetsch

7.1 Introduction

Many ill-posed problems of mathematical physics may be phrased as evaluating closed unbounded linear operators on Hilbert space. A well-known specific example occurs in potential theory. Consider the model Cauchy problem for Laplace's equation on an infinite strip. Here the unbounded operator maps one boundary distribution into another. Specifically, let $u(x,y)$ be a harmonic function in the strip

$$\Omega = \{(x,y) : 0 < x < 1, \ -\infty < y < \infty\},$$

satisfying a homogenous Neumann condition on the boundary $x = 0$, and suppose that one wishes to determine the boundary values $g(y) = u(1,y)$ given the boundary values $f(y) = u(0,y)$. That is, $u(x,y)$ satisfies

$$\begin{aligned} \Delta u &= 0 \quad \text{in } \Omega, \\ u(0,y) &= f(y), \\ u_x(0,y) &= 0, \\ u(1,y) &= g(y), \end{aligned}$$

for $-\infty < y < \infty$.

Applying the Fourier transform $\hat{u} = \mathcal{F}\{u\}$ with respect to the y variable,

$$\hat{u}(x:\omega) = \frac{1}{\sqrt{2\pi}} \int_{-\infty}^{\infty} u(x,y) e^{-i\omega y}\, dy,$$

results in the initial value problem involving the frequency parameter ω:

$$\begin{aligned} \frac{d^2}{dx^2}\hat{u} &= \omega^2 \hat{u}, \\ \hat{u}(0) &= \hat{f}, \\ \frac{d}{dx}\hat{u}(0) &= 0, \end{aligned}$$

This work was supported in part by the Charles Phelps Taft Foundation.

giving

$$\hat{g}(\omega) = \hat{u}(1, \omega) = \hat{f}(\omega) \cosh(\omega).$$

Therefore the linear operator connecting f to g is given by

$$g = Lf = \mathcal{F}^{-1}\{\hat{f}(\omega) \cosh(\omega)\}.$$

This linear operator is defined only on

$$\mathcal{D}(L) = \{f \in L^2(R) : \hat{f}(\omega) \cosh(\omega) = \hat{g}\}$$

for some function $g \in L^2(R)$, a condition that says the high frequency components of f must decay very rapidly. In particular, L is defined on band-limited functions and hence $\mathcal{D}(L)$ is dense in $L^2(R)$. Also, operator L is closed. For if $\{f_n\} \subset L^2(R)$, $\hat{f}_n(\omega) \cosh(\omega) \in L^2(R)$, and $\mathcal{F}^{-1}\{\hat{f}_n(\omega) \cosh(\omega)\} \to g \in L^2(R)$, then $\hat{f}_n(\omega) \cosh(\omega) \to \hat{g} \in L^2(R)$, and since $\hat{f}_n \to \hat{f}$, we have

$$\hat{f}(\omega) \cosh(\omega) = \hat{g} \in L^2(R),$$

and hence $f \in \mathcal{D}(L)$ and $Lf = g$, that is, L is closed.

However, L is unbounded. Indeed, if

$$\hat{f}_n(\omega) = \chi_{[n-1/2, n+1/2]}$$

is the characteristic function of $[n - 1/2, n + 1/2]$, then $f_n \in L^2(R)$ and $\|f_n\| = 1$, by the Parseval–Plancherel relation. However,

$$\|Lf_n\|^2 = \int_{n-1/2}^{n+1/2} (\cosh \omega)^2 d\omega \geq \frac{1}{4} \int_{n-1/2}^{n+1/2} e^{2\omega} d\omega \to \infty \quad \text{as} \quad n \to \infty,$$

showing that L is unbounded.

The basic problem in the evaluation of such an unbounded operator is that the data vector to which the operator is to be applied might be only approximately known, and this approximation may fail to be in the domain of the operator. Or worse, a given sequence of approximate data vectors, even if it lies within the domain of the operator, might, upon application of the operator, lead to a nonconvergent sequence since the operator is unbounded. What are needed are *bounded* approximations to the unbounded operator, whose values converge in an appropriate sense to the required value of the unbounded operator. Such schemes are called stabilization methods and the best known stabilization method is the Tikhonov–Morozov method.

In this note we answer a question, raised by M. Mitrea at IMSE2004 in Orlando, concerning a converse of a convergence theorem for the Tikhonov–Morozov method. In fact, we give two quite distinct proofs of the fact that if the method achieves a rate of convergence of the form $O(\alpha)$, where α is the stabilization parameter, then the true data lies in the space $\mathcal{D}(LL^*L)$.

7.2 The Tikhonov–Morozov Method

Suppose $L : \mathcal{D}(L) \subseteq H_1 \to H_2$ is a closed densely defined linear operator acting on a Hilbert space H_1 and taking values in a Hilbert space H_2. The Tikhonov–Morozov method (see [1]) consists of approximating the vector Lx by Lx_α where x_α is the minimizer over $\mathcal{D}(L)$ of the functional

$$\Phi_\alpha(z;x) = \|z - x\|^2 + \alpha\|Lz\|^2$$

in which α is a positive stabilization parameter. That is,

$$x_\alpha = \text{argmin}_{z\in\mathcal{D}(L)}\Phi_\alpha(z;x).$$

It is easy to see that the unique minimizer x_α of this functional in fact lies in $\mathcal{D}(L^*L)$ and is given by

$$x_\alpha = (I + \alpha L^*L)^{-1}x.$$

Also $L(I+\alpha L^*L)^{-1}$ is an everywhere defined *bounded* linear operator, that is, the Tikhonov–Morozov method is a stabilization method.

The expression for x_α may be conveniently reformulated by use of von Neumann's classic theorem which states that if L is a closed densely defined linear operator, then the operators $\widetilde{L}$ and $\widehat{L}$ defined by

$$\widetilde{L} = (I + L^*L)^{-1} \quad \text{and} \quad \widehat{L} = (I + LL^*)^{-1}$$

are everywhere defined bounded self-adjoint operators (see [2]). The approximation may then be written

$$x_\alpha = \widetilde{L}((1-\alpha)\widetilde{L} + \alpha I)^{-1}x$$

and one sees immediately that $Lx_\alpha = L\widetilde{L}((1-\alpha)\widetilde{L} + \alpha I)^{-1}x$ is a stable approximation to Lx. For example, the operator L defined in the previous section may be written $L = \mathcal{F}^{-1}M\mathcal{F}$ where M is the unbounded multiplication operator densely defined on $L^2(R)$ by

$$(M\varphi)(\omega) = (\cosh\omega)\varphi(\omega).$$

In this case one finds that

$$Lx_\alpha = \mathcal{F}^{-1}M(I + \alpha M^2)^{-1}\mathcal{F}x$$

and for each $\alpha > 0$, one has

$$\|\mathcal{F}^{-1}M(I + \alpha M^2)^{-1}\mathcal{F}\| \le 1/\alpha.$$

Therefore, Lx_α is stable with respect to perturbations in x. It is known (see [3]) that if $x \in \mathcal{D}(LL^*L)$, then

$$\|Lx_\alpha - Lx\| = O(\alpha).$$

In this note we give two proofs of converses of this fact, one based on spectral theory and the other relying on the weak topology.

7.3 Operators with Compact Resolvent

The special case of the converse in which $\widehat{L}$ is compact has an instructive proof and will be considered separately in this section. Such operators with compact resolvent are quite common. In fact, Kato [4] has noted that in many cases arising in mathematical physics the unbounded operator L^*L has compact resolvent, that is, the operator

$$\widehat{L} = (I + LL^*)^{-1}$$

is compact. In Kato's words:

"Operators with compact resolvent occur frequently in mathematical physics. It may be said that most differential operators that appear in *classical* boundary value problems are of this type."

A prototypical example of such a closed densely defined linear operator with compact resolvent is provided by the differentiation operator $Lf = f'$ defined on

$$\mathcal{D}(L) = \{f \in L^2[0,1] : f \text{ abs. cont.}, \ f' \in L^2[0,1]\}.$$

Then

$$\mathcal{D}(LL^*) = \{y \in L^2[0,1] : y' \text{ abs. cont.}, \ y'' \in L^2[0,1], \ y(0) = y(1) = 0\}.$$

In this case $\widehat{L} = (I + LL^*)^{-1}$ is the compact integral operator on $L^2[0,1]$ defined by

$$\widehat{L}h(s) = \int_0^1 g(s,t)h(t)\,dt,$$

where $g(s,t)$ is the continuous symmetric kernel

$$g(s,t) = \begin{cases} \dfrac{\sinh(1-s)\sinh t}{\sinh 1}, & t \le s, \\[2ex] \dfrac{\sinh(1-t)\sinh s}{\sinh 1}, & s \le t. \end{cases}$$

In this section we give a proof of the promised converse in this case. Before doing so, we note the following simple assertion.

Lemma. $R(\widehat{L}) = \mathcal{D}(LL^*)$.

Proof. Since $\widehat{L} = (I+LL^*)^{-1}$, it follows that $\widehat{L}z \in \mathcal{D}(LL^*)$ for all $z \in H_2$, that is, $R(\widehat{L}) \subseteq \mathcal{D}(LL^*)$. On the other hand, if $y \in \mathcal{D}(LL^*)$, then $y = \widehat{L}w$ where $w = y + LL^*y$, i.e., $\mathcal{D}(LL^*) \subseteq R(\widehat{L})$.

Theorem. *Suppose $\widehat{L}$ is compact and $x \in \mathcal{D}(L)$. If $\|Lx_\alpha - Lx\| = O(\alpha)$, then $x \in \mathcal{D}(LL^*L)$.*

Proof. Let $\{u_j; \lambda_j\}$ be a complete orthonormal eigensystem for H_2, generated by the self-adjoint bounded operator $\widehat{L}$. N ote that the eigenvalues $\{\lambda_j\}$ lie in the interval $(0, 1]$. Suppose they are ordered as

$$0 < \cdots \leq \lambda_{n+1} \leq \lambda_n \leq \cdots \leq \lambda_2 \leq \lambda_1.$$

Then

$$\begin{aligned} Lx_\alpha &= L\widetilde{L}(\alpha I + (1-\alpha)\widetilde{L})^{-1}x \\ &= \widehat{L}(\alpha I + (1-\alpha)\widehat{L})^{-1}Lx \\ &= \sum_{j=1}^{\infty} \frac{\lambda_j}{\alpha + (1-\alpha)\lambda_j} \langle Lx, u_j \rangle u_j, \end{aligned}$$

and hence,

$$\begin{aligned} \|Lx - Lx_\alpha\|^2 &= \alpha^2 \sum_{j=1}^{\infty} \frac{(1-\lambda_j)^2}{(\alpha + (1-\alpha)\lambda_j)^2} |\langle Lx, u_j \rangle|^2 \\ &\geq \alpha^2(1-\lambda_1)^2 \sum_{j=1}^{\infty} (\alpha + (1-\alpha)\lambda_j)^{-2} |\langle Lx, u_j \rangle|^2. \end{aligned}$$

Therefore, if $\|Lx - Lx_\alpha\| = O(\alpha)$, we have

$$\sum_{j=1}^{\infty} (\alpha + (1-\alpha)\lambda_j)^{-2} |\langle Lx, u_j \rangle|^2 \leq C$$

for some constant C and all $\alpha \in (0, 1]$. In particular, all of the partial sums of the above series are uniformly bounded by C. Letting $\alpha \to 0^+$ in each of the individual partial sums shows that

$$\sum_{j=1}^{n} \lambda_j^{-2} |\langle Lx, u_j \rangle|^2 \leq C$$

for each n, and hence the series $\sum_{j=1}^{\infty} \lambda_j^{-2} |\langle Lx, u_j \rangle|^2$ is convergent. The vector

$$z = \sum_{j=1}^{\infty} \lambda_j^{-1} \langle Lx, u_j \rangle u_j$$

is therefore well defined and

$$\widehat{L}z = \sum_{j=1}^{\infty} \langle Lx, u_j \rangle u_j = Lx,$$

that is, $Lx \in R(\widehat{L})$; hence, $x \in \mathcal{D}(LL^*L)$.

We note that the vector LL^*Lx is given in terms of the eigenexpansion of Lx by

$$LL^*Lx = \sum_{j=1}^{\infty} \frac{1-\lambda_j}{\lambda_j} \langle Lx, u_j \rangle u_j.$$

7.4 The General Case

We now drop the assumption that $\widehat{L}$ is compact and give a very different proof of the converse that is inspired by an argument of Neubauer [5]. First, we need a well-known result.

Lemma. *If L is closed and densely defined, then LL^* is closed.*

Proof. Suppose that $y_n \in \mathcal{D}(LL^*)$, $y_n \to y$, and $LL^*y_n \to u \in H_2$. Then $(I + LL^*)y_n \to y + u$ and hence, since $\widehat{L}$ is bounded, $y_n \to \widehat{L}(y+u)$. Therefore, $\widehat{L}(y+u) = y$, that is, $y \in \mathcal{D}(LL^*)$ and $y + LL^*y = y + u$, or $LL^*y = u$, so LL^* is closed.

Theorem. *If $x \in \mathcal{D}(L)$ and $\|Lx - Lx_\alpha\| = O(\alpha)$, then $x \in \mathcal{D}(LL^*L)$.*

Proof. First, note that

$$x_\alpha = (I + \alpha L^*L)^{-1}x \in \mathcal{D}(L^*L),$$

and hence,

$$Lx_\alpha = (I + \alpha LL^*)^{-1}Lx \in \mathcal{D}(LL^*).$$

Also,

$$Lx_\alpha - Lx = -\alpha LL^*Lx_\alpha.$$

Therefore, by the hypothesis,

$$\|LL^*Lx_\alpha\| = O(1).$$

By the lemma, we know that LL^* is closed. The graph of LL^* is therefore closed and convex, and hence weakly closed. Since $\{\|LL^*Lx_\alpha\|\}$ is bounded, there is a sequence $\alpha_n \to 0$ with

$$LL^*Lx_{\alpha_n} \rightharpoonup w$$

for some w (here $\rightharpoonup$ indicates weak convergence). But $Lx_{\alpha_n} \to Lx$. Since the graph of LL^* is weakly closed, it follows that $Lx \in \mathcal{D}(LL^*)$ and $LL^*Lx = w$. In particular, $x \in \mathcal{D}(LL^*L)$, as claimed.

References

1. V.A. Morozov, *Methods for Solving Incorrectly Posed Problems*, Springer-Verlag, New York, 1984.
2. F. Riesz and B. Sz.-Nagy, *Functional Analysis,* Ungar, New York, 1955.
3. C.W. Groetsch and O. Scherzer, The optimal order of convergence for stable evaluation of differential operators, *Electronic J. Diff. Equations* **3** (1993), 1–12.
4. T. Kato, *Perturbation Theory for Linear Operators*, Springer-Verlag, Berlin, 1966.
5. A. Neubauer, *Regularization of Ill-posed Linear Operator Equations on Closed Convex Sets,* dissertation, Linz, 1985.

8 A Weakly Singular Boundary Integral Formulation of the External Helmholtz Problem Valid for All Wavenumbers

Paul J. Harris, Ke Chen, and Jin Cheng

8.1 Introduction

Over the last forty or so years the boundary integral method has become established as one of the most widely used methods for solving the exterior Helmholtz problem. The underlying differential equation can be reformulated as a boundary integral equation either by applying Green's theorem directly to the solution or by representing the solution in terms of a layer potential function. It is well known that the integral equation formulation arising from either of these methods does not have a unique solution for all real and positive values of the wavenumber. Over the years a number of methods for overcoming this problem have been proposed, most notably the so-called CHIEF method [1] and the Burton and Miller method [2] for Green's theorem formulation, and methods similar to those proposed by Panich [3] for the layer potential formulation. A survey of different methods for overcoming these problems is given in Amini et al. [4].

In the work presented in this paper we shall only consider the Burton and Miller formulation for overcoming the nonuniqueness problem. This method has the advantage that it is guaranteed to have a unique solution for all real and positive wavenumbers, but has the disadvantage that it introduces an integral operator with a hypersingular kernel. In this paper we present a method for reformulating the hypersingular integral operator in terms of integral operators that have kernel functions which are at worst weakly singular and hence relatively straightforward to approximate by standard numerical methods. Further, we shall show that the numerical results obtained using the methods described here are considerably more accurate than those obtained by the most widely used existing methods.

8.2 Boundary Integral Formulation

Consider the Helmholtz (or reduced wave-scattering) equation

$$\nabla\phi + k^2\phi = 0 \tag{8.1}$$

in some domain D_+ exterior to some closed, finite region D_- with closed and piecewise smooth surface S, subject to the boundary condition that $\partial\phi/\partial n$ is known on S and to the radiation condition

$$\lim_{r\to\infty} r\left(\frac{\partial\phi}{\partial r} - ik\phi\right) = 0,$$

where k is the acoustic wavenumber. In this work we shall assume that k is real and positive.

An application of Green's second theorem leads to

$$\int_S \phi(q)\frac{\partial G_k(p,q)}{\partial n_q} - G_k(p,q)\frac{\partial\phi(q)}{\partial n_q} dS_q = \frac{1}{2}\phi(p) \tag{8.2}$$

for $p \in S$, where

$$G_k(p,q) = \frac{e^{ik|p-q|}}{4\pi|p-q|}$$

is the free-space Green's function for the Helmholtz equation. However, it is well known that (8.2) does not possess a unique solution for certain discrete values of the wavenumber k, although the underlying differential equation (8.1) does possess a unique solution for all real and positive k. Further, the exact values of k for which (8.2) fails will depend on the shape of the surface S. Burton and Miller [2] proposed the alternative integral equation

$$\begin{aligned} -\frac{1}{2}\phi(p) + \int_S \phi(q)\left(\frac{\partial G_k(p,q)}{\partial n_q} + \alpha\frac{\partial^2 G_k(p,q)}{\partial n_p \partial n_q}\right) dS_q \\ = \frac{\alpha}{2}\frac{\partial\phi(p)}{\partial n_p} + \int_S \frac{\partial\phi(q)}{\partial n_q}\left(G_k(p,q) + \alpha\frac{\partial G_k(p,q)}{\partial n_p}\right) dS_q \end{aligned} \tag{8.3}$$

and showed that (8.3) has a unique solution for all real and positive values of the wavenumber k provided the imaginary part of the coupling constant α is nonzero. Further, it is shown in [4,5] that the almost optimal choice of α is $\alpha = i/k$ as this almost minimizes the condition number of the integral operator. However, we note that this formulation has introduced the integral operator with kernel function

$$\frac{\partial^2 G_k(p,q)}{\partial n_p \partial n_q},$$

which contains a $1/r^3$ hypersingularity. It is worth noting here that all the other integral operators appearing in (8.3) have kernel functions which are at worst weakly singular and so can be numerically evaluated using appropriate quadrature rules. Methods for evaluating this hypersingular integral operator will be discussed in the next section.

8.3 Numerical Methods

We now describe the collocation method with high-order elements. To solve (8.3), we first approximate the solution ϕ by

$$\tilde{\phi}(q) = \sum_{i=1}^{N} \phi_j \psi_j(q), \tag{8.4}$$

where $\{\psi_j(q)\}$ are a set of known basis functions and $\{\phi_j\}$ are a set of constants to be determined. Substituting (8.4) into (8.3) yields

$$\sum_{j=1}^{n} \phi_j \left[-\frac{1}{2}\psi_j(p) + \int_S \psi_j(q) \left(\frac{\partial G_k(p,q)}{\partial n_q} + \alpha \frac{\partial^2 G_k(p,q)}{\partial n_p \partial n_q} \right) dS_q \right]$$
$$= \frac{\alpha}{2} \frac{\partial \phi(p)}{\partial n_p} + \int_S \frac{\partial \phi(q)}{\partial n_q} \left(G_k(p,q) + \alpha \frac{\partial G_k(p,q)}{\partial n_p} \right) dS_q + R(p), \tag{8.5}$$

where R is a residual function. A linear system of equations is obtained by picking N points $p_1, p_2, \ldots, p_N$ at which we force $R(p_i) = 0$.

In order to be able to make use of (8.5) we need a method for evaluating the hypersingular integral

$$\int_S \psi_j(q) \frac{\partial^2 G_k(p_i,q)}{\partial n_p \partial n_q} dS_q. \tag{8.6}$$

Let us first review the commonly used collocation method with piecewise constant elements. This consists of rewriting (8.6) as

$$\int_S \psi_j(q) \frac{\partial^2 G_k(p_i,q)}{\partial n_p \partial n_q} dS_q$$
$$= \int_S (\psi_j(q) - \psi_j(p_i)) \frac{\partial^2 G_k(p_i,q)}{\partial n_p \partial n_q} dS_q + \psi_j(p_i) \int_S \frac{\partial^2 G_k(p_i,q)}{\partial n_p \partial n_q} dS_q. \tag{8.7}$$

Using the result given in [6], the second integral on the right-hand side of (8.7) can be made weakly singular and hence be evaluated using an appropriate quadrature rule. This can seen by rewriting (8.7) as

$$\int_S \psi_j(q) \frac{\partial^2 G_k(p_i,q)}{\partial n_p \partial n_q} dS_q$$
$$= \int_S (\psi_j(q) - \psi_j(p_i)) \frac{\partial^2 G_k(p_i,q)}{\partial n_p \partial n_q} dS_q + \psi_j(p_i) k^2 \int_S G_k(p_i,q) n_p \cdot n_q dS_q. \tag{8.8}$$

If we now choose the basis functions ψ_j to be piecewise constant then whenever the collocation point p_i is in the same element as the integration point q then the first integral on the right-hand side of (8.8) is zero and the problems associated with the hypersingular kernel function have been avoided. However, this will not work if any other basis functions (such as higher-order piecewise polynomials) are used, as the first integral on the right-hand side of (8.8) is no longer zero. Hence the piecewise constant approximation to ϕ has been widely used in practise.

Next we discuss the more general case of using high order collocation methods. The nonconstant basis functions can be used with the help of the recent result in [7]

$$\begin{aligned}
&\int_S \psi_j(q) \frac{\partial^2 G_k(p,q)}{\partial n_p \partial n_q} dS_q \\
&= \int_S \left(\{\psi_j(q) - \psi_j(p) - \nabla\psi_j(p)\cdot(q-p)\} \frac{\partial^2 G_k(p,q)}{\partial n_p \partial n_q} \right) dS_q \\
&\quad + k^2 \psi_j(p) \int_S n_p \cdot n_q G_k(p,q) dS_q + \int_S \nabla\psi_j(p)\cdot n_q \frac{\partial G_k(p,q)}{\partial n_p} dS_q \\
&\quad - k^2 \int_{D_-} \nabla\psi_j(p)\cdot(q-p) \frac{\partial G_k(p,q)}{\partial n_p} dV_q - \frac{1}{2}\nabla\psi_j(p)\cdot n_p,
\end{aligned} \tag{8.9}$$

in which all of the integrals on the right-hand side are at worst weakly singular, but which has now introduced the volume integral over the interior D_- of S. Clearly we can avoid having to evaluate this volume integral if we only consider the case $k = 0$. This is possible by rewriting the hypersingular integral as

$$\begin{aligned}
&\int_S \psi_j(q) \frac{\partial^2 G_k(p,q)}{\partial n_p \partial n_q} dS_q \\
&= \int_S \psi_j(q) \left[\frac{\partial^2 G_k(p,q)}{\partial n_p \partial n_q} - \frac{\partial^2 G_0(p,q)}{\partial n_p \partial n_q} \right] dS_q + \int_S \psi_j(q) \frac{\partial^2 G_0(p,q)}{\partial n_p \partial n_q} dS_q.
\end{aligned} \tag{8.10}$$

The first integral on the right-hand side of (8.10) is at worst weakly singular whilst the second integral can be evaluated using (8.9) with $k = 0$ to yield

$$\begin{aligned}
&\int_S \psi_j(q) \frac{\partial^2 G_k(p,q)}{\partial n_p \partial n_q} dS_q \\
&= \int_S \psi_j(q) \left[\frac{\partial^2 G_k(p,q)}{\partial n_p \partial n_q} - \frac{\partial^2 G_0(p,q)}{\partial n_p \partial n_q} \right] dS_q \\
&\quad + \int_S \left(\{\psi_j(q) - \psi_j(p) - \nabla\psi_j(p)\cdot(q-p)\} \frac{\partial^2 G_0(p,q)}{\partial n_p \partial n_q} \right) dS_q \\
&\quad + \int_S \nabla\psi_j(p)\cdot n_q \frac{\partial G_0(p,q)}{\partial n_p} dS_q - \frac{1}{2}\nabla\psi_j(p)\cdot n_p.
\end{aligned} \tag{8.11}$$

Hence it is possible to solve (8.3) using (8.11) to evaluate the hypersingular integral operator by means of any type of high-order basis functions; in particular, high-order (i.e., nonconstant) polynomial basis functions.

8.4 Numerical Results

Here we present some results to illustrate the new formulation using high-order collocation methods. These are for the following test surfaces:

(i) A unit sphere with point sources at (0, 0, 0.5) and (0.25, 0.25, 0.25) and with strengths $2 + 3i$ and $4 - i$, respectively.

(ii) A cylinder of length 0.537 and radius 0.2685 with point sources at (0, 0, 0.15) and (0.25, 0.25, 0.25) and with strengths $2 + 3i$ and $4 - i$, respectively.

(iii) A 'peanut-shaped' surface defined by

$$x = \sqrt{\cos 2\theta + \sqrt{1.5 - \sin^2 2\theta}} \sin\theta \cos\gamma,$$

$$y = \sqrt{\cos 2\theta + \sqrt{1.5 - \sin^2 2\theta}} \sin\theta \sin\gamma,$$

$$z = \sqrt{\cos 2\theta + \sqrt{1.5 - \sin^2 2\theta}} \cos\theta$$

for $0 \le \theta \le \pi$ and $0 \le \gamma < 2\pi$, with point sources at (0.2, 0, 1) and (0, 0.2, −0.75) and with strengths $2 + 3i$ and $4 - i$, respectively. In each case the surface S has been approximated by a number of quadratically curved triangular elements.

The usual high-order piecewise polynomial basis functions for solving (8.3) on a surface made up of such elements are usually defined in terms of the nodal points of the surface elements (isoparametric elements). However, when trying to solve (8.3) there is a major problem with using the surface nodes as collocation points as we need to be able to calculate the normal at the collocation points (n_p in (8.3)), but the interpolated surface does not possess a well-defined normal at these points. One possible solution to this problem is to use the normal to the original surface, but this will not be possible if the original surface has an edge or a vertex. The alternative is to collocate at points which are interior to the elements and different from the surface interpolation nodes. This has the immediate advantage that the approximate surface will have a unique normal at each collocation point, but the disadvantage that the approximate solution will no longer be continuous at the element boundaries. The choice of the location of these interior points will have a significant effect on the overall accuracy of the numerical scheme. Here we have used discontinuous linear and quadratic schemes to calculate the solution. In the linear case, the location of the collocation points is determined by a parameter δ which governs the location of the three collocation points relative to the element centroid and each element vertex. In the quadratic case, the additional three collocation points are simply taken as the midpoints between each two consecutive collocation points of the linear case. Fig. 1 and 2 show the relative error in the computed surface pressure for each surface at the first characteristic wavenumber using discontinuous linear and quadratic elements, respectively. In the linear case, it is clear that the optimal value is $\delta = 0.4$, whilst in the quadratic case, the situation is not quite so clear-cut, but using $\delta = 0.25$ would seem to be appropriate. These are the values that have been used in the calculations described below.

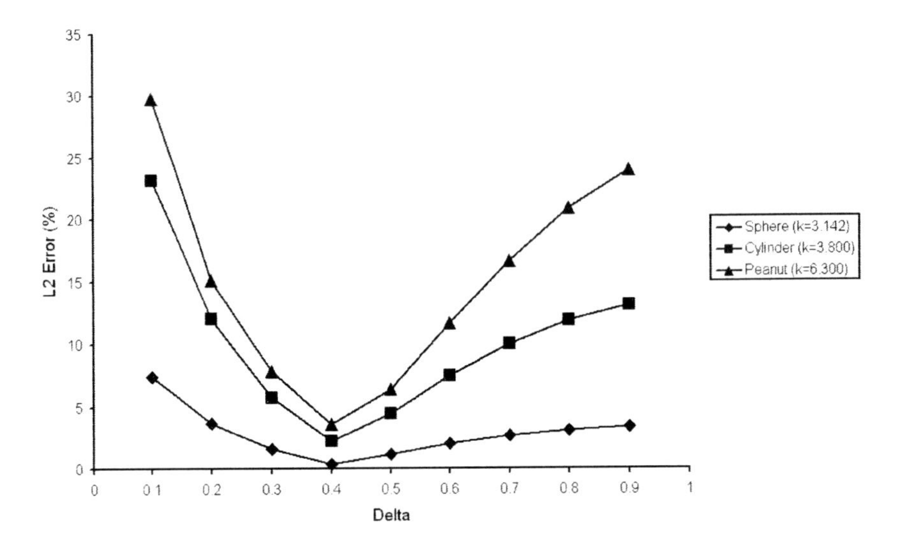

Fig. 1. The L_2 relative error on each test surface for different values of δ using the linear basis functions.

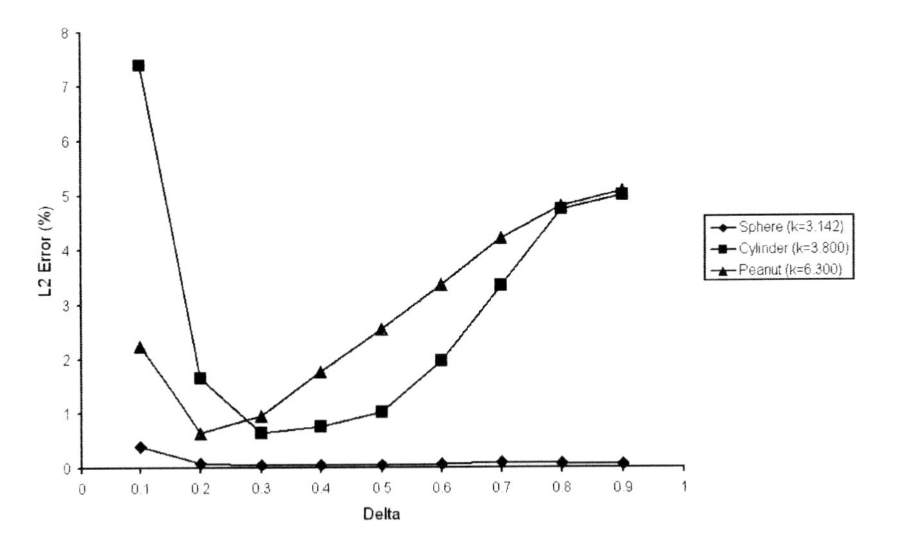

Fig. 2. The L_2 relative error on each test surface for different values of δ using the quadratic basis functions.

Fig. 3, 4, and 5 show the results of using the discontinuous constant, linear, and quadratic approximations for the sphere, cylinder, and peanut, respectively.

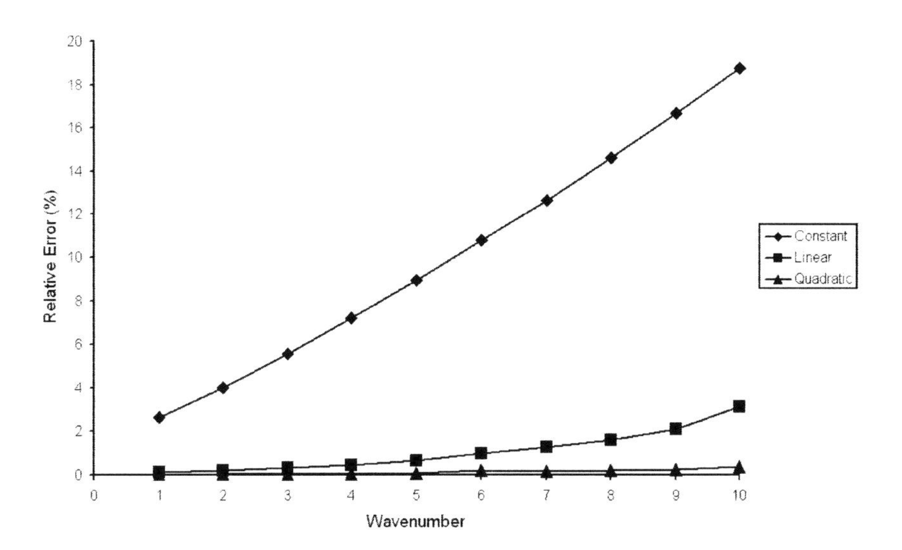

Fig. 3. The L_2 relative error on the unit sphere using discontinuous constant, linear, and quadratic basis functions.

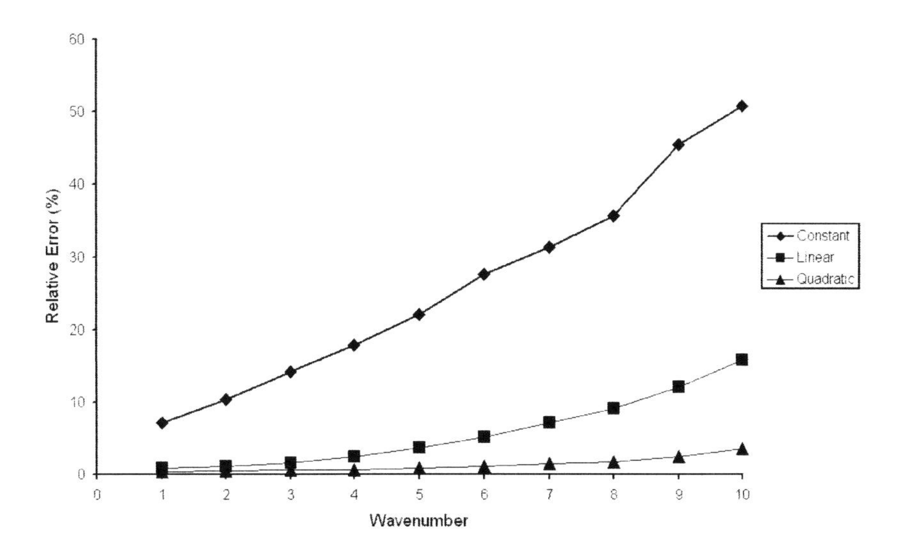

Fig. 4. The L_2 relative error on a cylinder using discontinuous constant, linear, and quadratic basis functions.

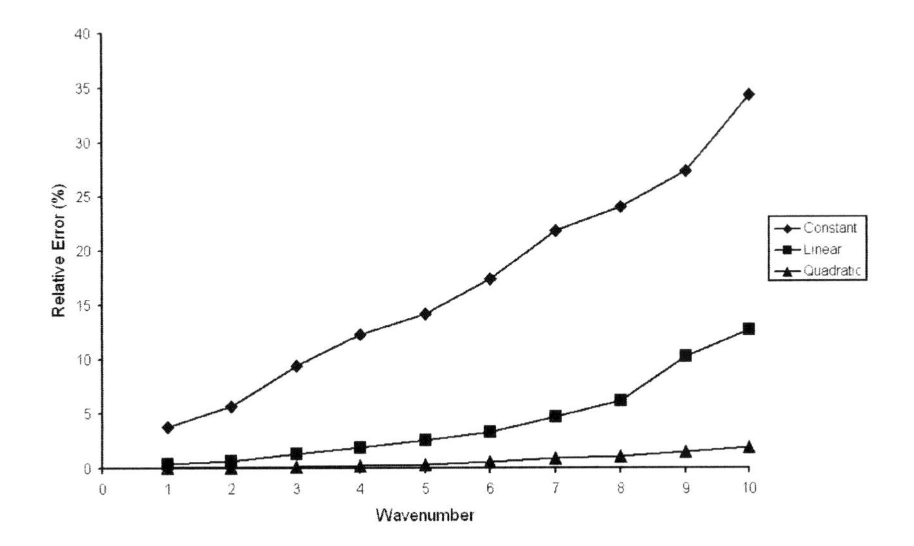

Fig. 5. The L_2 relative error on the peanut using discontinuous constant, linear, and quadratic basis functions.

The results of employing the commonly used piecewise constant approximation are also given as a comparison. In each case we see that the linear and quadratic discontinuous approximations are considerably more accurate than the usual piecewise constant approximation, and that the quadratic approximation is more accurate than the linear approximation.

8.5 Conclusions

Over the years, the Burton and Miller method has been shown to be a theoretically reliable method for determining the acoustic field radiated or scattered by an object. However, in practice, nearly all of the earlier work on this problem was restricted to using a piecewise constant approximation and there has always been the problem of how to evaluate the integral operator which involves the second derivative of the Green's function beyond the piecewise constants.

The work in the paper shows how this problem can be overcome; by using a singularity subtraction technique, one manages to avoid introducing any volume integrals. This new formulation allows a much wider class of basis functions to be considered. The numerical results show that the higher-order piecewise polynomials considered here give considerably more accurate results.

References

1. H.A. Schenck, Improver integral formulations for acoustic radiation problems, *J. Acoust. Soc. Amer.* **44** (1968), 41–58.

2. A.J. Burton and G.F. Miller, The application of integral equation methods for the numerical solution of boundary value problems, *Proc. Roy. Soc. London A* **232** (1971), 201–210.

3. O.I. Panich, On the question of solvability of the exterior boundary value problems for the wave equation and Maxwell's equations, *Russian Math. Surveys* **20**:A (1965), 221–226.

4. S. Amini, P.J. Harris, and D.T. Wilton, Coupled boundary and finite element methods for the solution of the dynamic fluid-structure interaction problem, in *Lecture Notes in Engng.* **77**, C.A. Brebbia and S.A. Orszag, eds. Springer-Verlag, London, 1992.

5. S. Amini, On the choice of coupling parameter in boundary integral formulations of the exterior acoustic problem, *Appl. Anal.* **35** (1989), 75–92.

6. W.L. Meyer, W.A. Bell, B.T. Zinn, and M.P. Stallybrass, Boundary integral solution of three dimensional acoustic radiation problems, *J. Sound Vibrations* **59** (1978), 245–262.

7. K. Chen, J. Cheng, and P.J. Harris, A new weakly-singular reformulation of the Burton-Miller method for solving the exterior Helmholtz problem in three dimensions, 2004 (submitted for publication).

9 Cross-referencing for Determining Regularization Parameters in Ill-Posed Imaging Problems

John W. Hilgers and Barbara S. Bertram

9.1 Introduction

Imaging problems offer some of the best examples of ill-posed Fredholm first kind integral equations. These problems are also very challenging because they consist of inherently two-dimensional integral operators with kernels, or point spread functions (psf), which amount to degraded identity operators. This means that the kernels resemble corrupted Dirac delta functions, or spikes that have been diffracted into peaks with finite diameters. Problems featuring such kernels are among the most pathological because the corresponding spectral falloff of eigenvalues with index is extremely rapid. This magnifies the ill-posed nature of the inverse problem [1].

The integral operator K in this case is defined by

$$Kf(x, y) = \int\int K(x - x', y - y')f(x', y')dx'dy', \quad 0 \leq x,\ y \leq 1.$$

Because K is a Hilbert–Schmidt kernel, operator K is completely continuous from $L_2([0, 1] \times [0, 1])$ into itself. As such, the inverse operator, K^{-1}, or generalized inverse, $K^{\dagger}$, is necessarily unbounded (unless K is degenerate)[2]. This is the source of the instability in the inverse problem.

The practical problem is usually encountered in a noise-contaminated form

$$Kf = \bar{g} = g + \epsilon, \tag{9.1}$$

where $g = Kf_0$, f_0 is the true solution and ϵ represents additive error. The inverse problem is to obtain f_0, or a good approximation of it, from the measured data $\bar{g}$ in spite of the noise ϵ. To this end, a least square approach is usually taken whereby one tries to solve

$$\min_f \|Kf - \bar{g}\|,$$

The second author wishes to thank Signature Research, Inc. for hosting her sabbatical.

any solution of which must solve the normal equation $K^*Kf = K^*\bar{g}$, which is even more ill posed than the original (9.1) (see [13]–[15]). There are a number of ways to avoid the instability of the normal equation by applying some form of regularization. Three general classes follow.

1. Solve
$$(K^*K + \alpha L^*L)f = K^*\bar{g}, \tag{9.2}$$
where L and α are the regularization operator and parameter, respectively. Here α is just the Lagrange multiplier in the constrained least squares (CLS) approach (see [13], [15], and [16]–[22]). For $\alpha > 0$, the operator in (9.2) is invertible when $N(K) \cap N(L) = \{0\}$ (N representing the null space).
2. Solve the normal equation by a spectral or singular value decomposition (SVD) expansion, and keep only the first terms. The number of terms retained becomes the de facto regularization parameter [2].
3. Use an iterative approach to solving (9.1) directly, or the normal equation, or even (9.2) (see [28], [25], and [26]). In this paper the conjugate gradient method is used. In all such methods the number of iterations is the equivalent regularization parameter.

What is clear is that all regularization techniques involve assigning values to one or more parameters which control the amount of smoothness to be forced on the approximate solution. How this is to be done is a problem of long standing and the subject of this paper.

9.2 The Parameter Choice Problem

How to determine the type and amount of regularization to be applied in an ill-posed inverse problem is an open question, probably dating back to the Russian A.N. Tikhonov and the work he did in the 1940's in inverting a solution to the heat equation in an effort to infer past climate from measured cores of Siberian permafrost. The papers [23]–[30] cited below, and hundreds of others, consider this problem in some way. The answer to the problem depends, among other things, on: the high frequency components of f_0, the spectrum of K^*K, the statistics of ϵ, and the size of $\|\epsilon\|/\|g\|$. Some specific results are given in [2]–[8] and [12] for regularization schemes like (9.2), and [1] discusses how the spectrum of K^*K can impact the stability of the optimal regularization parameter.

The methodology advanced herein, termed CREF for cross referencing different regularized families, has evolved over about 10 years (see [6]–[11]), and is based on three assumptions.

First, no matter what complex dependence exists between any particular regularized approximate to f_0, call it $\tilde{f}$, and the above-mentioned (as well as other) factors, the following behavior will always be noted. As the regularization parameter varies, whether it be continuous or discrete, $\tilde{f}$ will evolve from over-smoothed to something near f_0 to something increasingly unstable.

Second, when two different regularized approximates, say $\tilde{f}$ and $\tilde{h}$, both pass into their respective unstable regimes, the highly disordered nature of the instabilities means it is highly probable that $\|\tilde{f} - \tilde{h}\|$ increases rapidly with continued parameter variation, a phenomenon which is easily detected.

Third, quantities that get close to the same thing get close to each other (i.e., the triangle inequality).

Based on these three assumptions, the CREF method is simply defined as the solution to

$$\text{argmin} \parallel \tilde{f} - \tilde{h} \parallel, \tag{9.3}$$

where the arguments are the regularization parameters of the two families. In [7] the condition

$$\|\epsilon\|/\|g\| < 1$$

is shown to be a likely indicator of the existence of (9.3) for regularizers of the form (9.2). In cases run to date (see [6]–[11]), (9.3) has never failed to exist, and has always provided stable approximations of high acuity, even when $\|\epsilon\|/\|g\| \approx 0.5$.

9.3 Advantages of CREF

The major advantage of CREF is that the approximation inherent in (9.3) is undertaken in the domain space (of K). In contrast, the generalized cross-validation (GCV) method [5] minimizes a function which approximates $\|Kf_0 - K\tilde{f}\|$, thereby executing a comparison in the range space, where the action of the kernel can suppress very pathological features in $\tilde{f}$.

To state this another way, the optimal parameter(s) which minimize $\|\tilde{f} - f_0\|$ can fluctuate over several orders of magnitude for different error vectors, ϵ, even when $\|\epsilon\|$ is fixed. This phenomenon is driven by the rate of decrease of the spectrum of K^*K, and is a feature shared by the GCV approximation to the optimal parameter(s) [1].

However, the optimal regularizer, $\tilde{f}$, considered as a vector in the appropriate function space, is a very stable approximator to f_0, and the quality of the approximation appears to depend only on $\|\epsilon\|$ no matter how unstable the dependence of the optimal parameter(s) becomes. Using (9.3) accesses only stable approximations. Basically the only way CREF can fail is:

(1) If the regularization method is pathologically wrong for the particular problem so the optimal parameter is infinite or undefined, or if $\|\epsilon\|/\|g\|$ is excessive, in which cases none of the competing methods would perform well either.

(2) That the second assumption of the method fails, namely that both regularizers in (9.3) will become unstable in the very same way, so (9.3) remains small even when $\tilde{f}$ and $\tilde{h}$ become unstable.

While it may well be possible to engineer examples where (1) and (2) occur, in problems undertaken with standard regularization techniques, with $\|\epsilon\|/\|g\| < 1$, problems with CREF have never been experienced in practice (see [6]–[11]).

9.4 Examples

All images are 256 × 256, centered on 512 × 512 for zero padding. In the figures only the 256 × 256 image is displayed. The noise is additive and Gaussian. All convolutions and deconvolutions were done in the frequency domain using standard MATLAB FFT routines.

9.4.1 The Garage

The psf was computed by the MATLAB AIRY function with a support diameter of about 10 pixels. The object was the "garage" (see Fig. 1). Gaussian noise was added to the image with $\|\epsilon\|/\|g\| = .04$, and this is shown in Fig. 1.

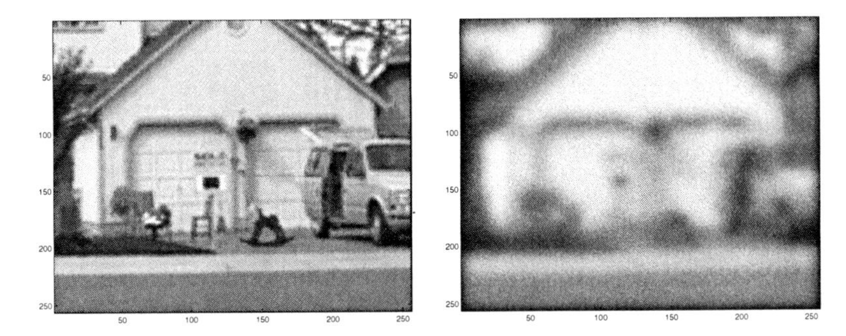

Fig. 1. The garage as object and its image with noise.

9.4.1.1 CLS Example

Two CLS regularizers with $L_1 = I$ and $L_2 = \Delta$, the identity and Laplacian, respectively, were compared. A power of ten grid search was executed with the result that $\alpha = 10$ and $\beta = 100$ where α and β are the parameters for L_1 and L_2, respectively. These values are identical to the true optimum parameter values (also taken to the nearest power of 10). The results are shown in Fig. 2.

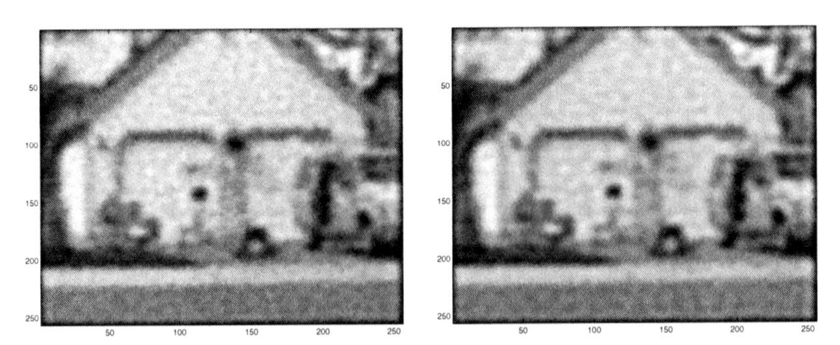

Fig. 2. Reconstructions of the garage with the identity and Laplacian, respectively.

9.4.1.2 Conjugate Gradient and CLS/Laplacian Example

This example is the same as 9.4.1.1 except that here CREF compared the conjugate gradient (CG) method with a positivity constraint invoked with CLS, $L = \Delta$. The search was done on a grid pitting multiples of ten iterations of CG against powers of ten for α with the result that a minimum was found for 20 iterations of CG and $\alpha = 1000$ for the CLS method. The resulting approximations are shown in Fig. 3.

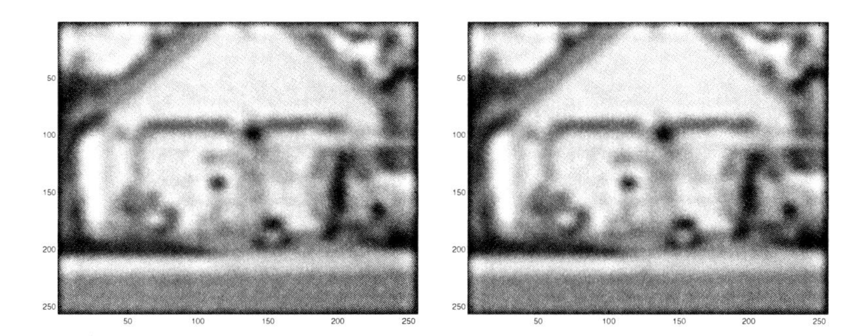

Fig. 3. Reconstructions of the garage using conjugate gradient and Laplacian, respectively.

9.4.1.3 Conjugate Gradient and CLS/Identity Example

This is like examples 9.4.1.1 and 9.4.1.2 except that here CG with a positivity constraint was used with CLS, $L = I$. The same grid was searched as in example 9.4.1.2 with the minimum occurring for 70 iterations of CG and $\alpha = 10$ for the CLS. This is the optimal α value for CLS with $L = I$. The reconstructions are shown in Fig. 4.

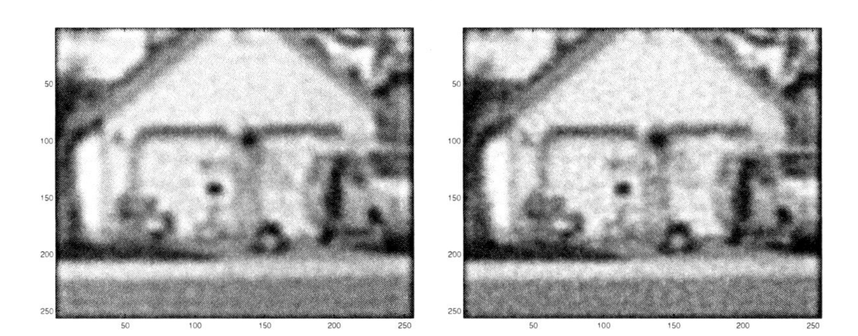

Fig. 4. Reconstructions of the garage using conjugate gradient and identity, respectively.

9.4.2 The Jeep

The psf is measured and is approximately Gaussian. The object was the "jeep" shown in Fig. 5. Gaussian noise was added to the image with the extreme noise level, $\|\epsilon\|/\|g\| = 0.42$, and this, too, is shown in Fig. 5.

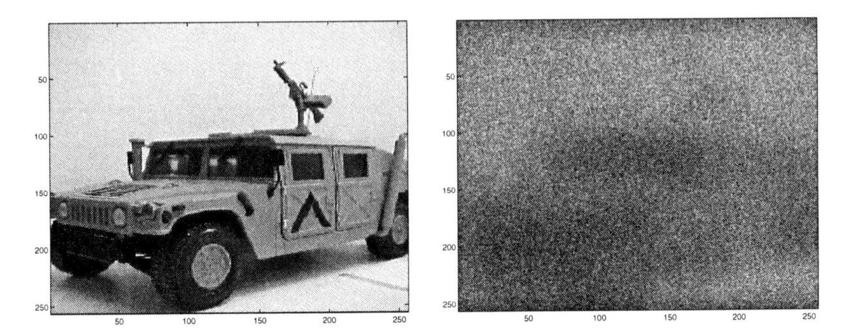

Fig. 5. The jeep as object and its image with noise.

9.4.2.1 CLS Example

Two CLS regularizers with $L_1 = I$ and $L_2 = \Delta$, the identity and Laplacian, respectively, were compared. A CREF power of ten grid search was executed with the result that $\alpha = 10^{-4}$ and $\beta = 0.1$, where α and β are the parameters for L_1 and L_2, respectively. The reconstructions are shown in Fig. 6.

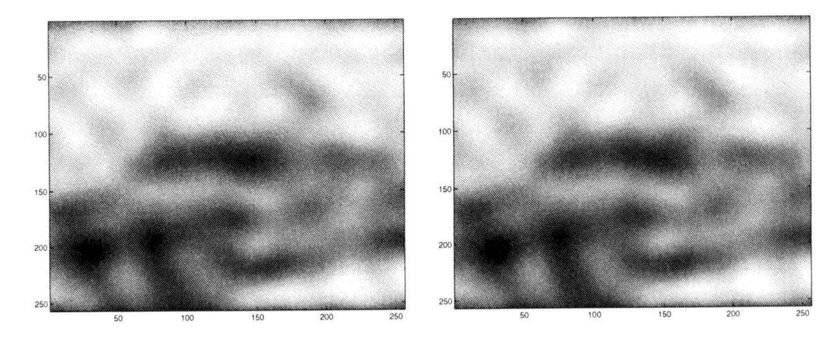

Fig. 6. Reconstructions of the jeep with the identity and Laplacian, respectively.

9.4.2.2 Conjugate Gradient and CLS/Laplacian Example

Example 9.4.2.1 was repeated using CG (with positivity constraint) and CLS with $L = \Delta$ as the two regularizers. The parameters are the number n of iterations for CG and β as in example 9.4.2.1. The search for solving (9.3) was conducted over negative powers of ten and nonnegative integers for β (and, of course, integers for n). The minimum square variation was attained for $n = 3$ iterations and $\beta = 5$. The results are shown in Fig. 7. Further comments are included in the summary.

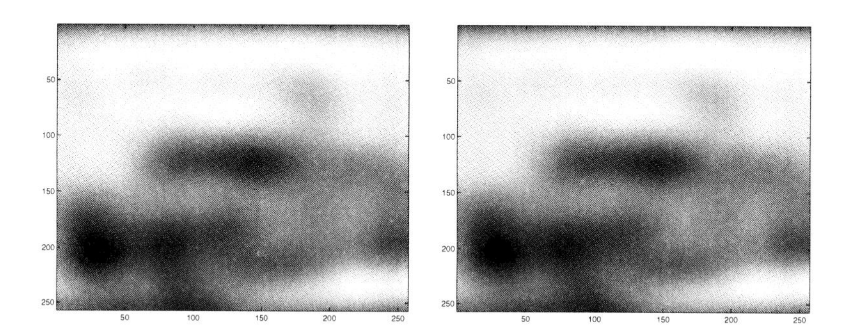

Fig. 7. Reconstructions with CG (with positivity constraint) and constrained least squares with Laplacian.

9.5 Summary

The CREF method continues to provide highly stable approximations to the solutions of ill-posed inverse problems, particularly in the presence of high noise levels, and also in the case of two-dimensional imaging problems.

It should be emphasized that in example 9.4.2.1, CREF provided both regularized approximates with their true optimum parameter values (to the nearest power of ten). The true parameter values are found by comparing each reconstruction with the true object shown in Fig. 1 and 5. Note that example 9.4.2.1 shows a bit less blur but more noise when compared with the solutions of example 9.4.2.2.

This is not surprising since β of example 9.4.2.1 is 0.1 while that of example 9.4.2.2 is 5. This, in turn, is due to the fact that in performing each iteration of CG, a nonnegativity constraint was also enforced by the simple expedient of zeroing out negative-going regions of the reconstruction. As a result the reconstruction is smoother and when used in CREF pulls β up to the larger value. In example 9.4.2.1 neither CLS regularizer included a nonnegativity constraint.

In general, whether one favors the deblur or the noise removal functions of the regularization method depends on the level of each degrading factor present. In this case Fig. 6 may appear preferable to the "optimal" solution of Fig. 7.

References

1. J.W. Hilgers and W.R. Reynolds, Instabilities in the optimization parameter relating to image recovery problems, *J. Opt. Soc. Amer. A* **9** (1992), 1273–1279.

2. C.W. Groetsch, *The Theory of Tikhonov Regularization for Fredholm Equations of the First Kind,* Res. Notes in Math. **105**, Pitman, London, 1977.

3. V.A. Morozov, *Methods for Solving Incorrectly Posed Problems,* Springer-Verlag, Berlin-Heidelberg, 1984.

4. S.J. Reeves and R.M. Mersereau, Optimal estimation of the regularization parameter and stabilizing functional for regularized image restoration, *Optical Engng.* **29** (1990), 446–454.

5. G.H. Golub, M. Heath, and G. Wahba, Generalized cross-validation as method for choosing a good ridge parameter, *Technometrics* **21** (1979), 215–223.

6. J. Hilgers, B. Bertram, and W. Reynolds, Cross-referencing different regularized families of approximate solutions to ill-posed problems for solving the parameter choice problem in generalized CLS, in *Integral Methods in Science and Engineering,* P. Schiavone, C. Constanda, and A. Mioduchowski, eds., Birkhauser, Boston, 2002, 105–110.

7. J. Hilgers and B. Bertram, Comparing different regularizers for choosing the parameters in CLS, *Computers and Math. with Appl.* (to appear).

8. J.W. Hilgers, B.S. Bertram, and W.R. Reynolds, Extension of constrained least squares for obtaining regularized solutions to first-kind integral equations, in *Integral Methods in Science and Engineering,* B. Bertram, C. Constanda, and A. Struthers, eds., Chapman & Hall/ CRC, Boca Raton, FL, 2000, 185–189.

9. E.M. Wilcheck, A comparison of cross-referencing and generalized cross-validation in the optimal regularization of first kind equations, Master's Thesis, Michigan Tech. Univ., 1992.

10. J.W. Hilgers, B.S. Bertram, M.M. Alger, and W.R. Reynolds, Extensions of the cross-referencing method for choosing good regularized solutions to image recovery problems, in *Proc. SPIE* **3171**, Barbour et al., eds., 1997, 234–237.

11. M.M. Alger, J.W. Hilgers, B.S. Bertram, and W.R. Reynolds, An extension of the Tikhonov regularization based on varying the singular values of the regularization operator, *Proc. SPIE* **3171**, Barbour et al., eds., (1997), 225–231.

12. H.W. Engl, M. Hanke, and A. Neubauer, *Regularization of Inverse Problems,* Kluwer, Dordrecht, 1996.

13. A.N. Tikhonov, Solution of incorrectly formulated problems and the regularization method, *Soviet Math. Dokl.* **4** (1963), 1035–1038.

14. C.K. Rushforth, Signal restoration, functional analysis and Fredholm integral equations of the first kind, in *Image Recovery: Theory and Application,* H. Stark, ed., Academic Press, New York, 1987.

15. A.N. Tikhonov and V.Y. Arsenin, *Solutions of Ill-posed Problems,* Winston and Sons, Washington, DC, 1977.

16. A.M. Thompson et. al., A Study of methods of choosing the smoothing parameter in image restoration by regularization, *IEEE Trans. on PAMI* **13** (1991), 326–339.

17. M. Bertero, *Regularization Methods for Linear Inverse Problems,* Lect. Notes in Math., Springer-Verlag, Berlin-Heidelberg, 1986.

18. A.K. Katsaggelos, Iterative image restoration algorithms, *Optical Engng.* **28** (1989), 735–738.

19. J.W. Hilgers, Non-iterative methods for solving operator equations of the first kind, TSR 1413 Math. Res. Center, Univ. of Wisconsin-Madison, 1974.

20. B. Bertram, On the use of wavelet expansions and the conjugate gradient method for solving first kind integral equations, in *Integral Methods in Science and Engineering,* B. Bertram, C. Constanda, and A. Struthers, eds., Chapman & Hall/CRC, Boca Raton, FL, 2000, 62–66.

21. B. Bertram and H. Cheng, On the use of the conjugate gradient method for the numerical solution of first kind integral equations in two variables, in *Integral Methods in Science and Engineering,* P. Shiavone, C. Constanda, and A. Mioduchowski, eds., Birkhauser, Boston, 2002, 51–56.

22. H. Cheng and B. Bertram, On the stopping criteria for conjugate gradient solutions of first kind integral equations in two variables, in *Integral Methods in Science and Engineering,* P. Schiavone, C. Constanda, and A. Mioduchowski, eds., Birkhauser, Boston, 2002, 57–62.

23. L. Desbat and D. Girard, The "minimum reconstruction error" choice of reconstruction parameters: some more efficient methods and the application to deconvolution problems, *SIAM J. Sci. Comput.* **16** (1995), 1387–1403.

24. M. Hanke and T. Raus, A general heuristic for choosing the regularization parameter in ill-posed problems, *SIAM J. Sci. Comput.* **17** (1996), 956–972.

25. A. Frommer and P. Maass, Fast CG-based methods for Tikhonov- Phillips regularization, *SIAM. J. Sci. Comput.* **20** (1999), 1831–1850.

26. M.E. Kilmer and D.P. O'Leary, Choosing regularization parameters in iterative methods for ill-posed problems, *SIAM J. Matrix Anal. Appl.* **22** (2001), 1204–1221.

27. D.P. O'Leary, Near-optimal parameters for Tikhonov and other regularization methods, *SIAM J. Sci. Comput.* **23** (2001), 1161–1171.

28. S. Saitoh, *Integral Transformations, Reproducing Kernels, and Their Applications,* Addison-Wesley Longman, London, 1997.

29. M. Rojas and D.C. Sorensen, A trust region approach to the regularization of large-scale discrete forms of ill-posed problems, *SIAM J. Sci. Comput.* **23** (2002), 1842–1860.

30. P.R. Johnston and R.M. Julrajani, An analysis of the zero-crossing method for choosing regularization parameters, *SIAM J. Sci. Comput.* **24** (2002), 428–442.

10 A Numerical Integration Method for Oscillatory Functions over an Infinite Interval by Substitution and Taylor Series

Hiroshi Hirayama

10.1 Introduction

The arithmetic operations and functions of power series can be defined easily in Fortran 90, C++ [1], and C#. The functions represented in these languages, which consist of arithmetic operations, predefined functions, and conditional statements, can be expanded in power series.

We consider the integral of an oscillatory function over an infinite interval, of the form

$$I = \int_0^\infty f(x) g(h(x))\, dx, \tag{10.1}$$

where $f(x)$ is a slowly decaying function, $g(x)$ is one of $\sin x$, $\cos x$, $J_n(x)$, and $Y_n(x)$ (the Bessel function of the first kind of integral order), and $h'(x) > 0$.

Let $t = h(x)$; then

$$I = \int_0^\infty f(h^{-1}(t))(\frac{d}{dt} h^{-1}(t)) g(t)\, dt.$$

The solution of an ordinary differential equation can be expanded in a Taylor series by Picard's method of successive approximations and the inverse function of $f(x)$ satisfies the ordinary differential equation

$$\frac{dy}{dx} = \frac{1}{f'(y)}.$$

Therefore, the integrand above, that is,

$$f(h^{-1}(t))(\frac{d}{dt} h^{-1}(t)),$$

can be expanded in a Taylor series. Using these Taylor series, we can give an effective numerical integration method for this type of integral.

10.2 Taylor Series

In this section, we explain the basic idea of expanding functions in Taylor series. The reader is referred to [2]–[4] for details.

An arithmetic program for Taylor series can easily be made. The following relations are valid not only at the origin, but also at any other point. Taylor series can be defined in the form

$$f(x) = f_0 + f_1 x + f_2 x^2 + f_3 x^3 + f_4 x^4 + \cdots, \tag{10.2}$$
$$g(x) = g_0 + g_1 x + g_2 x^2 + g_3 x^3 + g_4 x^4 + \cdots,$$
$$h(x) = h_0 + h_1 x + h_2 x^2 + h_3 x^3 + h_4 x^4 + \cdots. \tag{10.3}$$

Addition and subtraction. If $h(x) = f(x) \pm g(x)$, then

$$h_i = f_i \pm g_i.$$

Multiplication. If $h(x) = f(x)g(x)$, then

$$h_n = \sum_{k=0}^{n} f_i g_{n-i}.$$

Division. If $h(x) = f(x)/g(x)$, then

$$h_0 = \frac{f_0}{g_0}, \quad h_n = \frac{1}{g_0}\left(f_n - \sum_{k=0}^{n-1} h_k g_{n-k}\right), \quad n \geq 1.$$

Differentiation. If $h(x) = df(x)/dx$, then

$$h_m = 0, \quad h_n = (n+1) f_{n+1}, \quad n = 1, \ldots, m-1.$$

Integration. If $h(x) = \int f(x)dx$, then

$$h_0 = 0, \quad h_n = \frac{1}{n} f_{n-1}, \quad n \geq 1.$$

Exponential function. If $h(x) = e^{f(x)}$, then $dh/dx = hdf/dx$. Substituting (10.2) and (10.3) in this differential equation and comparing the coefficients, we find that

$$h_0 = e^{f_0}, \quad h_n = \frac{1}{n} \sum_{k=1}^{n} k h_{n-k} f_k, \quad n \geq 1.$$

We can easily get the same type of differential equation and the same type of relation between the coefficients of Taylor series for other elementary transcendental functions.

10.3 Integrals of Oscillatory Type

We consider the oscillatory-type integral

$$I = \int_0^\infty f(x) \sin x \, dx, \tag{10.4}$$

where $f(x)$ is $O(x^\alpha)$, $\alpha < 0$, and $f^{(n)}(x)$ is $O(x^{\alpha-n})$ as $x \to \infty$.

Integral (10.4) can be rewritten as

$$I = \int_0^a f(x) \sin x \, dx + \int_a^\infty f(x) \sin x \, dx.$$

The first integral on the right-hand side can be computed using ordinary numerical integration methods.

The second integral on the right-hand side is transformed through integration by parts into

$$\begin{aligned}\int_a^\infty f(x) \sin x \, dx &= [-f(x) \cos x]_a^\infty + \int_a^\infty f'(x) \cos x \, dx \\ &= f(a) \cos a + \int_a^\infty f'(x) \cos x \, dx.\end{aligned}$$

Repeating this operation M times, we arrive at

$$\begin{aligned}\int_a^\infty f(x) \sin x \, dx &= f(a) \cos a - f'(a) \sin a - f''(a) \cos a \\ &+ \cdots + f^{(M-1)}(a) \sin \left(x + \frac{M\pi}{2}\right) + \int_a^\infty f^{(M)}(x) \sin \left(x + \frac{M\pi}{2}\right) dx.\end{aligned}$$

Similarly, for $\cos x$,

$$\begin{aligned}\int_a^\infty f(x) \cos x \, dx &= -f(a) \sin a - f'(a) \cos a - f''(a) \sin a \\ &+ \cdots + f^{(M-1)}(a) \cos \left(x + \frac{M\pi}{2}\right) + \int_a^\infty f^{(M)}(x) \cos (x + \frac{M\pi}{2}) \, dx.\end{aligned}$$

The above results can be written as

$$\begin{aligned}\int_a^\infty f(x) \sin x dx &= (f(a) - f''(a) + f^{(4)}(a) - f^{(6)}(a) + \cdots) \cos a \\ &- (f'(a) - f_{(3)}(a) + f^{(5)}(a) + \cdots) \sin a, \qquad (10.5)\\ \int_a^\infty f(x) \cos x dx &= -(f(a) - f''(a) + f^{(4)}(a) - f^{(6)}(a) + \cdots) \sin a \\ &- (f'(a) - f_{(3)}(a) + f^{(5)}(a) + \cdots) \cos a. \qquad (10.6)\end{aligned}$$

These series are generally divergent. Calculation must be stopped when the minimum of the absolute value of the added terms is achieved.

Analogous results are obtained for

$$I = \int_0^\infty f(x) J_n(x)\, dx,$$

where $J_n(x)$ is the Bessel function of the first kind and order n, $f(x)$ is smooth and $O(x^\alpha)$, $\alpha < 1/2$, and $f^{(n)}(x)$ is $O(x^{\alpha-n})$ as $x \to \infty$.

The above integral can be split as

$$I = \int_0^a f(x) J_n(x)\, dx + \int_a^\infty f(x) J_n(x)\, dx.$$

The first integral on the right-hand side is calculated by means of ordinary numerical methods. The second integral is transformed through integration by parts. Using the equality

$$\int x^n J_{n-1}(x)\, dx = x^n J_n(x),$$

we obtain

$$\begin{aligned}
&\int_a^\infty f(x) J_n(x)\, dx \\
&\quad = [f(x) J_{n+1}(x)]_a^\infty + \int_a^\infty \frac{d}{dx}\left(x^{-n-1} f(x)\right) x^{n+1} J_{n+1}(x)\, dx \\
&\quad = -f(a) J_{n+1}(a) + \int_a^\infty \frac{d}{dx}\left(x^{-n-1} f(x)\right) x^{n+1} J_{n+1}(x)\, dx.
\end{aligned}$$

The Bessel function satisfies

$$J_n(x) \approx \sqrt{\frac{2}{x\pi}} \cos\left(x - \tfrac{1}{2} n - \tfrac{1}{4}\right) \quad \text{as } x \to \infty$$

and $f(x)$ is $O(x^\alpha)$, $\alpha < 1/2$; therefore, $f(x) J_n(x) \to 0$ as $x \to \infty$.

Repeating the above operation M times yields

$$\begin{aligned}
&\int_a^\infty f(x) J_n(x)\, dx \\
&\quad = \sum_{k=1}^{M} \left[(-1)^k \left(\frac{1}{x} \frac{d}{dx} \right)^{k-1} (x^{-n-1} f(x)) x^{n+k} J_{n+k}(x) \right]_{x=a} \\
&\qquad + (-1)^M \int_a^\infty \left(\frac{1}{x} \frac{d}{dx} \right)^M (x^{-n-1} f(x)) x^{n+M+1} J_{n+M}(x)\, dx. \qquad (10.7)
\end{aligned}$$

The Bessel function is easily evaluated by Miller's method. If the Taylor series of $f(x)$ is given, then the second integral on the right-hand side above can be computed without difficulty.

10.4 Numerical Examples

We use formulas (10.5)–(10.7) to compute numerically two simple examples of integrals of the type (10.1).

10.4.1 Example 1

Consider the integral

$$I = \int_0^\infty \frac{1}{x+1} \sin(x \log(1+x))\, dx.$$

As above, we can write

$$I = \int_0^a \frac{1}{x+1} \sin(x \log(1+x))\, dx + \int_a^\infty \frac{1}{x+1} \sin(x \log(1+x))\, dx. \quad (10.8)$$

Let $t = h(x) = x \log(1+x)$ in the second integral in (10.8); then this integral becomes

$$I_2 = \int_a^\infty \frac{1}{x+1} \sin(x \log(1+x))\, dx = \int_b^\infty v(t) \sin(t)\, dt,$$

where

$$v(t) = \frac{\frac{d}{dt} dh^{-1}(t)}{h^{-1}+1}, \qquad b = a \log(1+a).$$

Thus, for $a = 13$, we find that

$$\begin{aligned} h(x) &= 34.3077 + 3.56763(x-13) + 0.0382653(x-13)^2 \\ &\quad - 0.000971817(x-13)^3 + 3.6877 \times 10^{-5}(x-13)^4 + \cdots \end{aligned}$$

and that the inverse function of $h(x)$ is

$$\begin{aligned} h^{-1}(t) &= 13 + 0.280298(t-b) - 0.000842687(t-b)^2 \\ &\quad + 1.10657 \times 10^{-5}(t-b)^3 - 1.92062 \times 10^{-7}(t-b)^4 + \cdots, \end{aligned}$$

with $b = 13 \log(14)$; therefore,

$$\begin{aligned} v(t) &= 0.0200213 - 0.000521236(t-b) + 1.40122 \times 10^{-5}(t-b)^2 \\ &\quad - 3.82616 \times 10^{-7}(t-b)^3 + 1.05501 \times 10^{-8}(t-b)^4 + \cdots, \end{aligned}$$

so

$$I = \int_0^\infty \frac{1}{x+1} \sin(x \log(1+x))\, dx = 0.437992011202960.$$

10.4.2 Example 2

Consider the integral

$$I = \int_0^\infty \frac{x}{x^2+1} J_0(x \log(1+x))\, dx. \tag{10.9}$$

As in Example 1,

$$I = \int_0^a \frac{x J_0(x \log(1+x))}{x^2+1}\, dx + \int_a^\infty \frac{x J_0(x \log(1+x))}{x^2+1}\, dx.$$

Integral (10.9) can now be computed easily with, for example, $a = 30$, by means of the procedure explained above. The result is

$$I = \int_0^\infty \frac{x}{1+x^2} J_0(x \log(1+x))\, dx = 0.510502342713232.$$

10.5 Conclusion

Since Taylor series can be constructed for smooth functions, for the solution of an ordinary differential equation, and for an inverse function, we can approximate certain integrals in various ways by means of such series, in quite an effective manner.

References

1. M.A. Ellis and B. Stroustrup, *The Annotated C++ Reference Manual,* Addison-Wesley, New York, 1990.
2. P. Henrici, *Applied Computational Complex Analysis,* vol. 1, Wiley, New York, 1974.
3. H. Hirayama, Numerical technique for solving an ordinary differential equation by Picard's method, in *Integral Methods in Science and Engineering,* P. Schiavone, C. Constanda, and A. Mioduchowski (eds.), Birkhäuser, Boston, 2002, 111–116.
4. L.B. Rall, *Automatic Differentiation–Technique and Applications*, Lect. Notes in Comp. Sci. **120**, Springer-Verlag, Berlin-Heidelberg-New York, 1981.

11 On the Stability of Discrete Systems

Alexander O. Ignatyev and Oleksiy A. Ignatyev

11.1 Introduction

Difference equations have been studied in various branches of mathematics for a long time. The first results in the qualitative theory of such systems were obtained by Poincaré and Perron at the end of the nineteenth and the beginning of the twentieth centuries. A systematic description of the theory of difference equations can be found in [1]–[4]. Difference equations are a convenient model for the description of discrete dynamical systems and for the mathematical simulation of systems with impulse effects (see [5]–[9]). One of the directions arising from the applications of difference equations is linked with the qualitative investigation of their solutions (stability, boundedness, controllability, observability, oscillation, robustness, and so on; see [10]–[19]).

Consider a discrete system of the form

$$x_{n+1} = f_n(x_n), \quad f_n(0) = 0, \tag{11.1}$$

where $n = 0, 1, 2, \ldots$ is the discrete time, $x_n = (x_n^1, x_n^2, ..., x_n^p) \in \mathbb{R}^p$, and $f_n = (f_n^1, f_n^2, ..., f_n^p) \in \mathbb{R}^p$ satisfy Lipschitz conditions uniformly in n, that is, $\|f_n(x) - f_n(y)\| \leq L_r \|x - y\|$ for $\|x\| \leq r$, $\|y\| \leq r$. System (11.1) has the trivial (zero) solution

$$x_n \equiv 0. \tag{11.2}$$

We denote by $x_n(n_0, u)$ the solution of system (11.1) coinciding with u for $n = n_0$. We also make the notation $B_r = \{x \in \mathbb{R}^p : \ \|x\| \leq r\}$. We assume that the functions $f_n(x)$ are defined in B_H, where $H > 0$ is some fixed number. As in [4], we denote by $\mathbb{Z}_+$ the set of nonnegative integers.

11.2 Main Definitions and Preliminaries

By analogy with ordinary differential equations (see [20]–[22]), let us introduce the following definitions.

Definition 1. Solution (11.2) of system (11.1) is said to be stable if for any $\varepsilon > 0$ and $n_0 \in \mathbb{Z}_+$ there exists $\delta = \delta(\varepsilon, n_0) > 0$ such that $\|x_{n_0}\| \leq \delta$ implies that $\|x_n\| \leq \varepsilon$ for each $n > n_0$.

Definition 2. The trivial solution of system (11.1) is said to be uniformly stable if δ in Definition 1 can be chosen independent of n_0, i.e., $\delta = \delta(\varepsilon)$.

Definition 3. Solution (11.2) of system (11.1) is called attractive if for every $n_0 \in \mathbb{Z}_+$ there exists $\eta = \eta(n_0) > 0$ and for every $\varepsilon > 0$ and $x_{n_0} \in B_\eta$ there exists $\sigma = \sigma(\varepsilon, n_0, x_{n_0}) \in \mathbb{N}$ such that $\|x_n\| < \varepsilon$ for any $n \geq n_0 + \sigma$. Here $\mathbb{N}$ is the set of natural numbers.

In other words, the solution (11.2) of system (11.1) is attractive if

$$\lim_{n\to\infty} \|x_n(n_0, x_{n_0})\| = 0. \tag{11.3}$$

Definition 4. The zero solution of system (11.1) is called equiattractive if for every $n_0 \in \mathbb{Z}_+$ there exists $\eta = \eta(n_0) > 0$ and for any $\varepsilon > 0$ there is $\sigma = \sigma(\varepsilon, n_0) \in \mathbb{N}$ such that $\|x_n(n_0, x_{n_0})\| < \varepsilon$ for all $x_{n_0} \in B_\eta$ and $n \geq n_0 + \sigma$.

In other words, the zero solution of (11.1) is equiattractive if the limit relation (11.3) holds uniformly in $x_{n_0} \in B_\eta$.

Definition 5. Solution (11.2) of system (11.1) is said to be uniformly attractive if for some $\eta > 0$ and any $\varepsilon > 0$ there exists $\sigma = \sigma(\varepsilon) \in \mathbb{N}$ such that $\|x_n(n_0, x_{n_0})\| < \varepsilon$ for all $n_0 \in \mathbb{Z}_+$, $x_{n_0} \in B_\eta$, and $n \geq n_0 + \sigma$.

In other words, the solution (11.2) of system (11.1) is uniformly attractive if the limit relation (11.3) holds uniformly in $n_0 \in \mathbb{Z}_+$, $x_{n_0} \in B_\eta$.

Definition 6. The trivial solution (11.2) of system (11.1) is called
- asymptotically stable if it is stable and attractive;
- equiasymptotically stable if it is stable and equiattractive;
- uniformly asymptotically stable if it is uniformly stable and uniformly attractive.

Definition 7. A function $r : R_+ \to R_+$ belongs to the class of Hahn functions $\mathcal{K}$ ($r \in \mathcal{K}$) if r is a continuous increasing function and $r(0) = 0$ (see [20] and [21]).

The following assertion was proved in [6].

Theorem 1. *Solution* (11.2) *of system* (11.1) *is uniformly stable if there exists a sequence of functions* $\{V_n(x)\}$ *such that*

$$a(\|x\|) \leq V_n(x) \leq b(\|x\|), \quad a \in \mathcal{K},\ b \in \mathcal{K},\ n \in \mathbb{Z}_+, \tag{11.4}$$

$$V_n(x_n) \geq V_{n+1}(x_{n+1}) \quad \text{for every solution } x_n. \tag{11.5}$$

Theorem 2. *Suppose that there exists a sequence of functions* $\{V_n(x)\}$ *with properties* (11.4) *and such that*

$$V_{n+1}(x_{n+1}) - V_n(x_n) \leq -c(\|x_n\|), \quad c \in \mathcal{K}, \tag{11.6}$$

$$|V_n(x) - V_n(y)| \leq L\|x - y\|, \quad n \in Z_+,\ x \in B_H,\ y \in B_H,\ L > 0.$$

Then the zero solution of system (11.1) *is uniformly asymptotically stable.*

In the particular case when system (11.1) is autonomous, that is, $f_n(x) = f(x)$, the following theorem holds (see [6], p. 34).

Theorem 3. *If there exists a continuous function* $V(x)$ *such that* $a(\|x\|) \leq V(x) \leq b(\|x\|)$, $a \in \mathcal{K}$, $b \in \mathcal{K}$, *and*

$$V(x_{n+1}) - V(x_n) \leq 0 \tag{11.7}$$

for every solution x_n *of system* (11.1), *and if* (11.7) *holds as an equality in some set that does not contain entire semitrajectories, then solution* (11.2) *of system* (11.1) *is asymptotically stable.*

The purpose of this paper is to obtain conditions for the asymptotic stability of solution (11.2) of system (11.1) assuming that the sequences $\{f_n(x)\}$ are periodic or almost periodic.

11.3 Stability of Periodic Systems

Definition 8. System (11.1) is said to be periodic with period q if

$$f_n(x) \equiv f_{n+q}(x) \quad \text{for each } n \in Z_+, \; x \in B_H. \tag{11.8}$$

Throughout this section we assume that (11.1) is periodic with period q.

Theorem 4. *If solution* (11.2) *of system* (11.1) *is stable, then it is uniformly stable.*

Proof. Conditions (11.8) imply that

$$x_{n+q}(n_0 + q, x) \equiv x_n(n_0, x), \tag{11.9}$$

so it suffices to show that for any $\varepsilon > 0$ there exists $\delta = \delta(\varepsilon) > 0$ such that for each $n_0 = 0, 1, \ldots, q-1$ and $x_{n_0} \in B_\delta$ the inequality $\|x_n(n_0, x_{n_0})\| \leq \varepsilon$ holds for $n \geq n_0$. According to the assumption, for any $\varepsilon > 0$ there exists $\delta_1 > 0$ such that if $x_q = x_q(0, x_{n_0})$ satisfies the condition $x_q \in B_{\delta_1}$, then $x_n(q, x_q) \in B_\varepsilon$ for $n \geq q$. The functions f_n satisfy the Lipschitz condition with Lipschitz constant L, uniformly in n. Let us choose $\delta = L^{-q}\delta_1$. If for any $0 \leq n_0 \leq q-1$ the condition $x_{n_0} \in B_\delta$ holds, then $x_n(n_0, x_{n_0}) \in B_\varepsilon$. This completes the proof.

Theorem 5. *If the zero solution of system* (11.1) *is asymptotically stable, then it is uniformly asymptotically stable.*

Proof. Since solution (11.2) of system (11.1) is asymptotically stable, it follows that (11.3) holds in the set

$$n_0 \in Z_+, \quad x_{n_0} \in B_\lambda, \tag{11.10}$$

where λ is a sufficiently small positive number. Because of the periodicity of system (11.1), we assume that n_0 satisfies the condition $0 \leq n_0 \leq q-1$. First, we define the number $\eta = \eta(\varepsilon)$ from the condition

$$\|x_n(n_0, x_{n_0})\| \leq \varepsilon \;\; \text{for} \;\; x_{n_0} \in B_\eta, \;\; n > n_0. \tag{11.11}$$

This is always possible because of the uniform stability of the zero solution. Let us show that the limit relation (11.3) holds uniformly in n_0 and x_{n_0} from (11.10), that is, we show that for every $\varepsilon > 0$, there is $\sigma = \sigma(\varepsilon) \in \mathbb{N}$ such that the inequality $\|x_n(n_0, x_{n_0})\| \le \varepsilon$ holds for all $n \ge n_0 + \sigma$. We assume the opposite, namely, that there is no such $\sigma = \sigma(\varepsilon)$. Then for any large natural number m, there is $n_m \in \mathbb{N}$ such that $n_m > mq$, and there are initial data $(n_{0m}, x_{n_{0m}})$ such that $0 \le n_{0m} \le q - 1$, $x_{n_{0m}} \in B_\lambda$, and

$$\|x_{n_m}(n_{0m}, x_{n_{0m}})\| > \varepsilon. \tag{11.12}$$

Since the sequence $\{n_{0m}\}$ is finite and $\{x_{n_{0m}}\}$ lies in a compact set, the sequence $\{n_{0m}, x_{0m}\}$ contains a subsequence that converges to (n_*, x_*), where $0 \le n_* \le q - 1$ and $x_* \in B_\lambda$. Without loss of generality, we may assume that the sequence $\{n_{0m}\}$ coincides with n_* and that $\{x_{n_{0m}}\}$ itself converges to x_*. Hence, (11.3) is valid for values $n_0 = n_*$ and $x_{n_0} = x_*$, so there is a sufficiently large $k = k(\varepsilon) \in \mathbb{N}$ such that

$$\|x_{n_*+kq}(n_*, x_*)\| < \tfrac{1}{2}\,\eta(\varepsilon).$$

Then, by virtue of the continuous dependence of solutions on the initial data, there are sufficiently large values of m for which

$$\|x^{(k)}\| < \eta(\varepsilon), \tag{11.13}$$

where $x^{(k)} = x_{n_{0m}+kq}(n_{0m}, x_{0m})$. Inequalities (11.13) and (11.11) imply that $\|x_n(n_{0m}, x^{(k)})\| \le \varepsilon$ for all $n > n_{0m}$. According to (11.9) and the property of uniqueness of the solution, this implies that

$$\varepsilon \ge \|x_n(n_{0m}, x^{(k)})\| \equiv \|x_{n+kq}(n_{0m} + kq, x^{(k)})\| \equiv \|x_{n+kq}(n_{0m}, x_{n_{0m}})\|.$$

This inequality contradicts assumption (11.12) because there exists n_m such that $n_m > kq$. The contradiction proves that limit relation (11.3) holds uniformly in n_0 and x_{n_0}, as required.

Definition 9. A sequence of numbers $\{u_k\}_{k=1}^{\infty}$ is said to be ultimately nonzero if for any natural number M there exists $k > M$ such that $u_k \ne 0$.

Theorem 6. *Suppose that there exists a periodic sequence of functions* $\{V_n(x)\}$ *with period* q*, each term of which satisfies* (11.4), (11.5), *and a Lipschitz condition, and that the sequence* $\{V_n(x_n) - V_{n+1}(x_{n+1})\}$ *is ultimately nonzero for each nonzero solution of system* (11.1). *Then the zero solution of* (11.1) *is uniformly asymptotically stable.*

Proof. Theorem 1 implies that solution (11.2) of system (11.1) is uniformly stable, that is, for every $\varepsilon > 0$ there exists $\delta = \delta(\varepsilon) > 0$ such that for any $n_0 \in Z_+$ and $x_{n_0} \in B_\delta$, $n > n_0$, the inequality $\|x_n(n_0, x_{n_0})\| \le \varepsilon$ holds. Let us show that each trajectory $x_n = x_n(n_0, x_{n_0})$ with such initial conditions has property (11.3).

Consider the sequence $\{v_n\}$, where

$$v_n = V_n(x_n(n_0, x_{n_0})).$$

This sequence does not increase and is bounded below, so there exists $\lim_{n\to\infty} v_n = \eta \geq 0$. Let us show that $\eta = 0$. We assume the opposite, namely, that

$$\eta = \lim_{n\to\infty} V_n(x_n(n_0, x_{n_0})) > 0. \tag{11.14}$$

Consider the sequence $\{x^{(k)}\}$, where $x^{(k)} = x_{n_0+kq}(n_0, x_{n_0})$. Since $\|x^{(k)}\| \leq \varepsilon < H$, we conclude that there exists a subsequence that converges to $x_* \in B_\varepsilon$. Without loss of generality, suppose that the sequence $\{x^{(k)}\}$ itself converges to $x_* \neq 0$. Since the $V_n(x)$ are periodic in n and each $V_n(x)$ is continuous in x, it follows that $V_{n_0}(x_*) = \eta$. Consider the semitrajectory $x_n(n_0, x_*)$ of system (11.1) for $n \geq n_0$, and the sequence $\{v_n^*\}$, where $v_n^* = V_n(x_n(n_0, x_*))$. This sequence does not increase, and $\{v_n^* - v_{n+1}^*\}$ is ultimately nonzero. Hence, there is $n_* \in \mathbb{N}$, $n_* > n_0$, such that

$$V_{n_*}(x_{n_*}(n_0, x_*)) = \eta_1 < \eta.$$

Since $\{x^{(k)}\}$ tends to x_* as $k \to \infty$ and solutions are continuously dependent on the initial conditions, we have

$$\|x_{n_*}(n_0, x_*) - x_{n_*}(n_0, x^{(k)})\| < \gamma$$

for all $k > M(\gamma) \in \mathbb{N}$ with any small $\gamma > 0$; hence,

$$\lim_{k\to\infty} V_{n_*}(x_{n_*}(n_0, x^{(k)}) = \eta_1. \tag{11.15}$$

Taking into account the periodicity of system (11.1), we can write

$$x_{n_*}(n_0, x^{(k)}) = x_{n_*}(n_0, x_{n_0+kq}(n_0, x_{n_0})) = x_{n_*+kq}(n_0, x_{n_0}). \tag{11.16}$$

In fact, trajectories I and II of (11.1) with initial conditions $(n_0, x^{(k)})$ and $(n_0+kq, x^{(k)})$ for the discrete time $\Delta n = n_* - n_0$ pass through $x_{n_*}(n_0, x^{(k)})$ and $x_{n_*+kq}(n_0, x_{n_0})$, respectively. This proves (11.16). The periodicity of V_n in n implies that $V_{n_*}(x) = V_{n_*+kq}(x)$; hence, in view of (11.16), condition (11.15) can be written as

$$\lim_{k\to\infty} V_{n_*+kq}(x_{n_*+kq}(n_0, x_{n_0})) = \eta_1. \tag{11.17}$$

But (11.17) contradicts the inequality $V_n(x_n(n_0, x_{n_0})) \geq \eta_1$ because $\eta_1 < \eta$. This contradiction proves that assumption (11.14) was incorrect; therefore, $\eta = 0$. Condition (11.4) implies limit relation (11.3). Using Theorem 5, we deduce that the zero solution of system (11.1) is uniformly asymptotically stable.

11.4 Stability of Almost Periodic Systems

Definition 10. A sequence $\{u_n\}_{-\infty}^{+\infty}$ is said to be almost periodic if for every $\varepsilon > 0$ there is $l = l(\varepsilon) \in \mathbb{N}$ such that each segment $[sl, (s+1)l]$, $s \in \mathbb{Z}$, contains an integer m such that $\|u_{n+m} - u_n\| < \varepsilon$ for all $n \in \mathbb{Z}$. Here $\mathbb{Z}$ is the set of integers. Numbers m with such properties are called ε-almost periods of the sequence $\{u_n\}$.

Definition 11. A sequence of functions $\{f_n(x)\}$ is called uniformly almost periodic if for every $\varepsilon > 0$ there exists $l = l(\varepsilon, r) \in \mathbb{N}$ such that each segment of the form $[sl, (s+1)l]$, $s \in \mathbb{Z}$, contains an integer m such that $\|f_{n+m}(x) - f_n(x)\| < \varepsilon$ for all $n \in \mathbb{Z}$ and $\|x\| < r$.

Lemma 1. [23, p. 125] *Let sequences $\{u_n^1\}, \{u_n^2\}, \dots, \{u_n^M\}$ be almost periodic. Then for every $\varepsilon > 0$, there exists $l = l(\varepsilon) \in \mathbb{N}$ such that each segment of the form $[sl, (s+1)l]$, $s \in \mathbb{Z}$, contains at least one ε-almost period that is common to all these sequences.*

Lemma 2. *If for every $x \in B_H$ a sequence $\{F_n(x)\}$ is almost periodic, and if each function $F_n(x)$ satisfies a Lipschitz condition uniformly in $n \in \mathbb{Z}$ and $x \in B_H$, then this sequence is uniformly almost periodic.*

Proof. The functions $F_n(x)$ satisfy the Lipschitz condition, so

$$\|F_n(x) - F_n(y)\| \le L_1 \|x - y\|, \tag{11.18}$$

where L_1 is the Lipschitz constant. Let ε be any positive number. B_H is bounded and closed, therefore it is compact. Consequently, there exists a finite set of points $z_1, \dots, z_M$ such that $z_j \in B_H$, $j = 1, \dots, M$, and for any $x \in B_H$ there exists a natural number i, $1 \le i \le M$, such that

$$\|x - z_i\| < \frac{\varepsilon}{3L_1}.$$

By Lemma 1, there is $l = l(\varepsilon) \in \mathbb{N}$ such that every segment $[sl, (s+1)l]$, $s \in \mathbb{Z}$, contains a number $m \in \mathbb{Z}$ such that

$$\|F_n(z_i) - F_{n+m}(z_i)\| < \frac{\varepsilon}{3} \tag{11.19}$$

for all $1 \le i \le M$ and $n \in \mathbb{Z}$.

We now show that for every $x \in B_H$, any integer m satisfying (11.19) is an ε-almost period of the sequence $\{F_n(x)\}$. Let z_k be the same element of the set $z_1, \dots, z_M$ for which $\|x - z_k\| < \varepsilon/(3L_1)$. Then, by (11.18) and (11.19),

$$\begin{aligned} &\|F_{n+m}(x) - F_n(x)\| \\ &\quad \le \|F_{n+m}(x) - F_{n+m}(z_k)\| + \|F_{n+m}(z_k) - F_n(z_k)\| + \|F_n(z_k) - F_n(x)\| \\ &\quad \le \frac{\varepsilon}{3} + 2L_1 \cdot \frac{\varepsilon}{3L_1} = \varepsilon. \end{aligned} \tag{11.20}$$

Inequality (11.20) completes the proof of the lemma.

Theorem 7. *Suppose that a sequence of continuous functions $\{V_n(x)\}$ satisfies*

$$a(\|x\|) \leq V_n(x), \quad a \in \mathcal{K}, \ x \in B_H, \quad V_n(0) = 0, \tag{11.21}$$

and that for every $n_0 \in \mathbb{Z}_+$ there exists $\Delta(n_0) > 0$ such that $\|x_{n_0}\| < \Delta$ implies that the sequence $\{V_n(x_n(n_0, x_{n_0}))\}$ is monotonically nonincreasing and converges to zero. Then the zero solution of system (11.1) is equiattractive.

Proof. Choose an arbitrary $\delta = \delta(n_0) \in (0, \Delta)$. According to the conditions of the theorem, for any $\varepsilon > 0$, $n_0 \in \mathbb{Z}_+$, and $x_{n_0} \in B_\delta$ there exists $\sigma = \sigma(\varepsilon, n_0, x_{n_0}) \in \mathbb{N}$ such that

$$V_{n_0+\sigma}(x_{n_0+\sigma}(n_0, x_{n_0})) < \tfrac{1}{2}\,\varepsilon.$$

Because of the continuity of the $V_n(x)$ and the continuous dependence of solutions on the initial data, there is a neighborhood $Q(x_{n_0})$ of x_{n_0} where

$$V_{n_0+\sigma}(x_{n_0+\sigma}(n_0, y)) < \varepsilon \ \text{ for } \ y \in Q(x_{n_0}). \tag{11.22}$$

Since the sequence $\{V_n\}$ is monotonically nonincreasing along the solutions of system (11.1), from (11.22) it follows that

$$V_n(x_n(n_0, y)) < \varepsilon \ \text{ for } \ n \geq n_0 + \sigma(\varepsilon, n_0, x_{n_0}), \ y \in Q(x_{n_0}).$$

So, the compact set B_δ is covered by the system of neighborhoods $\{Q(x_{n_0})\}$ from which, by the Heine–Borel lemma, it is possible to select a finite subcovering $Q_1, ..., Q_j$ with corresponding numbers $\sigma_1, \dots, \sigma_j$. Let

$$\sigma(\varepsilon, n_0) = \max\{\sigma_1, \dots, \sigma_j\},$$

where σ depends only on ε and n_0. Then $V_n(x_n(n_0, x_{n_0})) < \varepsilon$ for all $n \geq n_0 + \sigma(\varepsilon, n_0)$ if $\|x_{n_0}\| \leq \delta(n_0)$. This inequality proves that solution (11.2) of system (11.1) is equiattractive.

In the rest of this section, we assume that the sequence $\{f_n(x)\}$ on the right-hand side of (11.1) is almost periodic for every fixed $x \in B_H$, and that the functions $f_n(x)$ satisfy a Lipschitz condition uniformly in n.

Lemma 3. *Consider a solution $x_n(n_0, x_{n_0})$ of system (11.1) and suppose that $x_n(n_0, x_{n_0})$ belongs to B_r, $0 < r < H$, for $n \geq n_0$. Let $\{\varepsilon_k\}$ be a monotonic sequence of positive numbers converging to zero, and let $\{m_k\}$ be a sequence of ε_k-almost periods of $\{f_n(x)\}$ (for every ε_k there corresponds an ε_k-almost period m_k). Then*

$$\lim_{k \to \infty} \|x_{n_*}(n_0, x^{(k)}) - x_{n_*+m_k}(n_0, x_{n_0})\| = 0, \tag{11.23}$$

where $x^{(k)} = x_{n_0+m_k}(n_0, x_{n_0})$ and n_* is a fixed natural number greater than n_0, that is, $n_* > n_0$.

Proof. Consider the solutions

$$x_n(n_0, x^{(k)}) \tag{11.24}$$

and

$$x_n(n_0 + m_k, x^{(k)}) \tag{11.25}$$

of (11.1). After $\Delta n = n_* - n_0$ steps, the point $x^{(k)}$ passes to $x_{n_*}(n_0, x_{n_0})$ along the solution (11.24), and $x^{(k)}$ passes to $x_{n_*+m_k}(n_0 + m_k, x^{(k)}) = x_{n_*+m_k}(n_0, x_{n_0})$ along (11.25). Solution (11.25) of (11.1) with initial condition $(n_0 + m_k, x^{(k)})$ can be interpreted as the solution of the system

$$x_{n+1} = f_{n+m_k}(x_n) \tag{11.26}$$

with initial data $(n_0, x^{(k)})$. The sequence $\{f_n(x)\}$ is almost periodic, and every function $f_n(x)$ satisfies a Lipschitz condition; hence, the difference between the right-hand sides of (11.1) and (11.26) is arbitrarily small for k large enough. This implies limit relation (11.23).

Theorem 8. *Suppose that there exists a sequence of functions $\{V_n(x)\}$ such that*

(a) *for every fixed $x \in B_H$, the sequence $\{V_n(x)\}$ is almost periodic;*
(b) *each $V_n(x)$ satisfies* (11.21) *and a Lipschitz condition uniformly in n;*
(c) *$V_n(x_n) \geq V_{n+1}(x_{n+1})$ along any solution of* (11.1)*;*
(d) *the sequence $\{V_n(x_n)\}$ is ultimately nonzero along any nonzero solution of* (11.1).

Then the zero solution of system (11.1) *is equiasymptotically stable.*

Proof. First, we show that solution (11.2) of system (11.1) is stable. We choose arbitrary $\varepsilon \in (0, H)$ and $n_0 \in \mathbb{Z}_+$. Let $\delta = \delta(\varepsilon, n_0) > 0$ be such that $V_{n_0}(x) < a(\varepsilon)$ for $x \in B_\delta$; then

$$a(\|x_n\|) \leq V_n(x_n) \leq V_{n_0}(x_{n_0}) < a(\varepsilon),$$

so we have $\|x_n\| < \varepsilon$ for $n > n_0$.

We now show that solution (11.2) is equiattractive. We choose an arbitrary $x_{n_0} \in B_\delta$. The sequence $\{V_n(x_n(n_0, x_{n_0}))\}$ is monotonically nonincreasing, therefore there exists

$$\lim_{n\to\infty} V_n(x_n(n_0, x_{n_0})) = \eta \geq 0,$$

and $V_n(x_n(n_0, x_{n_0})) \geq \eta$ for $n \geq n_0$. We claim that $\eta = 0$. We assume the opposite, namely, that $\eta > 0$, and consider a monotonic sequence of positive numbers $\{\varepsilon_k\}$ converging to zero, where ε_1 is sufficiently small. By Lemmas 2 and 1, for every ε_i, for the sequences $\{f_n(x)\}$ and $\{V_n(x)\}$ there exists a sequence of ε_i-almost periods $m_{i,1}, m_{i,2}, \ldots, m_{i,k}, \ldots$ with $m_{i,k} < m_{i,k+1}$ and $\lim_{k\to+\infty} m_{i,k} = +\infty$, such that

$$|V_{n+m_{i,k}}(x) - V_n(x)| < \varepsilon_i,$$
$$\|f_{n+m_{i,k}}(x) - f_n(x)\| < \varepsilon_i$$

for any $n \in \mathbb{Z}$ and $x \in B_\varepsilon$. Without loss of generality, suppose that $m_{i,k} < m_{i+1,k}$ for all $i \in \mathbb{N}$, $k \in \mathbb{N}$, and write $m_k = m_{k,k}$.

Consider the sequence $\{x^{(k)}\}$, $x^{(k)} = x_{n_0+m_k}(n_0, x_{n_0})$, $k = 1, 2, \ldots$. This sequence is bounded, therefore there exists a subsequence that converges to some point x^*. Without loss of generality, suppose that the sequence $\{x^{(k)}\}$ itself converges to x^*. The sequence $\{V_n(x)\}$ is almost periodic for every fixed $x \in B_H$ and each function $V_n(x)$ is continuous; hence,

$$V_{n_0}(x_*) = \lim_{n\to\infty} V_{n_0}(x_n) = \lim_{k\to\infty} \lim_{n\to\infty} V_{n_0+m_k}(x_n)$$
$$= \lim_{n\to\infty} V_{n_0+m_n}(x_n) = \lim_{n\to\infty} V_{n_0+m_n}(x_{n_0+m_n}(n_0, x_{n_0})) = \eta.$$

Consider the sequence $\{x_n(n_0, x^*)\}$. From the conditions of the theorem, there exists $n_* > n_0$ ($n_* \in \mathbb{N}$) such that

$$V_{n_*}(x_{n_*}(n_0, x^*)) = \eta_1 < \eta.$$

The functions $f_n(x)$ satisfy a Lipschitz condition, hence,

$$\lim_{k\to\infty} \|x_{n_*}(n_0, x^{(k)}) - x_{n_*}(n_0, x^*)\| = 0$$

because

$$\lim_{k\to\infty} \|x^{(k)} - x^*\| = 0.$$

This implies that

$$\lim_{k\to\infty} V_{n_*}(x_{n_*}(n_0, x^{(k)})) = \eta_1. \tag{11.27}$$

The almost periodicity of the sequence $\{f_n(x)\}$ and limit relation (11.23) yield

$$\|x_{n_*}(n_0, x^{(k)}) - x_{n_*+m_k}(n_0, x_{n_0})\| \le \gamma_k, \tag{11.28}$$

where $\gamma_k \to 0$ as $k \to \infty$. Since the sequence $\{V_n\}$ is almost periodic, we have

$$|V_{n_*}(x) - V_{n_*+m_k}(x)| < \varepsilon_k \tag{11.29}$$

for every $x \in B_H$, and conditions (11.27) and (11.28) imply that

$$|V_{n_*}(x_{n_*+m_k}(n_0, x_{n_0})) - \eta_1| < \xi_k, \tag{11.30}$$

where $\xi_k \to 0$ as $k \to \infty$. From (11.29) it follows that

$$|V_{n_*}(x_{n_*+m_k}(n_0, x_{n_0})) - V_{n_*+m_k}(x_{n_*+m_k}(n_0, x_{n_0}))| < \varepsilon_k. \tag{11.31}$$

By (11.30) and (11.31),

$$|V_{n_*+m_k}(x_{n_*+m_k}(n_0, x_{n_0})) - \eta_1| < \xi_k + \varepsilon_k, \tag{11.32}$$

where $\xi_k + \varepsilon_k \to 0$ as $k \to \infty$.

On the other hand,

$$\lim_{k\to\infty} V_{n_*+m_k}(x_{n_*+m_k}(n_0, x_{n_0})) = \eta. \tag{11.33}$$

Inequality (11.32) and (11.33) contradict the inequality $\eta_1 < \eta$, which proves that $\eta = 0$; hence, by Theorem 7, solution (11.2) of system (11.1) is equiasymptotically stable.

Theorem 9. *Suppose that there exists a sequence of functions $\{V_n(x)\}$ such that for every $x \in B_H$, the sequence $\{V_n(x)\}$ is almost periodic, and that each function $V_n(x)$ satisfies a Lipschitz condition uniformly in n and is such that*

(a) $|V_n(x)| \le b(\|x\|)$, $b \in \mathcal{K}$, $n \in \mathbb{Z}_+$, *for* $x \in B_H$;
(b) *for any* $n \in \mathbb{Z}_+$ *and* $\delta > 0$, *there is* $x \in B_\delta$ *such that* $V_n(x) > 0$;
(c) $V_{n+1}(x_{n+1}) \ge V_n(x_n)$ *along any solution* x_n.

Then solution (11.2) *of system* (11.1) *is unstable.*

Proof. Let $\varepsilon \in (0, H)$ be an arbitrary number. Choose any $n_0 \in \mathbb{Z}_+$ and any sufficiently small $\delta > 0$. Also, choose $x_0 \in B_\delta$ so that $V_{n_0}(x_{n_0}) > 0$. From the conditions of the theorem it follows that there exists $\eta > 0$ such that $|V_n(x)| < V_{n_0}(x_{n_0})$ for every $x \in B_\eta$. Consider the sequence $\{V_n\}$, where $V_n = V_n(x_n(n_0, x_{n_0}))$. This sequence is nonincreasing, that is, $V_n(x_n(n_0, x_{n_0})) \ge V_{n_0}(x_{n_0})$ for $n \ge n_0$. This means that $\|x_n(n_0, x_{n_0})\| \ge \eta$ for every $n \ge n_0$. We show that there is $N_0 \in \mathbb{N}$, $N_0 > n_0$, such that $\|x_{N_0}(n_0, x_{n_0})\| > \varepsilon$. Assume the opposite, namely, that

$$\eta \le \|x_n(n_0, x_{n_0})\| \le \varepsilon \tag{11.34}$$

for all $n > n_0$. Using the conditions of the theorem and inequality (11.34), we arrive at a contradiction just as in the proof of Theorem 8, so we omit details. This contradiction shows that the solution $x_n(n_0, x_{n_0})$ leaves B_ε, which completes the proof.

Example. Consider the system

$$x_{n+1} = y_n \cos(\pi n), \quad y_{n+1} = -x_n \cos n \tag{11.35}$$

and the function $V_n(x_n, y_n) = x_n^2 + y_n^2$; then

$$V_{n+1}(x_{n+1}, y_{n+1}) - V_n(x_n, y_n) = -(\sin^2 n)\, x_n^2 - (\sin^2 \pi n)\, y_n^2. \tag{11.36}$$

In [23] it is shown that for any sufficiently small $\varepsilon > 0$, there exists a sequence $n_1, n_2, \ldots, n_k, \ldots \to \infty$ such that

$$0 < \sin^2 n_k < \varepsilon, \quad 0 < \sin^2(\pi n_k) < \varepsilon, \quad k = 1, 2, \ldots.$$

This means that there is no $c \in \mathcal{K}$ such that the left-hand side of (11.36) satisfies inequality (11.6), so Theorem 2 cannot be applied to this system. System (11.35) is not autonomous, therefore Theorem 3 cannot be applied to the study of the stability of its zero solution. But this system is almost periodic, and the right-hand side of (11.36) is negative for each nonzero solution of (11.35). Hence, by Theorem 8, the zero solution of system (11.35) is equiasymptotically stable.

References

1. R.P. Agarwal, *Difference Equations and Inequalities,* Marcel Dekker, New York, 1992.
2. S. Elaydi, *An Introduction to Difference Equations,* 2nd ed., Springer-Verlag, New York, 1999.
3. I.V. Gaishun, *Systems with Discrete Time,* Institute of Mathematics of Belarus, Minsk, 2001 (Russian).
4. V. Lakshmikantham and D. Trigiante, *Theory of Difference Equations: Numerical Methods and Applications*, Academic Press, New York-London, 1998.
5. R.I. Gladilina and A.O. Ignatyev, On necessary and sufficient conditions for the asymptotic stability of impulsive systems, *Ukrain. Math. J.* **55** (2003), 1254–1264.
6. A. Halanay and D. Wexler, *Qualitative Theory of Impulsive Systems*, Mir, Moscow, 1971 (Russian).
7. A.O. Ignatyev, Method of Lyapunov functions in problems of stability of solutions of systems of differential equations with impulse action, *Sb. Mat.* **194** (2003), 1543–1558.
8. V. Lakshmikantham, D.D. Bainov, and P.S. Simeonov, *Theory of Impulsive Differential Equations*, Wiley, Singapore-London, 1989.
9. A.M. Samoilenko and N.A. Perestyuk, *Impulsive Differential Equations,* World Scientific, Singapore, 1995.
10. R. Abu-Saris, S. Elaydi, and S. Jang, Poincaré type solutions of systems of difference equations, *J. Math. Anal. Appl.* **275** (2002), 69–83.
11. R.P. Agarwal, W.-T. Li, and P.Y.H. Pang, Asymptotic behavior of nonlinear difference systems, *Appl. Math. Comput.* **140** (2003), 307–316.
12. C. Corduneanu, Discrete qualitative inequalities and applications, *Nonlinear Anal. TMA* **25** (1995), 933–939.
13. I. Győri, G. Ladas, and P.N. Vlahos, Global attractivity in a delay difference equation, *Nonlinear Anal. TMA* **17** (1991), 473–479.
14. I. Győri and M. Pituk, The converse of the theorem on stability by the first approximation for difference equations, *Nonlinear Anal.* **47** (2001), 4635–4640.

15. J.W. Hooker, M.K. Kwong, and W.T. Patula, Oscillatory second order linear difference equations and Riccati equations, *SIAM J. Math. Anal.* **18** (1987), 54–63.

16. P. Marzulli and D. Trigiante, Stability and convergence of boundary value methods for solving ODE, *J. Difference Equations Appl.* **1** (1995), 45–55.

17. A. Bacciotti and A. Biglio, Some remarks about stability of nonlinear discrete-time control systems, *Nonlinear Differential Equations Appl.* **8** (2001), 425–438.

18. A.O. Ignatyev, Stability of the zero solution of an almost periodic system of finite-difference equations, *Differential Equations* **40** (2004), 105–110.

19. R. Bouyekhf and L.T. Gruyitch, Novel development of the Lyapunov stability theory for discrete-time systems, *Nonlinear Anal.* **42** (2000), 463–485.

20. W. Hahn, *Stability of Motion*, Springer-Verlag, Berlin-Heidelberg-New York, 1967.

21. N. Rouche, P. Habets, and M. Laloy, *Stability Theory by Liapunov's Direct Method*, Springer-Verlag, New York, 1977.

22. A.Ya. Savchenko and A.O. Ignatyev, *Some Problems of the Stability Theory*, Naukova Dumka, Kiev, 1989 (Russian).

23. C. Corduneanu, *Almost Periodic Functions*, 2nd ed., Chelsea, New York, 1989.

12 Parallel Domain Decomposition Boundary Element Method for Large-scale Heat Transfer Problems

Alain J. Kassab and Eduardo A. Divo

12.1 Introduction

Numerical solutions of engineering problems often require large complex systems of equations to be set up and solved. For any system of equations, the amount of computer memory required for storage is proportional to the square of the number of unknowns, which for large problems can exceed machine limitations. For this reason, almost all kinds of computational software use some type of problem decomposition for large-scale problems. For methods that result in sparse matrices, the storage alone can be decomposed to save memory, but techniques such as the boundary element method (BEM) generally yield fully populated matrices, so another approach is needed.

The BEM requires only a surface mesh to solve a large class of field equations; furthermore, the nodal unknowns appearing in the BEM equations are the surface values of the field variable and its normal derivative. Thus, the BEM lends itself ideally not only to the analysis of field problems, but also to modeling coupled field problems such as those arising in conjugate heat transfer (CHT). However, in implementing the BEM for intricate 3D structures, the number of surface unknowns required to resolve the temperature field can readily number in the 10s to 100s of thousands. Since the ensuing matrix equation is fully populated, this poses a serious problem regarding both the storage requirements and the need to solve a large set of nonsymmetric equations.

The BEM community has generally approached this problem by (i) artificially sub-sectioning the 3D model into a multi-region model—an idea that originated in the treatment of piecewise nonhomogeneous media (see [1]–[3]), in conjunction with block solvers reminiscent of finite element method (FEM) frontal solvers (see [4] and [5]) or iterative methods (see [6]–[9])—and (ii) using fast-multipole methods adapted to BEM coupled to a generalized minimal residuals (GMRES) nonsymmetric iterative solver (see [10] and [11]). The first approach is readily adapted to existing BEM codes, whereas the multipole approach, although very efficient, requires the rewriting of existing BEM codes. Recently, a technique using wavelet decomposition has been proposed to compress the BEM matrix once it is formed and stored in order to accelerate the solution phase without major alteration of traditional BEM codes [12].

In this paper, a particular domain decomposition or artificial multi-region sub-sectioning technique is presented along with a region-by-region iteration algorithm tailored for parallel computation [13]. The domain decomposition is applied to two problems. The first one addresses BEM modeling of large-scale, three-dimensional, steady-state, nonlinear heat conduction problems, which allows for multiple regions of different nonlinear conductivities. A nonsymmetric update of the interfacial fluxes to ensure equality of fluxes at the subdomain interfaces is formulated. The second application considers the problem of large-scale transient heat conduction. Here, the transient heat conduction equation is transformed into a modified Helmholtz equation using the Laplace transformation. The time-domain solution is retrieved with a Stehfest numerical inversion routine.

The domain decomposition technique described below employs an iteration scheme, which is used to ensure the continuity of both the temperature and heat flux at the region interfaces. In order to provide a sufficiently accurate initial guess for the iterative process, a physically based initial guess for the temperatures at the domain interfaces is derived, and a coarse grid solution obtained with constant elements is employed. The results of the constant grid model serve as an initial guess for finer discretizations obtained with linear and quadratic boundary element models. The process converges very efficiently, offers substantial savings in memory, and does not demand the complex data-structure preparation required by the block-solvers or multipole approaches. Moreover, the process is shown to converge for steady-state linear and nonlinear problems, as well as for transient problems. The nonlinear problems are treated using the classical Kirchhoff transformation. Results from two numerical examples are presented. The first example considers the 3D steady-state solution in a composite linear and nonlinear conducting rod. The second example deals with a transient case in a cooled conducting blade. The solution of the second example is compared to that of a commercial code.

12.2 Applications in Heat Transfer

Below we give the BEM formulations for steady-state 3D nonlinear heat conduction and transient 2D linear conduction.

12.2.1 Three-dimensional Nonlinear Heat Conduction

The initial discussion will focus primarily on nonlinear 3D heat transfer, governed by the steady-state nonlinear heat conduction equation

$$\nabla \cdot [k(T) \nabla T] = 0, \tag{12.1}$$

where T is the temperature and k is the thermal conductivity of the material. If the thermal conductivity is assumed constant, then the above reduces to the Laplace equation for the temperature

$$\nabla^2 T = 0. \tag{12.2}$$

When the dependence of the thermal conductivity on temperature is an important concern, the nonlinearity in the steady-state heat conduction equation can readily be removed by introducing the classical Kirchhoff transform $U(T)$ [14], defined by

$$U(T) = \frac{1}{k_0} \int_{T_0}^{T} k(T)\, dT, \tag{12.3}$$

where T_0 is the reference temperature and k_0 is the reference thermal conductivity. The transform and its inverse are readily evaluated, either analytically or numerically. Since U is nothing but the area under the k vs. T curve, it is a monotonically increasing function of T, and the back-transform $T(U)$ is unique. The heat conduction equation then transforms to the Laplace equation for the transformed parameter $U(T)$:

$$\nabla^2 U = 0.$$

The boundary conditions are transformed linearly as long as they are of the first or second kind, that is,

$$\begin{aligned} T|_{r_s} = T_s \quad &\Rightarrow \quad U|_{r_s} = U(T_s) = U_s, \\ -k \frac{\partial T}{\partial n}\bigg|_{r_s} = q_s \quad &\Rightarrow \quad -k_0 \frac{\partial U}{\partial n}\bigg|_{r_s} = q_s. \end{aligned}$$

Here r_s denotes a point on the surface. After transformation, the boundary conditions of the third kind become nonlinear:

$$-k \frac{\partial T}{\partial n}\bigg|_{r_s} = h_s[T|_{r_s} - T_\infty] \quad \Rightarrow \quad -k_0 \frac{\partial U}{\partial n}\bigg|_{r_s} = h_s[T(U|_{r_s}) - T_\infty].$$

In this case, iteration is required. This is accomplished by rewriting the convective boundary condition as

$$-k_0 \frac{\partial U}{\partial n}\bigg|_{r_s} = h_s[U|_{r_s} - T_\infty] + h_s[T(U|_{r_s}) - U|_{r_s}]$$

and first solving the problem with the linearized boundary condition

$$-k_0 \frac{\partial U}{\partial n}\bigg|_{r_s} = h_s[U|_{r_s} - T_\infty]$$

to provide an initial guess for iteration.

Thus, the heat conduction equation can always be reduced to the Laplace equation. Therefore, for simplicity, from now on we use the symbol T for the dependent variable with the understanding that when dealing with a nonlinear problem T is interpreted as U.

The Laplace equation is readily solved by first converting it into a boundary integral equation (BIE) of the form (see [14] and [15])

$$C(\xi)\,T\,(\xi) + \oint_S q(x)G\,(x,\xi)\,dS(x) = \oint_S T(x)H\,(x,\xi)\,dS(x), \qquad (12.4)$$

where $S(x)$ is the surface bounding the domain of interest, ξ is the source point, x is the field point, $q(x) = -k\partial T(x)/\partial n$ is the heat flux, $G\,(x,\xi)$ is the fundamental solution, and $H\,(x,\xi) = -k\partial G\,(x,\xi)\,/\partial n$.

The fundamental solution is the response of the adjoint governing differential operator at any field point x due to a Dirac delta function acting at the source point ξ, and is given by $G\,(x,\xi) = \big(1/(4\pi k)\big)r(x,\xi)$ in 3D, where $r(x,\xi)$ is the Euclidean distance from the source point ξ to x. The free term $C(\xi) = \oint_{S(x)} H\,(x,\xi)\,dS(x)$ can be shown analytically to be the internal angle subtended at the source point, divided by 4π, when ξ is on the boundary, and equal to one when ξ is at the interior. In the standard BEM, polynomials are employed to discretize the boundary geometry and the distribution of the temperature and heat flux on the boundary. The discretized BIE is usually collocated at the boundary points, leading to the algebraic analog of (12.4), that is,

$$[H]\{T\} = [G]\{q\}.$$

These equations are readily solved upon imposition of boundary conditions. Sub-parametric constant, isoparametric bilinear, and super-parametric biquadratic, discontinuous boundary elements are used as the basic elements in the 3D BEM codes developed to implement these algorithms; these are illustrated in Fig. 1 (a)–(c). Such elements avoid the so-called star-point issue and allow for discontinuous fluxes. Moreover, the biquadratic elements used here are super-parametric, with a bilinear model of the geometry and a biquadratic model of the temperature and heat flux. This type of element provides compatibility of geometric models with grids generated by structured finite-volume grid generators.

12.2.2 Transient Heat Conduction

The BEM has been traditionally used to solve transient heat conduction problems via three different approaches: (i) using the convolution scheme, where a time-dependent Green's function is introduced to build a transient boundary integral equation model, (ii) using the dual reciprocity method (DRM) to expand the spatial portion of the governing equation by means of radial-basis functions and a finite difference scheme to march in time, and (iii) using the Laplace transformation of the governing equation to eliminate the time derivative and arrive at a modified Helmholtz equation that can be solved using a steady-state BEM approach, and then inverting the BEM solution back into real space-time by means of a numerical Laplace inversion scheme (see [14] and [16]–[25]).

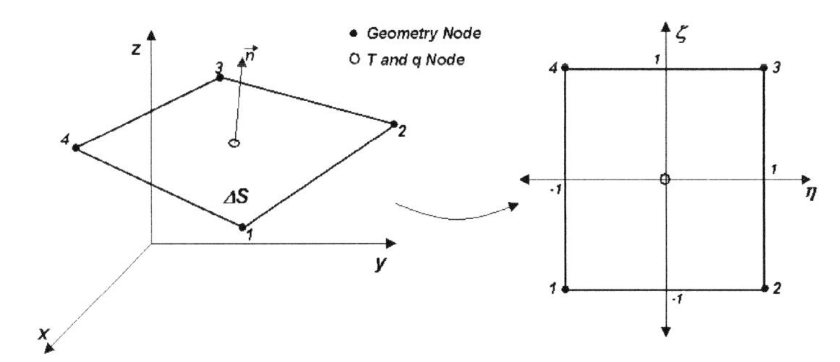

(a) Discontinuous sub-parametric constant element.

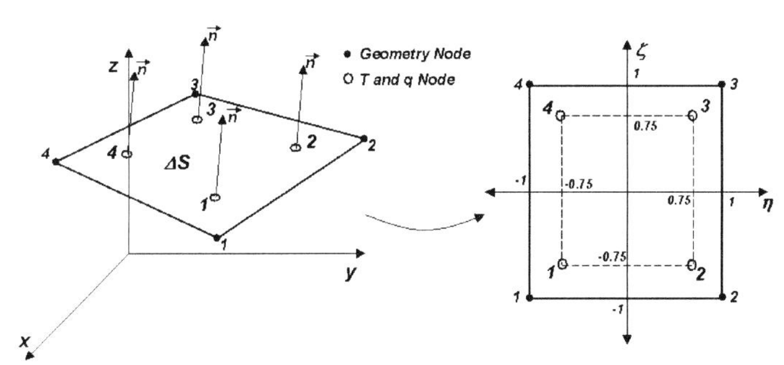

(b) Discontinuous isoparametric bilinear element.

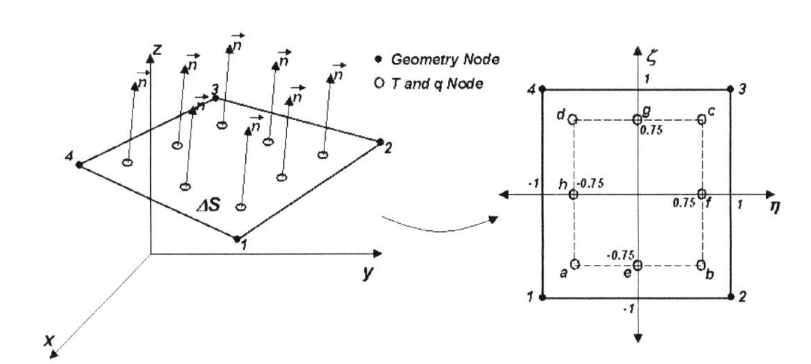

(c) Discontinuous super-parametric biquadratic element.

Fig. 1. Discontinuous elements.

The first approach will require the generation and storage of BEM influence coefficient matrices at every time step of the convolution scheme, making the technique unfeasible for medium or large problems, particularly in 3D applications, as the computational and storage requirements become unrealistically high. The second approach raises a different issue, because the global interpolation functions for the DRM, such as the widely used radial-basis functions (RBF), lack convergence and error estimation approximations, can at times lead to unwanted behavior, and significantly increase the conditioning number of the resulting algebraic system. The third approach, which originated in BEM applications by Rizzo and Shippy [17], does not require time marching or any type of interpolation, but it requires fine-tuning of the BEM solution of the modified Helmholtz equation and a numerical Laplace inversion of the results. Real variable-based numerical Laplace inversion techniques such as the Stehfest transformation (see [22] and [23]) provide very accurate results for nonoscillatory types of functions, such as those expected to result from transient heat conduction applications, as all poles of the transformed solution are real and distributed along the negative part of the real axis. One type of parallelization has been discussed by Davies and Crann [24], where individual solutions, as required for the numerical Laplace inversion, are obtained simultaneously in multiple processors. This type of parallelization reduces the computational time, but does not help with the storage requirements since the entire domain must be handled by each processor, thus leaving room for efficiency improvements.

In what follows, we briefly formulate a Laplace-transformed BEM algorithm which lends itself to the iterative parallel domain decomposition scheme to be discussed for both steady-state and transient heat conduction.

12.2.2.1 Governing Equation and the Laplace Transformation

Transient heat conduction is governed by the well-known diffusion equation, which for a 2D rectangular coordinate system is

$$\nabla \cdot \Big[k \, \nabla T(x,y,t) \Big] = \rho c \, \frac{\partial T}{\partial t}(x,y,t).$$

Applying the Laplace transformation, we arrive at

$$\nabla \cdot \Big[k \, \nabla \bar{T}(x,y,s) \Big] = \rho c s \, \bar{T}(x,y,s) \; - \; \rho c \, T(x,y,0), \tag{12.5}$$

where $\bar{T}(x,y,s)$ is the Laplace-transformed temperature and new dependent variable. The above expression can be further simplified by imposing the initial condition $T(x,y,0) = 0$, which is true for any case of uniform initial condition with a proper superposition. Equation (12.5) is also no longer time dependent; it now contains the Laplace transformation parameter s. This parameter can simply be treated as a constant in all further considerations. The dependence of the temperature field on s is thus eliminated, and we apply the initial condition to find that

$$\nabla \cdot \Big[k \, \nabla \bar{T}(x,y) \Big] - \rho c s \, \bar{T}(x,y) = 0.$$

Assuming that the thermal conductivity is independent of temperature, we see that the above is a modified Helmholtz equation. The solution to this equation is well known, since many other physical problems are governed by it [25]. Finally, the boundary conditions must be transformed in order to shift the entire problem to the proper Laplace transform space. Assuming time-independent boundary conditions, we obtain

$$\bar{T}(x,y,s)|_\Gamma = \left.\frac{T(x,y)}{s}\right|_\Gamma, \quad \bar{q}(x,y,s)|_\Gamma = \left.\frac{q(x,y)}{s}\right|_\Gamma,$$

where Γ is the boundary (control surface).

12.2.2.2 BEM for the Modified Helmholtz Equation

The development of a BEM solution begins by reducing the governing equation to a boundary-only integral equation. The current form of the Laplace-transformed transient heat conduction problem can be expressed in integral form by pre-multiplying the equation by a generalized function $G(x,y,\xi)$, integrating over the domain Ω of interest in the problem (control volume), identifying $G(x,y,\xi)$ as the fundamental solution, and using the sifting property of the Dirac delta function to obtain

$$\rho c\, C(\xi)\, \bar{T}(\xi) = \oint_\Gamma H(x,y,\xi)\, \bar{T}(x,y)\, d\Gamma - \oint_\Gamma G(x,y,\xi)\, \bar{q}(x,y)\, d\Gamma.$$

For this case, the fundamental solution G in 2D is [26]

$$G(x,y,\xi) = \frac{1}{2\pi\alpha} K_0\left(\sqrt{\frac{s}{\alpha}}\, r\right).$$

Here, $\alpha = k/\rho c$, K_0 is a modified Bessel function of the second kind of order zero, and $r = \left[(x-x_i)^2 + (y-y_i)^2\right]^{1/2}$. The normal derivative $H(x,y,\xi)$ of the fundamental solution is

$$H(x,y,\xi) = \frac{-\rho c}{2\pi r}\sqrt{\frac{s}{\alpha}}\, K_1\left(\sqrt{\frac{s}{\alpha}}\, r\right)\left[(x-x_i)n_x + (y-y_i)n_y\right],$$

where n_x and n_y are the x- and y-components of the unit outward normal n. Using standard BEM discretization leads to

$$\sum_{j=1}^{N} H_{ij}\bar{T}_j = \sum_{j=1}^{N} G_{ij}\bar{q}_j.$$

Here, $H_{ij} = \hat{H}_{ij} - \frac{1}{2}\rho c \delta_{ij}$ and δ_{ij} is the Kronecker delta. Boundary conditions can be further applied to reduce the above system of equations to the standard algebraic form $[A]\{x\} = \{b\}$. Once the system is solved by standard linear algebra methods, the solution must be inverted numerically from the Laplace space to the real transient space.

12.2.2.3 Numerical Inversion of the Laplace-transformed Solution

The final step of the overall numerical solution is the inversion of the Laplace-transformed BEM solution. While many techniques exist for such an inversion, the Stehfest transformation has the advantages of being quite stable, very accurate, and simple to implement. The Stehfest transformation works by computing a sample of solutions at a specified number of times and predicting the solution based on this sample [27]. Owing to the nonoscillatory behavior of the transient heat conduction equation, the Stehfest transformation works exceptionally well. The Stehfest inversion is considered the best attempt at an improvement using extrapolation methods on the result of an asymptotic expansion resulting from a specific delta sequence first proposed by Garver in 1966 [27]. The Stehfest inverse of the Laplace transform $\bar{f}(s)$ of a function of time $f(t)$ is given by

$$f(t) = \ln\left(\frac{2}{t}\right) \sum_{n=1}^{N} K_n \bar{f}(s_n),$$

where the sequence of s-values is

$$s_n = n\,\frac{\ln 2}{t}$$

and the series coefficients are

$$K_n = (-1)^{n+N/2} \sum_{k=(n+1)/2}^{\min(n,N/2)} \frac{k^{N/2}(2k)!}{(N/2-k)!k!(k-1)!(n-k)!(2k-n)!}.$$

The coefficients K_n are computed once and stored. Double precision arithmetic is mandatory to obtain accurate solutions. This method has been shown to provide accurate inversion for heat conduction problems in the BEM literature and is adopted in this study as the method to invert Laplace-transformed BEM solutions. Typically, the upper limit in the series is taken as $N = 12 \sim 14$, as cited by Stehfest [22]; however, for these types of BEM solution inversions, Moridis and Reddell [28] reported little gain in accuracy for $N = 6 \sim 10$ and demonstrated accurate results using $N = 6$. Davies and Crann [29] also report accurate results using $N = 8$ for BEM problems with periodic boundary conditions. In this work we have used $N = 12$, following the original results of Stehfest, and for maximum accuracy. It is also notable that owing to amplification effects of the large factorial coefficients K_n, on both round off and truncation errors BEM solutions must be carried to very high levels of precision. For this reason very accurate integration, linear solver, and iteration routines are necessary in the BEM solution. This requirement acts to further increase the computational power and time needed for accurate transient results. This inversion method still remains advantageous because of its consistent requirements for any-time solutions. The computation is independent of the given time value, which is a major advantage over time-marching schemes, which require much longer run-times for large-time solutions compared to small-time solutions.

12.3 Explicit Domain Decomposition

In the standard BEM solution process, if N is the number of boundary nodes used to discretize the problem, the number of floating-point operations (FLOPS) required to generate the algebraic system is proportional to N^2. Direct memory allocation is also proportional to N^2. Enforcing imposed boundary conditions yields

$$[H]\{T\} = [G]\{q\} \quad \Rightarrow \quad [A]\{x\} = \{b\},$$

where $\{x\}$ contains the nodal unknowns T or q, whichever is not specified in the boundary conditions. The solution of the algebraic system for the boundary unknowns can be performed using a direct solution method, such as LU decomposition, requiring FLOPS proportional to N^3 or an iterative method such as the biconjugate gradient or general minimization of residuals which, in general, require FLOPS proportional to N^2 to achieve convergence. In 3D problems of any appreciable size, the solution becomes computationally prohibitive and leads to enormous memory demands.

A domain decomposition solution process is adopted instead, where the domain is decomposed by artificially sub-sectioning the single domain of interest into K subdomains. Each of these is independently discretized and solved by a standard BEM, with enforcement of the continuity of temperature and heat flux at the interfaces. It is worth mentioning that the discretization of neighboring subdomains in this method of decomposition does not have to be coincident, that is, at the connecting interface, the boundary elements and nodes from the two adjoining subdomains are not required to be structured following a sequence or particular position. The only requirement at the connecting interface is that it forms a closed boundary with the same path on both sides. The information between neighboring subdomains separated by an interface can be effectively passed through an interpolation, for example, by compactly supported radial-basis functions.

The process is illustrated in 2D in Fig. 2, with a decomposition of four ($K = 4$) subdomains. The conduction problem is solved independently over each subdomain, where initially a guessed boundary condition is imposed over the interfaces in order to create a well-posed problem for each subdomain. The problem in the subdomain Ω_1 is transformed as follows:

$$\nabla^2 T_{\Omega_1}(x,y) = 0 \quad \Rightarrow \quad [H_{\Omega_1}]\{T_{\Omega_1}\} = [G_{\Omega_1}]\{q_{\Omega_1}\}.$$

The composition of this algebraic system requires n^2 FLOPS, where n is the number of boundary nodes in the subdomain, as well as n^2 for direct memory allocation. This new proportionality number n is roughly equivalent to $2N/(K+1)$, as long as the discretization along the interfaces has the same level of resolution as the discretization along the boundaries. The direct memory allocation requirement for the algebraic manipulation of the latter is now reduced to n^2, since the influence coefficient matrices can easily be stored in ROM memory for later use after the boundary value problems on the remaining subdomains have been effectively solved. For the example shown here, where the number of subdomains is $K = 4$, the new proportionality value n is approximately equal to $2N/5$. This simple multi-region example reduces the memory requirements to about $n^2/N^2 = 4/25 = 16\%$ of the standard BEM approach.

The algebraic system for the subdomain Ω_1 is rearranged, with the aid of the given and guessed boundary conditions, as

$$[H_{\Omega_1}]\{T_{\Omega_1}\} = [G_{\Omega_1}]\{q_{\Omega_1}\} \quad \Rightarrow \quad [A_{\Omega_1}]\{x_{\Omega_1}\} = \{b_{\Omega_1}\}.$$

The solution of the new algebraic system for Ω_1 requires that the number of FLOPS be proportional to $n^3/N^3 = 8/125 = 6.4\%$ of the standard BEM approach if a direct algebraic solution method is employed, or a number of floating-point operations proportional to $n^2/N^2 = 4/25 = 16\%$ of the standard BEM approach if an indirect algebraic solution method is employed. For both FLOPS count and direct memory requirement, this reduction is dramatic. However, as the first set of solutions for the subdomains was obtained using guessed boundary conditions along the interfaces, the global solution needs to follow an iteration process and satisfy a convergence criterion.

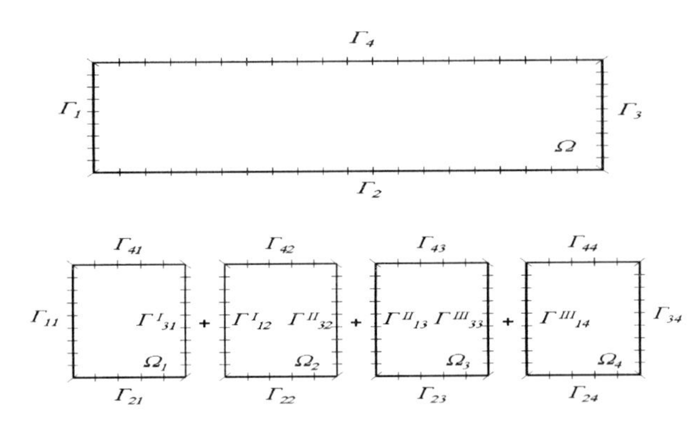

Fig. 2. BEM single region discretization and four domain BEM decomposition.

Globally, the FLOPS count for the formation of the algebraic setup for all K subdomains must be multiplied by K, therefore, the total operation count for the computation of the coefficient matrices is $K\frac{n^2}{N^2} \approx \frac{4K}{(K+1)^2}$. For this particular case with $K = 4$, this corresponds to $Kn^2/N^2 = 16/25 = 64\%$ of the standard BEM approach. Moreover, a more significant reduction is revealed in the RAM memory requirements since only the memory needs for one of the subdomains must be allocated at a time; the others can temporarily be stored into ROM, and when a parallel strategy is adopted, the matrices for each subdomain are stored by its assigned processor. Therefore, for this case of $K = 4$, the memory requirements are reduced to only $n^2/N^2 = 4/25 = 16\%$ of the standard single region case.

In order to reduce the computational efforts needed with respect to the algebraic solution of the system, a direct approach LU factorization is employed for all subdomains. The LU factors of the coefficient matrices for all subdomains are constant, as they are independent of the right-hand side vector and need to be computed only once, at the first iteration step, and stored on disc for later use during the iteration process. Therefore, at each iteration, only a forward and a backward substitution will be required for the algebraic solution. This feature allows a significant reduction in the operational count through the iteration process since only a number of floating-point operations proportional to n, as opposed to n^3, is required at each iteration step. The access to memory at each iteration step must also be added to this computation time. Typically, however, the overall convergence of the problem requires few iterations, and this access to memory is not a significant addition. Additionally, iterative solvers such as GMRES may offer a more efficient alternative.

12.4 Iterative Solution Algorithm

The initial guess is crucial to the success of any iteration scheme. In order to provide an adequate initial guess for the 3D case, the problem is first solved using a coarse-grid constant-element model, obtained by collapsing the nodes of the discontinuous bilinear element to the centroid and supplying that model with a physically based initial guess for interface temperatures. This converged solution then serves as the initial guess for a finer-grid solution, obtained using isoparametric bilinear elements; the latter, in turn, may be used to provide the starting point to a super-parametric biquadratic model (see Fig. 1 (a)–(c), where these three elements are illustrated).

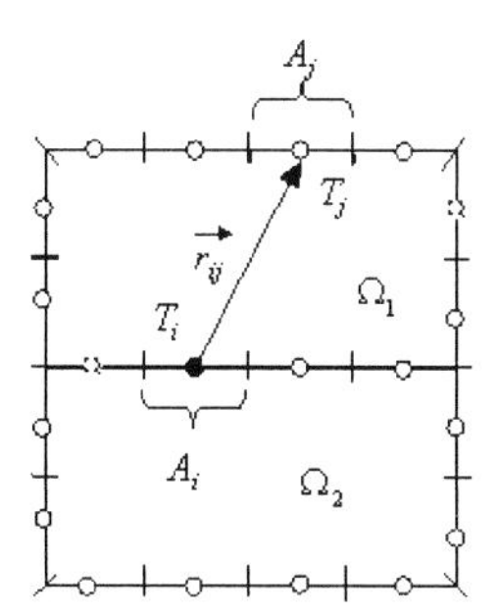

Fig. 3. Initial guess at the interface node i illustrated in 2D for a 2-region subdomain decomposition.

While the constant-element solution can be used as an initial guess for the later runs, an initial guess is still required for the solution of the constant-element case. An efficient initial guess can be made using a 1D heat conduction argument for every node on the external surfaces to every node at the interface of each subdomain. An "area over distance" argument is then used to weight the contribution of an external temperature node to an interface node (see Fig. 3).

Relating any interface node i to any exterior node j, we estimate that

$$T_i = \frac{\displaystyle\sum_{j=1}^{N_3} \frac{A_j T_j}{r_{ij}}}{\displaystyle\sum_{j=1}^{N_e} \frac{A_j}{r_{ij}}},$$

where $r_{ij} = |\vec{r}_{ij}|$ is the magnitude of the position vector from interfacial node i to surface node j and the area of element j is denoted by A_j. There are N_e exterior nodes imposed by the boundary conditions, N_T exterior nodes imposed by temperature, N_q exterior nodes subjected to heat flux conditions, and N_h exterior boundary nodes subjected to convective boundary conditions. The use of a 1D conduction argument for flux and convective nodes is shown in Fig. 4.

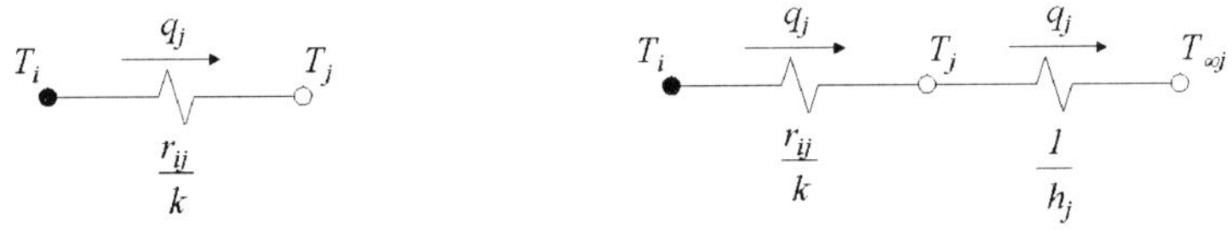

(a) Heat flux node j.

(b) Convective node j.

Fig. 4. Electric circuit analogy to 1D heat conduction from node i to node j.

Using these arguments, we readily conclude that the initial guess for any interfacial node is given by the simple algebraic expression

$$T_i = \frac{\displaystyle\sum_{j=1}^{N_T} B_{ij} T_j - \sum_{j=1}^{N_q} B_{ij} R_{ij} q_j + \sum_{j=1}^{N_h} \frac{B_{ij} H_{ij} T_{\infty j}}{H_{ij}+1}}{\displaystyle S_i - \sum_{j=1}^{N_T} B_{ij} + \sum_{j=1}^{N_h} \frac{B_{ij} H_{ij}}{H_{ij}+1}},$$

where

$$B_{ij} = A_j / r_{ij}, \quad R_{ij} = r_{ij}/k, \quad H_{ij} = h_j / k r_{ij}, \quad S_i = \sum_{j=1}^{N} A_j / r_{ij}.$$

The thermal conductivity of the medium is k, and the film coefficient at the jth convective surface is h_j. For a nonlinear problem, the conductivity of the medium is taken at a mean reference temperature. Once the initial temperatures are imposed as boundary conditions at the interfaces, a resulting set of normal heat fluxes along the interfaces will be computed. These are then nonsymmetrically averaged in an effort to match the heat flux from neighboring subdomains.

In a two-domain substructure, the averaging at the interface is explicitly given by

$$q_{\Omega_1}^I = q_{\Omega_1}^I - \frac{q_{\Omega_1}^I + q_{\Omega_2}^I}{2} \quad \text{and} \quad q_{\Omega_2}^I = q_{\Omega_2}^I - \frac{q_{\Omega_2}^I + q_{\Omega_1}^I}{2},$$

to ensure the flux continuity condition $q_{\Omega_1}^I = -q_{\Omega_2}^I$ after averaging. Compactly supported radial-basis interpolation can be employed in the flux averaging process in order to account for unstructured grids along the interface from neighboring subdomains. Using these fluxes, we solve the BEM equations again to obtain mismatched temperatures along the interfaces for neighboring subdomains. These temperatures are interpolated, if necessary, from one side of the interface to the other using compactly supported radial-basis functions, to account for the possibility of interface mismatch between the adjoining substructure grids. Once this is accomplished, the temperature is averaged out at each interface. Thus, in a two-domain substructure, the interface temperatures for regions 1 and 2 are

$$T_{\Omega_1}^I = \frac{T_{\Omega_1}^I + T_{\Omega_2}^I}{2} + \frac{R'' q_{\Omega_1}^I}{2} \quad \text{and} \quad T_{\Omega_2}^I = \frac{T_{\Omega_1}^I + T_{\Omega_2}^I}{2} + \frac{R'' q_{\Omega_2}^I}{2},$$

to account, in general, for a case where a physical interface exists and a thermal contact resistance is present between the connecting subdomains; here, R'' is the thermal contact resistance that imposes a jump on the interface temperature values. These temperatures, now matched along the interfaces, are used as the next set of boundary conditions.

It is important to note that when dealing with the nonlinear problem, the interfacial temperature update is performed in terms of the temperature T and not in terms of the Kirchhoff transform variable U. That is, given the current values of the transform variable from either side of the subdomain interface at the current iteration, these are both inverted to provide the actual temperatures, and it is these temperatures that are averaged. This is an important point, as the Kirchhoff transform amplifies the jump in temperature at the interface, leading to the divergence of the iterative process, as reported in the literature (see [30]–[32]). Also, if a convective boundary condition is imposed at the exposed surface of a subdomain, a sublevel iteration is carried out for that subdomain. However, as the solution for such a subdomain is part of the overall iterative process, the sublevel iterations are not carried out to convergence, but are limited to only a few. For such cases, the number of sublevel iterations is set to a default number of 5, with an option for the user to increase that number as needed. The overall iteration is continued until a convergence criterion is satisfied. A measure of convergence may be defined as the L_2-norm of the mismatched temperatures along all interfaces, that is,

$$L_2 = \left\{ \frac{1}{KN^I} \sum_{k=1}^{K} \sum_{i=1}^{N^I} \left(T^I - T_u^I\right)^2 \right\}^{1/2}.$$

This norm measures the standard deviation of the BEM-computed interface temperatures T^I and the averaged-out updated interface temperatures T_u^I. The iteration routine can be stopped once this standard deviation reaches a small fraction ϵ of ΔT_{max}, where ΔT_{max} is the maximum temperature span of the global field. We remark that an iteration means the process by which an iterative sweep is carried out to update both the interfacial fluxes and temperatures so that the above norm may be computed. Here it is important to note that for the steady-state problems, a value of $\epsilon = 5 \cdot 10^{-4}$ is sufficient for accurate solutions; however, owing to the amplification effects of the Stehfest transformation, the transient cases require values as small as $\epsilon = 10^{-15}$.

12.5 Parallel Implementation on a PC Cluster

The above domain decomposition BEM formulation is ideally suited to parallel computing. We ran our BEM solutions on a Windows XP-based cluster consisting of 10 Intel-based P3 and P4 CPUs (1.7GHz $\sim$ 2GHz), equipped with RAMBUS memory ranging from 768MB to 1,024MB. This small cluster is interconnected through a local workgroup in a 100 base-T Ethernet network with full duplex switches. A parallel version of the code is implemented under MPICH libraries, which conform with MPI and MPI2 standards (see [33]–[35]) and using the COMPAQ Visual FORTRAN compiler. The parallel code collapses to serial computation if a single processor is assigned to the cluster. Static load balancing is implemented for all computations.

12.6 Numerical Validation and Examples

We present a 3D steady-state nonlinear heat conduction example and a 2D transient conduction example. A cylinder of radius 1 and length 10 is considered. The cylinder is decomposed into 10 equal subdomains corresponding to a discretization of 2,080 elements and 2,080 degrees of freedom (DOF) for the constant-element discretization, 8,320 DOF for the bilinear discretization, and 16,640 DOF for the biquadratic discretization. Two cases are considered here, namely, (i) a rod with nonlinear conductivity

$$k(T) = 1.93[1 + 9.07 \times 10^{-4}\,(T - 720)],$$

and (ii) a composite rod with end-caps comprising 10% of the geometry and having a low nonlinear conductivity

$$k(T) = 7.51[1 + 4.49 \times 10^{-4}\,(T - 1420)],$$

with the same conductivity for the remainder of the rod as in (i), or $k(T) = 19.33[1 + 4.53 \times 10^{-4}(T - 1420)]$ over 80% of the interior. Convective boundary conditions are imposed everywhere on the cylinder walls, with the ends cooled by convection with $T_\infty = 0$ and $h = 10$, while the perimeter is heated by convection with $h = 1$ and T_∞, varying from 1,000 to 4,000. The timings and total iterations for convergence of the solutions are shown in Table 1.

Table 1. The number of iterations and timings for the rod problem.

P4 cluster∼2,080 elements	Case 1	Case 2
constant elements (2,080DOF)	5 iterations	9 iterations
bilinear elements (8,320 DOF)	1 iteration	1 iteration
biquadratic elements (16,640 DOF)	1 iteration	1 iteration
Total time to solution	284 seconds	292 seconds

The second example considers transient heat conduction in a laminar airfoil with three cooling passages. The entire model consisted of about 1,600 degrees of freedom which were split over eight separate subdomains. Constant convective boundary conditions were applied at all surfaces. The problem was also modeled using the commercial code Fluent 6.1 with a similar level of discretization (see Fig. 5 for each mesh). Since an analytic solution to this problem is not available, a time step convergence study was completed for the finite difference model to ensure stable, accurate results for the finite difference analysis. At a time step of 0.04 sec, the change in solutions becomes negligible and the problem was solved with that converged time step. The temperature solutions at two points in each model were recorded and are displayed over time in Fig. 6. Contour plots at a single representative time (40 sec) are also presented (see Fig. 7) to show the agreement of the entire field temperatures. These results show almost perfect agreement between the BEM and FD solutions.

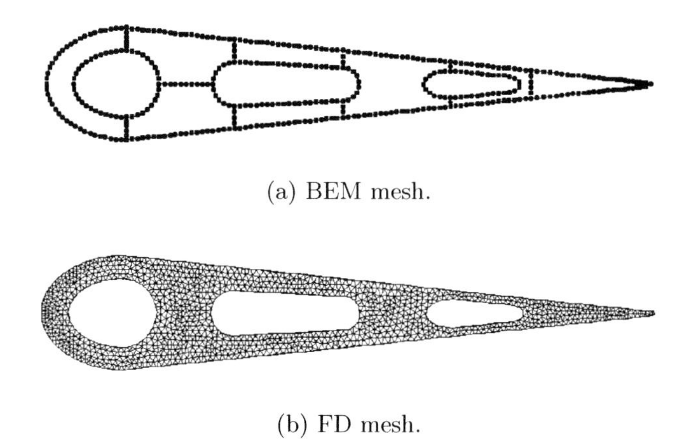

(a) BEM mesh.

(b) FD mesh.

Fig. 5. Laminar airfoil meshes (convective BCs imposed on all boundaries).

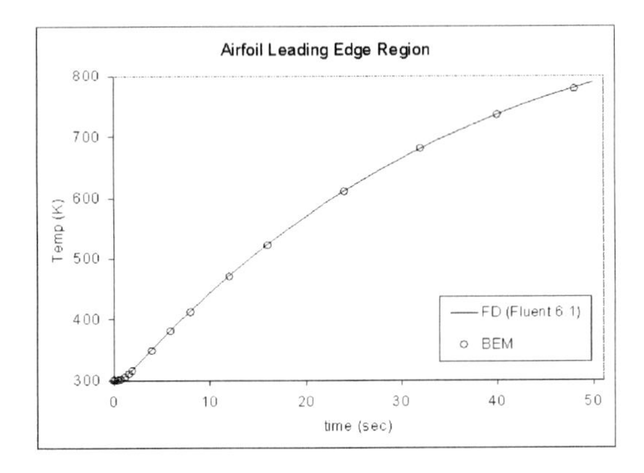

(a) Temperature at a point between the leading edge and the first passage.

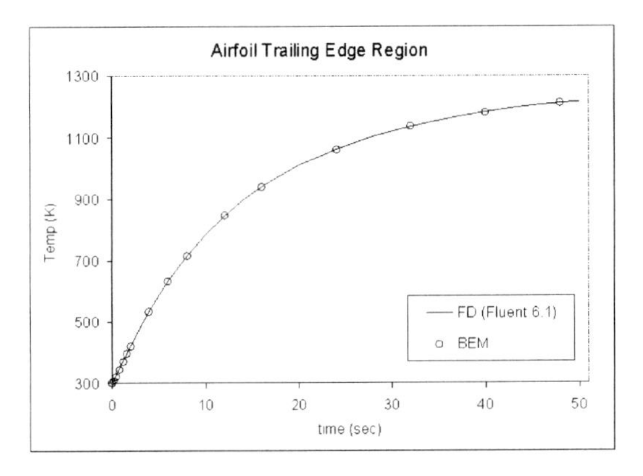

(b) Temperature at the trailing edge.

Fig. 6. Temperature solutions at two points.

12.7 Conclusions

The boundary element method (BEM) is often an efficient choice for the solution of various engineering field problems as it acts to decrease the dimensionality of the problem. However, the solution of large problems is still prohibitive since the BEM coefficient matrices are typically fully populated and difficult to subdivide or compress. We have presented an efficient iterative domain decomposition method to reduce the storage requirement and allow the solution of such large-scale problems. The decomposition approach lends itself ideally to parallel message-passing-type computing due to the independence of each of the BEM subregion solutions. With this approach, large-scale problems can be readily solved on small PC clusters. The iterative domain decomposition approach is general and can be applied to any type of BEM problem arising in such diverse fields as elasticity, thermoelasticity, and acoustics.

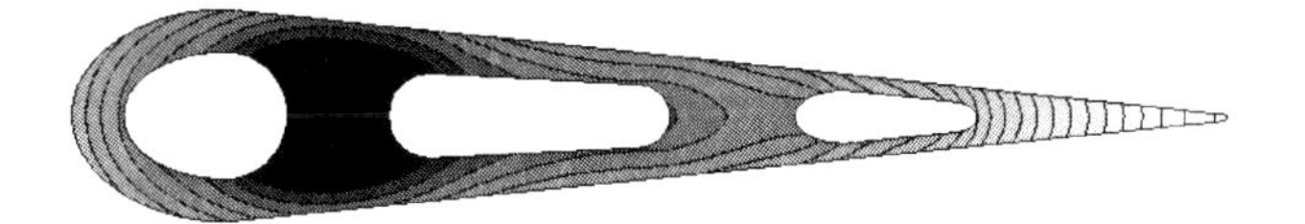

(a) BEM temperature field at $t = 40$ sec.

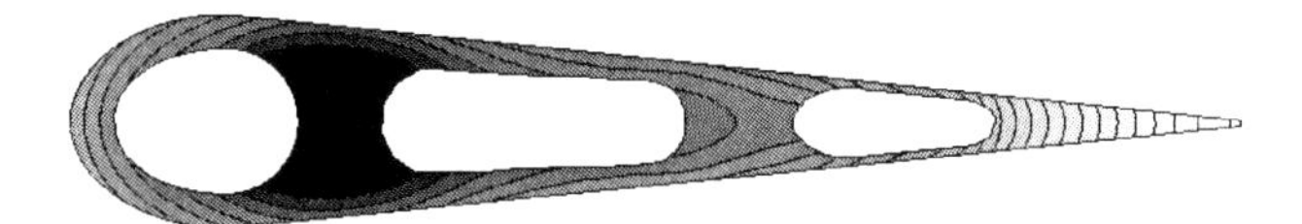

(b) Finite difference temperature field at $t = 40$ sec.

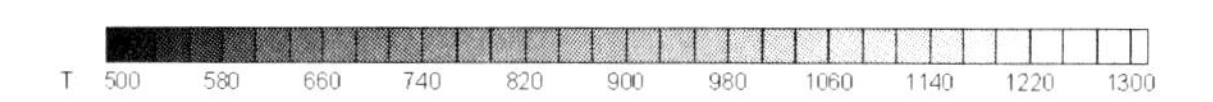

(c) Temperature scale for both solution fields.

Fig. 7. Comparison between BEM and a finite difference solver (Fluent 6.1).

References

1. A. Kassab, E. Divo, J. Heidmann, E. Steinthorsson, and F. Rodriguez, BEM/FVM conjugate heat transfer analysis of a three-dimensional film cooled turbine blade, *Internat. J. Numer. Methods Heat and Fluid Flow* **13** (2003), 581–610.
2. J.D. Heidmann, A.J. Kassab, E.A. Divo, F. Rodriguez, and E. Steinthorsson, Conjugate heat transfer effects on a realistic film-cooled turbine vane, ASME Paper GT2003-38553, 2003.
3. F. Rizzo and D.J. Shippy, A formulation and solution procedure for the general non-homogeneous elastic inclusion problem, *Internat J. Solids Structures* **4** (1968), 1161–1179.
4. R.A. Bialecki, M. Merkel, H. Mews, and G. Kuhn, In- and out-of-core BEM equation solver with parallel and nonlinear options, *Internat J. Numer. Methods Engng.* **39** (1996), 4215–4242.
5. J.H. Kane, B.L. Kashava-Kumar, and S. Saigal, An arbitrary condensing, non-condensing strategy for large scale, multi-zone boundary element analysis, *Comput. Methods Appl. Mech. Engng.* **79** (1990), 219–244.
6. B. Baltz and M.S. Ingber, A parallel implementation of the boundary element method for heat conduction analysis in heterogeneous media, *Engng. Anal.* **19** (1997), 3–11.

7. N. Kamiya, H. Iwase, and E. Kita, Parallel implementation of boundary element method with domain decomposition, *Engng. Anal.* **18** (1996), 209–216.
8. A.J. Davies and J. Mushtaq, The domain decomposition boundary element method on a network of transputers, Ertekin, BETECHXI, in *Proc. 11th Conf. Boundary Element Technology,* Computational Mechanics Publications, Southampton, 1996, 397–406.
9. N. Mai-Duy, P. Nguyen-Hong, and T. Tran-Cong, A fast convergent iterative boundary element method on PVM cluster, *Engng. Anal.* **22** (1998), 307–316.
10. L. Greengard and J. Strain, A fast algorithm for the evaluation of heat potentials, *Comm. Pure Appl. Math.* **43** (1990), 949–963.
11. W. Hackbush and Z.P. Nowak, On the fast multiplication in the boundary element method by panel clustering, *Numer. Math.* **54** (1989), 463–491.
12. H. Bucher and L.C. Wrobel, A novel approach to applying wavelet transforms in boundary element method, BETEQII, in *Advances in Boundary Element Techniques, II,* Hogaar Press, Switzerland, 2000, 3–13.
13. F. Rodriguez, E. Divo, and A.J. Kassab, A strategy for BEM modeling of large-scale three-dimensional heat transfer problems, in *Recent Advances in Theoretical and Applied Mechanics,* vol. XXI, A.J. Kassab, D.W. Nicholson, and I. Ionescu (eds.), Rivercross Publishing, Orlando, FL, 2002, 645–654.
14. C.A. Brebbia, J.C.F. Telles, and L.C. Wrobel, *Boundary Element Techniques,* Springer-Verlag, Berlin, 1984.
15. L.C. Wrobel, *The Boundary Element Method—Applications in Thermofluids and Acoustics,* vol. 1, Wiley, New York, 2002.
16. A.J. Kassab and L.C. Wrobel, Boundary element methods in heat conduction, in *Recent Advances in Numerical Heat Transfer,* vol. 2, W.J. Mincowycz and E.M. Sparrow (eds.), Taylor and Francis, New York, 2000, 143–188.
17. F.J. Rizzo and D.J. Shippy, A method of solution for certain problems of transient heat conduction, *AIAA J.* **8** (1970), 2004–2009.
18. E. Divo and A.J. Kassab, A boundary integral equation for steady heat conduction in anisotropic and heterogeneous media, *Numer. Heat Transfer B: Fundamentals* **32** (1997), 37–61.
19. E. Divo and A.J. Kassab, A generalized BIE for transient heat conduction in heterogeneous media, *J. Thermophysics and Heat Transfer* **12** (1998), 364–373.
20. E. Divo, A.J. Kassab, and F. Rodriguez, A parallelized iterative domain decomposition approach for 3D boundary elements in nonlinear heat conduction, *Numer. Heat Transfer B: Fundamentals* **44** (2003), 417–437.

21. A.H.-D. Cheng and K. Ou, An efficient Laplace transform solution for multiaquifer systems, *Water Resources Research* **25** (1989), 742–748.
22. H. Stehfest, Numerical inversion of Laplace transforms, *Comm. ACM* **13** (1970), 47–49.
23. H. Stehfest, Remarks on algorithm 368: numerical inversion of Laplace transforms, *Comm. ACM* **13** (1970), 624.
24. A.J. Davies and D. Crann, Parallel Laplace transform methods for boundary element solutions of diffusion-type problems, *J. Boundary Elements* BETEQ 2001, No. 2 (2002), 231–238.
25. E. Divo, A.J. Kassab, and M.S. Ingber, Shape optimization of acoustic scattering bodies, *Engng. Anal. Boundary Elements* **27** (2003), 695–704.
26. M. Greenberg, *Applications of Green's Functions in Engineering and Science,* Prentice-Hall, Englewood Cliffs, NJ, 1971.
27. B. Davies and B. Martin, Numerical inversion of the Laplace transform: a survey and comparison of methods, *J. Comput. Phys.* **33** (1979), 1–32.
28. G.J. Moridis and D.L. Reddell, The Laplace transform boundary element (LTBE) method for the solution of diffusion-type equations, in *Boundary Elements,* vol. XIII, WIT Press, Southampton, U.K., 1991, 83–97.
29. A.J. Davies and D. Crann, The Laplace transform boundary element methods for diffusion problems with periodic boundary conditions, in *Boundary Elements,* vol. XXVI, WIT Press, Southampton, U.K., 2004, 393–402.
30. J.P.S. Azevedo and L.C. Wrobel, Non-linear heat conduction in composite bodies: a boundary element formulation, *Internat. J. Numer. Methods Engng.* **26** (1988), 19–38.
31. R. Bialecki and R. Nahlik, Solving nonlinear steady-state potential problems in non-homogeneous bodies using the boundary element method, *Numer. Heat Transfer B* **16** (1989), 79–96.
32. R. Bialecki and G. Kuhn, Boundary element solution of heat conduction problems in multi-zone bodies of non-linear materials, *Internat. J. Numer. Methods Engng.* **36** (1993), 799–809.
33. W. Gropp, E. Lusk, and R. Thakur, *Using MPI: Portable Parallel Programming with the Message-passing Interface,* MIT Press, Cambridge, MA, 1999.
34. W. Gropp, E. Lusk, and R. Thakur, *Using MPI-2: Advanced Features of the Message-passing Interface,* MIT Press, Cambridge, MA, 1999.
35. T.E. Sterling, *Beowulf Cluster Computing with Windows,* MIT Press, Cambridge, MA, 2001.

13 The Poisson Problem for the Lamé System on Low-dimensional Lipschitz Domains

Svitlana Mayboroda and Marius Mitrea

13.1 Introduction and Statement of the Main Results

Consider the Lamé operator of linear elastostatics in $\mathbb{R}^3$,

$$\mathcal{L}\vec{u} := \mu\Delta\vec{u} + (\lambda+\mu)\nabla(\operatorname{div}\vec{u}), \quad \mu > 0,\ \lambda > -\tfrac{2}{3}\mu. \tag{13.1}$$

In this paper we study the well-posedness of the Poisson problems for the system of elastostatics equipped with either Dirichlet or Neumann-type boundary conditions in a bounded Lipschitz domain $\Omega \subset \mathbb{R}^3$ (see Section 13.6 for the two-dimensional case):

$$\mathcal{L}\vec{u} = \vec{f} \text{ in } \Omega, \quad \operatorname{Tr}\vec{u} = \vec{g} \text{ on } \partial\Omega, \tag{13.2}$$

$$\mathcal{L}\vec{u} = \vec{f} \text{ in } \Omega, \quad \partial_\nu\vec{u} = \vec{g} \text{ on } \partial\Omega. \tag{13.3}$$

Hereafter Tr stands for the *trace* map on $\partial\Omega$, and ∂_ν is the *traction conormal* defined by

$$\partial_\nu\vec{u} := \lambda(\operatorname{div}\vec{u})\nu + \mu[\nabla\vec{u} + \nabla\vec{u}^t]\nu, \tag{13.4}$$

where ν is the unit normal to $\partial\Omega$ and the superscript t indicates transposition (in this case, of the matrix $\nabla\vec{u} = (\partial_j u^\alpha)_{j,\alpha}$).

Relying on the method of layer potentials and suitable Rellich–Nečas–Payne–Weinberger formulas, the boundary value problems (13.2)–(13.3) with $\vec{f} = 0$ and $\vec{g} \in L^p(\partial\Omega)$, $2-\varepsilon < p < 2+\varepsilon$, have been treated (in all space dimensions) by B. Dahlberg, C. Kenig, and G. Verchota [7]. In the three-dimensional setting, these results have been subsequently extended to optimal ranges of p's ($2-\varepsilon < p \le \infty$ for the Dirichlet boundary condition, and $1 < p < 2+\varepsilon$ for the traction boundary condition, with $\varepsilon = \varepsilon(\partial\Omega) > 0$) in [6]. More recently, the results for the Dirichlet problem (i.e., (13.2) with $\vec{f} = 0$ and $\vec{g} \in L^p(\partial\Omega)$) have been further extended in dimension $n \ge 4$ to the range $2-\varepsilon < p < \frac{2(n-1)}{n-3} + \varepsilon$ by Z. Shen in [20] (though determining

The work of M. Mitrea was supported in part by grants from the NSF and the UM Office of Research.

the optimal range of p's remains an open problem at the moment). In all these works, nontangential maximal function estimates are sought for the solution $\vec{u}$.

Following the breakthrough in the case of the Dirichlet Laplacian in [11] (which further builds on [5]), as well as the subsequent developments in [8], a study of (13.2)–(13.3) on Sobolev–Besov spaces over low-dimensional Lipschitz domains has been initiated in [17], where the optimal ranges of indices have been identified under the assumption that all spaces involved are *Banach* (roughly speaking, this amounts to the requirement that all the integrability exponents are greater than one). Here we take the next natural step and extend the scope of this study to allow the consideration of Besov ($B^{p,q}_s$) and Triebel–Lizorkin ($F^{p,q}_s$) spaces for the full range of indices $0 < p, q \leq \infty$. This builds on the work in [13] where we have recently dealt with the case of the Laplacian. Our current goals are also motivated by questions regarding the regularity of the Green potentials associated with the Lamé system and here we prove new mapping properties of these potentials on L^p and Hardy spaces. For example, we show that (any) two derivatives on the elastic Green potential with a Dirichlet boundary condition yield an operator bounded on $L^p(\Omega)$, $1 < p \leq 2$, if $\Omega \subset \mathbb{R}^2$ is a convex domain and $|\lambda| < \sqrt{3}\mu$, and on the Hardy space $h^1(\Omega)$ if $\Omega \subset \mathbb{R}^3$ is a convex polyhedron.

Before stating our main results we briefly elaborate on notation and terminology. We refer to, e.g., [23] for definitions and basic properties of the Besov and Triebel–Lizorkin scales

$$B^{p,q}_\alpha(\mathbb{R}^n), \quad F^{p,q}_\alpha(\mathbb{R}^n), \quad n \geq 2.$$

An open, connected set $\Omega \subset \mathbb{R}^3$ is called a *Lipschitz domain* if its boundary can be locally described by means of Lipschitz graphs (in appropriate systems of coordinates); see, e.g., [21] for a more detailed discussion. Given $\Omega \subset \mathbb{R}^3$ Lipschitz and $0 < p, q \leq \infty$, $\alpha \in \mathbb{R}$, we set

$$B^{p,q}_\alpha(\Omega) := \{u \in D'(\Omega) : \exists\, v \in B^{p,q}_\alpha(\mathbb{R}^3) \text{ with } v|_\Omega = u\},$$

$$B^{p,q}_{\alpha,0}(\Omega) := \{u \in B^{p,q}_\alpha(\mathbb{R}^3) : \operatorname{supp} u \subseteq \overline{\Omega}\},$$

with similar definitions for $F^{p,q}_\alpha(\Omega)$ and $F^{p,q}_{\alpha,0}(\Omega)$. Here, $D(\Omega)$ denotes the collection of test functions in Ω (equipped with the usual inductive limit topology), while $D'(\Omega)$ stands for the space of distributions in Ω. Finally, if $|s| < 1$, $B^{p,q}_s(\partial\Omega)$ stands for the Besov class on the Lipschitz manifold $\partial\Omega$, obtained by transporting (via a partition of unity and pull-back) the standard scale $B^{p,q}_s(\mathbb{R}^2)$. Typically, we shall work with vector-valued distributions, such as $\vec{u} = (u^1, u^2, u^3)$, etc.; however, even when warranted, our notation for the various function spaces employed in this paper does not emphasize the vector nature of the objects involved (this will eventually be clear from the context).

To state our first result, for each $a \in \mathbb{R}$ set

$$(a)_+ := \max\{a, 0\}.$$

Theorem 1. *Suppose that Ω is a bounded Lipschitz domain in $\mathbb{R}^3$, and, for $\frac{2}{3} < p \leq \infty$, $0 < q \leq \infty$, $2\big(\frac{1}{p} - 1\big)_+ < s < 1$, consider the boundary value problem*

$$\mathcal{L}\vec{u} = \vec{f} \in B^{p,q}_{s+\frac{1}{p}-2}(\Omega), \quad \vec{u} \in B^{p,q}_{s+\frac{1}{p}}(\Omega), \quad \operatorname{Tr}\vec{u} = \vec{g} \in B^{p,q}_{s}(\partial\Omega). \tag{13.5}$$

Then there exists $\varepsilon = \varepsilon(\Omega) \in (0,1]$ such that (13.5) is well posed if the pair (s,p) satisfies one of the following three conditions:

$$\begin{array}{rl} (I): & \frac{2}{2+\varepsilon} < p \leq \frac{2}{1+\varepsilon} \ \ \text{and} \ \ \frac{2}{p} - 1 - \varepsilon < s < 1; \\ (II): & \frac{2}{1+\varepsilon} \leq p \leq \frac{2}{1-\varepsilon} \ \ \text{and} \ \ 0 < s < 1; \\ (III): & \frac{2}{1-\varepsilon} \leq p \leq \infty \ \ \text{and} \ \ 0 < s < \frac{2}{p} + \varepsilon. \end{array} \tag{13.6}$$

Furthermore, the solution has an integral representation formula in terms of (elastic) layer potential operators and satisfies the natural estimate

$$\|\vec{u}\|_{B^{p,q}_{s+\frac{1}{p}}(\Omega)} \leq C\|\vec{f}\|_{B^{p,q}_{s+\frac{1}{p}-2}(\Omega)} + C\|\vec{g}\|_{B^{p,q}_{s}(\partial\Omega)}.$$

Finally, an analogous well-posedness result holds on the Triebel–Lizorkin scale, i.e., for the problem

$$\mathcal{L}\vec{u} = \vec{f} \in F^{p,q}_{s+\frac{1}{p}-2}(\Omega), \quad \vec{u} \in F^{p,q}_{s+\frac{1}{p}}(\Omega), \quad \operatorname{Tr}\vec{u} = \vec{g} \in B^{p,p}_{s}(\partial\Omega). \tag{13.7}$$

This time, it is assumed that $p \neq \infty$ and

$$\Big(\min\big\{\tfrac{1}{3p} + 1,\ \tfrac{s}{2} + 1\big\}\Big)^{-1} < q \leq \infty. \tag{13.8}$$

In particular, the above inequality is true when $\min\{p,1\} \leq q \leq \infty$.

In the class of Lipschitz domains this result is sharp, but one may take $\varepsilon = 1$ if $\partial\Omega \in C^1$.

Fig. 1 depicts the two-dimensional pentagonal region consisting of all points with coordinates $(s, 1/p)$ satisfying (13.6).

To formulate our main result regarding the Poisson problem with traction boundary conditions we introduce one more piece of notation. Let Ψ be the six-dimensional linear space of vector-valued functions $\vec{\psi}$ in $\mathbb{R}^3$ satisfying

$$\partial_i\psi^j + \partial_j\psi^i = 0, \quad 1 \leq i,j \leq 3,$$

(note that each $\vec{\psi} \in \Psi$ is a null solution of $\mathcal{L}$ and $\partial_\nu\vec{\psi} = 0$), and for a space of vector-valued distributions $\mathcal{X}$ on a set $\mathcal{O}$, define

$$\Psi^\perp \cap \mathcal{X} := \{\vec{u} \in \mathcal{X} : \langle \vec{u}, \vec{\psi}|_{\mathcal{O}}\rangle = 0,\ \forall\, \vec{\psi} \in \Psi\}. \tag{13.9}$$

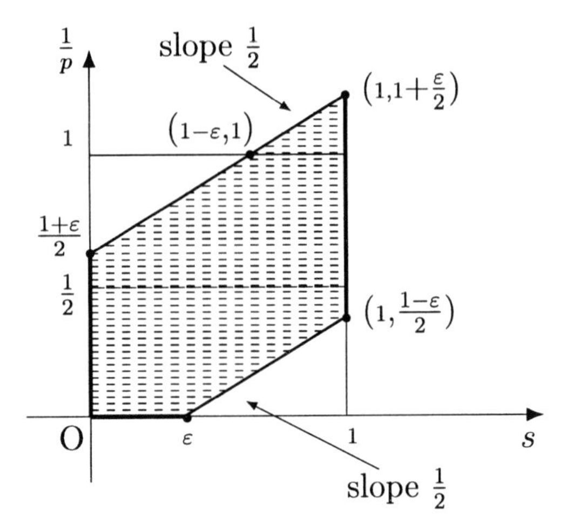

Fig. 1.

Theorem 2. *Let Ω be a bounded, connected Lipschitz domain in $\mathbb{R}^3$ and, for $\frac{2}{3} < p \le \infty$, $0 < q \le \infty$, and $2\big(\frac{1}{p}-1\big)_+ < s < 1$, consider the boundary value problem*

$$\mathcal{L}\vec{u} = \vec{f}\Big|_{\Omega}, \quad \vec{f} \in B^{p,q}_{s+\frac{1}{p}-2,0}(\Omega), \quad \vec{u} \in \Psi^{\perp} \cap B^{p,q}_{s+\frac{1}{p}}(\Omega), \quad \partial^{\vec{f}}_{\nu}\vec{u} = \vec{g} \in B^{p,q}_{s-1}(\partial\Omega), \tag{13.10}$$

where the data are assumed to satisfy the necessary compatibility condition

$$\langle \vec{f}, \vec{\psi}\rangle = \langle \vec{g}, \vec{\psi}\rangle, \quad \forall\, \vec{\psi} \in \Psi. \tag{13.11}$$

Then there exists $\varepsilon = \varepsilon(\Omega) \in (0,1]$ such that (13.10) *has a unique solution if the pair s, p satisfies one of the three conditions in* (13.6)*. In addition, the solution has an integral representation formula in terms of (elastic) layer potential operators and satisfies the estimate*

$$\|\vec{u}\|_{B^{p,q}_{s+\frac{1}{p}}(\Omega)} \le C\|\vec{f}\|_{B^{p,q}_{s+\frac{1}{p}-2,0}(\Omega)} + C\|\vec{g}\|_{B^{p,q}_{s-1}(\partial\Omega)}.$$

An analogous well-posedness result holds for the problem

$$\mathcal{L}\vec{u} = \vec{f}\Big|_{\Omega}, \quad \vec{f} \in F^{p,q}_{s+\frac{1}{p}-2,0}(\Omega), \quad \vec{u} \in \Psi^{\perp} \cap F^{p,q}_{s+\frac{1}{p}}(\Omega), \quad \partial^{\vec{f}}_{\nu}\vec{u} = \vec{g} \in B^{p,p}_{s-1}(\partial\Omega), \tag{13.12}$$

assuming (13.11), (13.8), *and $p \ne \infty$. When $\partial\Omega \in C^1$, one can take $\varepsilon = 1$ in each case.*

Above, $\partial^{\vec{f}}_{\nu}\vec{u}$ stands for a suitable concept of conormal derivative which takes into account both the field $\vec{u}$ and $\vec{f}$. See Section 13.3 for details.

Our strategy for dealing with the problems (13.5), (13.7), (13.10), and (13.12), is to rely on integral representation formulas of the solution in terms of elastic layer potential operators. This approach brings into focus the properties of singular integral operators involved in such representations. In particular, we study the boundedness of the elastic layer potentials on Besov and Triebel–Lizorkin scales in Section 13.2, as well as the invertibility of some of their boundary traces in Section 13.4. As a key prerequisite, in Section 13.3 we include a discussion of sharp results for traces and traction derivatives on Lipschitz domains. The properties of the Green potentials are then analyzed in Section 13.5. Finally, in Section 13.6, the two-dimensional setting is briefly discussed.

13.2 Estimates for Singular Integral Operators

Even though we shall work in the three-dimensional setting, all results proved in this section have natural versions in $\mathbb{R}^n$ with $n \geq 2$. Fix a Lipschitz domain $\Omega \subset \mathbb{R}^3$ and denote by $d\sigma$ and ν the surface measure and the unit normal on $\partial\Omega$, respectively. We start by reviewing the definitions of the elastic layer potentials associated with Ω. Recall that the standard fundamental solution for the system of elastostatics is given by the Kelvin matrix $\Gamma = (\Gamma_{\alpha\beta})_{\alpha,\beta}$, where for each $1 \leq \alpha, \beta \leq 3$, $x = (x_\alpha)_\alpha \in \mathbb{R}^3$,

$$\Gamma_{\alpha\beta}(x) := \frac{3}{8\pi}\left(\frac{1}{\mu} + \frac{1}{2\mu+\lambda}\right)\frac{\delta_{\alpha\beta}}{|x|} + \frac{3}{8\pi}\left(\frac{1}{\mu} - \frac{1}{2\mu+\lambda}\right)\frac{x_\alpha x_\beta}{|x|^3}.$$

Here and elsewhere, $\delta_{\alpha\beta}$ stands for the standard Kronecker symbol. For vector-valued functions on $\partial\Omega$, the *single (elastic) layer potential* operator is defined by

$$\mathcal{S}\vec{f}(x) := \int_{\partial\Omega} \Gamma(x-y)\,\vec{f}(y)\,d\sigma(y), \quad x \in \mathbb{R}^3 \setminus \partial\Omega, \tag{13.13}$$

where $d\sigma$ stands for the canonical surface measure on $\partial\Omega$. Also, the *double (elastic) layer potential* operator is given by

$$\mathcal{D}\vec{f}(x) := \int_{\partial\Omega} \left(\partial_{\nu_y}\Gamma(x-y)\right)^t \vec{f}(y)\,d\sigma(y), \quad x \in \mathbb{R}^3 \setminus \partial\Omega. \tag{13.14}$$

Here the operator ∂_ν (defined in (13.4)) applies to each column of the matrix Γ. The well-known jump relations for the operators (13.13)–(13.14) then read:

$$\begin{aligned}\lim_{\substack{x\to z\\ x\in\gamma_\pm(z)}} \mathcal{D}\vec{f}(x) &= \pm\tfrac{1}{2}\vec{f}(z) + \text{p.v.}\int_{\partial\Omega}\left(\partial_{\nu_y}\Gamma(z-y)\right)^t \vec{f}(y)\,d\sigma(y)\\ &=: \left(\pm\tfrac{1}{2}I + K\right)\vec{f}(z), \quad z \in \partial\Omega,\end{aligned} \tag{13.15}$$

and

$$\lim_{\substack{x\to z\\ x\in\gamma_\pm(z)}} \partial_\nu\mathcal{S}\vec{f}(x) = \left(\mp\tfrac{1}{2}I + K^*\right)\vec{f}(z), \quad z \in \partial\Omega, \tag{13.16}$$

where K^* denotes the formal adjoint of K. In (13.15)–(13.16), $\gamma_\pm(z)$ are suitable nontangential approach regions with vertex at $z \in \partial\Omega$, which are contained in $\Omega_+ := \Omega$ and $\Omega_- := \mathbb{R}^3 \setminus \bar{\Omega}$, respectively. See, e.g., [7] for more details.

Finally, the elastic Newtonian potential is defined as

$$\Pi\vec{f}(x) := \int_{\mathbb{R}^n} \Gamma(x-y)\vec{f}(y)\,dy, \quad x \in \mathbb{R}^3. \tag{13.17}$$

With regard to this operator, we first note the following consequence of known results about the mapping properties of pseudo-differential operators on Besov and Triebel–Lizorkin scales.

Proposition. *Suppose that*

$$0 < p < \infty, \quad 0 < q \leq \infty, \quad 3\Big(\tfrac{1}{\min\{1,p,q\}} - 1\Big) < \alpha,$$

and fix two arbitrary scalar-valued functions $\phi, \psi \in C_c^\infty(\mathbb{R}^3)$. *Then*

$$\phi\Pi\psi : F^{p,q}_{\alpha-2}(\mathbb{R}^3) \longrightarrow F^{p,q}_{\alpha}(\mathbb{R}^3)$$

is a bounded operator. Moreover, for $0 < p, q \leq \infty$ *and* $3\big(\frac{1}{\min\{1,p\}} - 1\big) < \alpha$,

$$\phi\Pi\psi : B^{p,q}_{\alpha-2}(\mathbb{R}^3) \longrightarrow B^{p,q}_{\alpha}(\mathbb{R}^3)$$

is also bounded.

We shall now state two theorems whose proofs are intertwined and will be presented together.

Theorem 3. *Consider a bounded Lipschitz domain* $\Omega \subset \mathbb{R}^3$ *and suppose that* p, q, s *are fixed numbers such that* $\frac{2}{3} < p \leq \infty$, $2(\frac{1}{p} - 1)_+ < s < 1$, *and* $0 < q \leq \infty$. *Then*

$$\begin{aligned} \partial_\nu\Pi : B^{p,q}_{s+1/p-2,0}(\Omega) &\longrightarrow B^{p,q}_{s-1}(\partial\Omega), \\ \partial_\nu\Pi : F^{p,q}_{s+1/p-2,0}(\Omega) &\longrightarrow B^{p,p}_{s-1}(\partial\Omega), \quad \text{if } p \neq \infty, \end{aligned} \tag{13.18}$$

are well-defined, linear, and bounded operators.

Theorem 4. *Let* Ω *be a bounded Lipschitz domain in* $\mathbb{R}^3$ *and assume that the indices* p, s *satisfy* $\frac{2}{3} < p \leq \infty$ *and* $2(\frac{1}{p} - 1)_+ < s < 1$. *Then*

$$\mathcal{D} : B^{p,q}_s(\partial\Omega) \longrightarrow B^{p,q}_{s+\frac{1}{p}}(\Omega), \quad \mathcal{S} : B^{p,q}_{s-1}(\partial\Omega) \longrightarrow B^{p,q}_{s+\frac{1}{p}}(\Omega)$$

are well-defined, bounded, linear operators for each $0 < q \leq \infty$.

The same holds for

$$\mathcal{D}: B^{p,p}_s(\partial\Omega) \longrightarrow F^{p,q}_{s+\frac{1}{p}}(\Omega), \quad \mathcal{S}: B^{p,p}_{s-1}(\partial\Omega) \longrightarrow F^{p,q}_{s+\frac{1}{p}}(\Omega), \tag{13.19}$$

provided $p \neq \infty$ *and* (13.8) *is satisfied.*

Proof of Theorems 4 and 3. This parallels arguments developed for harmonic layer potentials in [14], so we shall present a sketch in which only the novel aspects are emphasized. We proceed in a series of steps.

Step I. Consider first the case when $1 \le p = q \le \infty$ for Besov spaces, and $1 < p < \infty$, $q = 2$, for Triebel–Lizorkin spaces. For a proof of Theorems 4 and 3 under these assumptions, see [17].

Step II. For $0 < p < \infty$, $\alpha \in \mathbb{R}$, introduce the space

$$\mathbf{L}^p_\alpha(\Omega) := \{\vec{u} \in L^p(\Omega) : \mathcal{L}\vec{u} = 0 \text{ and } \|\vec{u}\|_{\mathbf{L}^p_\alpha(\Omega)} < \infty\},$$

$$\|\vec{u}\|_{\mathbf{L}^p_\alpha(\Omega)} := \|\delta^{\langle\alpha\rangle-\alpha}|\nabla^{\langle\alpha\rangle}\vec{u}|\|_{L^p(\Omega)} + \sum_{j=0}^{\langle\alpha\rangle-1} \|\nabla^j \vec{u}\|_{L^p(\Omega)}.$$

Throughout the paper, $\langle\alpha\rangle$ will denote the smallest nonnegative integer greater than or equal to α, and $\nabla^j\vec{u}$ will stand for the vector of all mixed partial derivatives of order j of the components of $\vec{u}$. Also, here and elsewhere, we set $\delta(x) := \mathrm{dist}(x, \partial\Omega)$ for $x \in \Omega$.

The claim we make in this scenario is that

$$\mathcal{D}: B^{p,p}_s(\partial\Omega) \longrightarrow \mathbf{L}^p_{s+\frac{1}{p}}(\Omega), \quad \mathcal{S}: B^{p,p}_{s-1}(\partial\Omega) \longrightarrow \mathbf{L}^p_{s+\frac{1}{p}}(\Omega)$$

are bounded operators provided $\frac{2}{3} < p \le \infty$ and $2(\frac{1}{p} - 1)_+ < s < 1$. This is proved by means of atomic decomposition results for Besov spaces and pointwise estimates on the kernel of the elastic layer potentials in question. The reader is referred to [14] for details.

Step III. The following embedding holds:

$$\mathbf{L}^p_\alpha(\Omega) \hookrightarrow F^{p,2}_\alpha(\Omega) \quad \text{if} \quad 0 < p < \infty, \quad \alpha \ge 0. \tag{13.20}$$

This has been established in [11] for $1 < p < \infty$, so the novelty here is the consideration of the range $0 < p \le 1$, on which we shall focus subsequently. First, we verify that, if $0 < p \le 1$, then

$$\mathcal{L}\vec{u} = 0 \text{ in } \Omega \quad \text{and} \quad \vec{u} \in L^p(\Omega) \quad \Longrightarrow \quad \vec{u} \in h^p(\Omega), \tag{13.21}$$

where membership to the local Hardy space $h^p(\Omega)$ can be characterized as follows. Fix $\psi \in C^\infty_c(B(0,1))$ such that $\int_{B(0,1)} \psi(x)\,dx = 1$ and set $\psi_t(x) := t^{-n}\psi(x/t)$. Then for any $0 < p \le 1$ and any $\vec{u} \in D'(\Omega)$,

$$\vec{u} \in h^p(\Omega) \iff (\vec{u})^+(x) := \sup_{0<t<\delta(x)} |(\psi_t * \vec{u})(x)| \in L^p(\Omega). \tag{13.22}$$

With an eye on (13.21) we first observe that the null solutions of $\mathcal{L}$ satisfy a sub-averaging property. That is, there exists a finite, positive constant C such that if $\mathcal{L}\vec{u}=0$ in Ω, then

$$|\vec{u}(x)| \leq \frac{C}{r^3}\int_{B_r(x)} |\vec{u}(y)|\,dy \quad \forall\, x\in\Omega,\ 0<r<\delta(x). \tag{13.23}$$

Indeed, it was proved in [12] that there exist two constants C_1 and C_2 such that for every null solution $\vec{u}=(u^\alpha)_\alpha$ of the Lamé system in an open domain Ω in $\mathbb{R}^3$,

$$u^\alpha(x)=\frac{1}{4\pi\rho^2}\int_{|y-x|=\rho}\sum_\beta\Big(C_1\delta_{\alpha\beta}+C_2\frac{x_\alpha-y_\alpha}{|x-y|}\frac{x_\beta-y_\beta}{|x-y|}\Big)u^\beta(y)\,d\sigma_y$$

for $x\in\Omega$ and $0<\rho<\delta(x)$, $1\leq\alpha\leq 3$. Multiplying both sides above by ρ^2 and integrating with respect to ρ for $0\leq\rho\leq r$, we get

$$u^\alpha(x)=\frac{3}{4\pi r^3}\int_{B_r(x)}\sum_\beta\Big(C_1\delta_{\alpha\beta}+C_2\frac{x_\alpha-y_\alpha}{|x-y|}\frac{x_\beta-y_\beta}{|x-y|}\Big)u^\beta(y)\,dy$$

for $x\in\Omega$ and $0<r<\delta(x)$, $1\leq\alpha\leq 3$, which readily yields (13.23). Making a dilation (to reduce matters to $r=1$) and differentiating under the integral sign, we also obtain, in a similar fashion,

$$|\nabla^k\vec{u}(x)|\leq\frac{C_k}{r^{3+k}}\int_{B_r(x)}|\vec{u}(y)|\,dy,\quad x\in\Omega,\ 0<r<\delta(x),\ k=0,1,2,\ldots.$$

It is known (going back to an old proof of Hardy and Littlewood) that the sub-averaging property (13.23) entails

$$|\vec{u}(x)|^p\leq\frac{C_p}{r^3}\int_{B_r(x)}|\vec{u}(y)|^p\,dy,\quad x\in\Omega,\ 0<r<\delta(x),\ 0<p<\infty.$$

Thus, for each $0<p<\infty$, $k=0,1,2,\ldots$, $x\in\Omega$ and $0<r<\delta(x)$ we may write

$$\begin{aligned}
|\nabla^k\vec{u}(x)| &\leq \frac{C_k}{r^{3+k}}\int_{B_{r/2}(x)}|\vec{u}(y)|\,dy\\
&\leq \frac{C_k}{r^{3+k}}\int_{B_{r/2}(x)}\Big(\frac{C_p}{r^3}\int_{B_{r/2}(y)}|\vec{u}(z)|^p\,dz\Big)^{1/p}dy\\
&\leq \frac{C_k}{r^{3+k}}\int_{B_{r/2}(x)}\Big(\frac{C_p}{r^3}\int_{B_r(x)}|\vec{u}(z)|^p\,dz\Big)^{1/p}dy\\
&\leq \frac{C_k}{r^k}\Big(\frac{C_p}{r^3}\int_{B_r(x)}|\vec{u}(y)|^p\,dy\Big)^{1/p}
\end{aligned}$$

so that for each $0 < p < \infty$ and $k = 0, 1, \dots,$

$$|\nabla^k \vec{u}(x)|^p \leq \frac{C_{p,k}}{r^{3+kp}} \int_{B_r(x)} |\vec{u}(y)|^p \, dy, \qquad x \in \Omega, \; 0 < r < \delta(x). \tag{13.24}$$

In particular, by (13.24) and Fubini's theorem,

$$\begin{aligned} &\int_\Omega |\delta(x)^k \nabla^k \vec{u}(x)|^p \, dx \\ &\qquad \leq C \int_\Omega \Big(\delta(x)^{-3} \int_{B_{\delta(x)/2}(x)} |\vec{u}(y)|^p \, dy \Big) dx \leq C \int_\Omega |\vec{u}(x)|^p \, dx \end{aligned}$$

since $\delta(x) \approx \delta(y)$ uniformly for $y \in B_{\delta(x)/2}(x)$.

Next, observe that $\mathcal{L}\vec{u} = 0$ in Ω implies $\Delta^2 \vec{u} = 0$ in Ω. Bring in the mean value theorem for biharmonic functions in $\mathbb{R}^3$, namely

$$\vec{u}(x) = \frac{1}{4\pi\rho^2} \int_{|y-x|=\rho} \vec{u}(y) \, d\sigma_y - \frac{\rho^2}{6} (\Delta \vec{u})(x), \qquad x \in \Omega, \; 0 < \rho < \delta(x),$$

and pick a nonnegative, radial function $\psi \in C_c^\infty(B(0,1))$ such that $\int_{B_1(0)} \psi(x) \, dx = 1$. We then write

$$\begin{aligned} (\vec{u})^+(x) &= \sup_{0<t<\delta(x)} \frac{1}{t^3} \Big| \int_{B_t(x)} \psi\Big(\frac{|x-y|}{t}\Big) \vec{u}(y) \, dy \Big| \\ &= \sup_{0<t<\delta(x)} \frac{1}{t^3} \Big| \int_0^t \psi\Big(\frac{\rho}{t}\Big) \int_{|x-z|=t} \vec{u}(z) \, d\sigma_z d\rho \Big| \\ &\leq C \sup_{0<t<\delta(x)} \frac{1}{t} \int_0^t \psi\Big(\frac{\rho}{t}\Big) \big(|\vec{u}(x)| + t^2 |\nabla^2 \vec{u}(x)|\big) \, d\rho \\ &\leq C\big(|\vec{u}(x)| + \delta(x)^2 |\nabla^2 \vec{u}(x)|\big), \qquad x \in \Omega. \end{aligned}$$

Therefore,

$$\|(\vec{u})^+\|_{L^p(\Omega)} \leq C\|\vec{u}\|_{L^p(\Omega)} + C\|\delta^2 \nabla^2 \vec{u}\|_{L^p(\Omega)} \leq C\|\vec{u}\|_{L^p(\Omega)},$$

which, in view of (13.22), proves (13.21). Given that $F_0^{p,2}(\Omega) = h^p(\Omega)$ for $0 < p \leq 1$, this justifies (13.20) when $\alpha = 0$. This argument can also be carried out when $\alpha \in \mathbf{N}$ is large, because of the Hardy space extension results of A. Miyachi [18].

There remains to interpolate between these partial results. To this end, we note that with $0 < p < \infty$ fixed, the spaces $\mathbf{L}^p_\alpha(\Omega)$, indexed by $\alpha \in \mathbb{R}$, form a complex interpolation scale. With analytic functions (of several complex variables) in place of elastic fields, this has been proved in [22] but essentially the argument goes through since it only uses the sub-averaging properties of the elastic fields. This finishes the proof of (13.20).

Step IV. In concert, Steps II and III justify the boundedness of the operators (13.19) for $q = 2$, provided $\frac{2}{3} < p \leq \infty$ and $2(\frac{1}{p} - 1)_+ < s < 1$. With this in hand, interpolation with the results in Step I allows us to further extend the range of q's for which the operators (13.19) are bounded.

Step V. The fact that the operator in (13.18) is well defined and bounded when $\frac{2}{3} < p \leq 1$, $2(\frac{1}{p} - 1) < s < 1$, and $0 < q \leq \infty$ can be proved using an atomic decomposition result for the space $F^{p,q}_{s+1/p-2,0}(\Omega)$ in which the atoms are supported in interior Whitney cubes for Ω. This, in turn, is proved in [14] (cf. also the decomposition results for Hardy spaces from [3]).

Step VI. Formally, the operators $\partial_\nu \Pi$ and $\mathcal{D}$ are dual to each other (as an inspection of their kernels shows). In fact, so are $\mathrm{Tr} \circ \Pi$ and $\mathcal{S}$, where the trace operator is analyzed in the next section. Dualize the results in Step IV (when $1 < p, q < \infty$) and interpolate with the results in Step V, then finally dualize this new round of results (again, when $1 < p, q < \infty$) and interpolate with the results in Step IV. Close inspection of these arguments shows that this procedure yields precisely the range of indices advertised in Theorems 4 and 3 for the operators (13.18) and (13.19).

Step VII. Finally, the results on Besov scales follow from what we have proved so far and real interpolation.

13.3 Traces and Conormal Derivatives

The issue of traces is central in the context of boundary value problems. The following result, proved in [14], is going to play an important role for us here.

Theorem 5. *Let $\Omega \subset \mathbb{R}^3$ be Lipschitz and assume that the indices p, s satisfy*

$$\frac{2}{3} < p \leq \infty, \quad 2\Big(\tfrac{1}{p} - 1\Big)_+ < s < 1.$$

Then the restriction to the boundary extends to a linear, bounded operator

$$\mathrm{Tr} : B^{p,q}_{s+\frac{1}{p}}(\Omega) \longrightarrow B^{p,q}_s(\partial\Omega) \quad \textit{for} \quad 0 < q \leq \infty.$$

Furthermore, for this range of indices, Tr is onto and has a bounded right inverse

$$\mathrm{Ex} : B^{p,q}_s(\partial\Omega) \longrightarrow B^{p,q}_{s+\frac{1}{p}}(\Omega).$$

Moreover, similar considerations hold for

$$\mathrm{Tr} : F^{p,q}_{s+\frac{1}{p}}(\Omega) \longrightarrow B^{p,p}_s(\partial\Omega)$$

when it is understood that $q = \infty$ *if* $p = \infty$. *In this situation, there exists a linear, bounded right inverse*

$$\mathrm{Ex} : B^{p,p}_{s}(\partial\Omega) \longrightarrow F^{p,q}_{s+\frac{1}{p}}(\Omega), \tag{13.25}$$

provided that (13.8) *also holds.*

Turning to conormal derivatives, we adopt a slightly more general point of view and assume that the Lamé system (13.1) is written in the form

$$(\mathcal{L}\vec{u})^{\alpha} = \sum_{\beta}\sum_{j,k} a^{\alpha\beta}_{jk}\partial_j\partial_k u^{\beta}, \quad \alpha = 1,2,3, \tag{13.26}$$

for some (real-valued, constant) coefficient tensor, i.e., some (9×9)-matrix $A := \left(a^{\alpha\beta}_{jk}\right)_{j,k,\alpha,\beta}$. The coefficient tensor A is not uniquely determined by the initial system of partial differential equations and each choice gives rise to a conormal derivative, i.e.,

$$(\partial_{\nu}\vec{u})^{\alpha} = \sum_{\beta}\sum_{j,k} a^{\alpha\beta}_{jk}\nu_j\partial_k u^{\beta}, \quad \alpha,\beta,j,k = 1,2,3. \tag{13.27}$$

Evidently, the traction conormal (13.4) is of this type, by choosing

$$a^{\alpha\beta}_{jk} := \lambda\delta_{\alpha j}\delta_{\beta k} + \mu(\delta_{\alpha\beta}\delta_{jk} + \delta_{\alpha k}\delta_{\beta j}), \quad \alpha,\beta,j,k = 1,2,3,$$

which we shall assume from now on.

Our goal is to make sense of (13.27) on the boundary of a Lipschitz domain $\Omega \subset \mathbb{R}^3$ when the field $\vec{u}$ is not regular enough in Ω to make sense of $\mathrm{Tr}\,(\partial_k u^{\beta})$. The approach we propose will take into account not only the smoothness of the field $\vec{u}$ itself but the smoothness of $\mathcal{L}\vec{u}$ as well. For starters, when $1 < p,q < \infty$, $0 < s < 1$, $\vec{u} \in F^{p,q}_{s+\frac{1}{p}}(\Omega)$, and $\vec{f} \in F^{p,q}_{s+\frac{1}{p}-2,0}(\Omega)$ are such that $\mathcal{L}\vec{u} = \vec{f}|_{\Omega}$, then we define $\partial^{\vec{f}}_{\nu}\vec{u} \in F^{p,q}_{s-1}(\partial\Omega) = \big(F^{p',q'}_{1-s}(\partial\Omega)\big)^{*}$, $1/p + 1/p' = 1$, $1/q + 1/q' = 1$, by

$$\begin{aligned}\langle\partial^{\vec{f}}_{\nu}\vec{u},\vec{\psi}\rangle &:= \langle\vec{f},\mathrm{Ex}(\vec{\psi})\rangle + \mu\langle\nabla\vec{u} + \nabla\vec{u}^{t}, \nabla\mathrm{Ex}(\vec{\psi}) + \nabla(\mathrm{Ex}(\vec{\psi}))^{t}\rangle \\ &\quad + \lambda\langle\mathrm{div}\,\vec{u}\,,\,\mathrm{div}\,\mathrm{Ex}(\vec{\psi})\rangle \quad \forall\,\vec{\psi} \in B^{p',q'}_{1-s}(\partial\Omega),\end{aligned} \tag{13.28}$$

where Ex is the (vector-valued version of the) extension operator (13.25). Note that all pairings of distributions in Ω are meaningful, so this procedure does yield a definition of the traction conormal, though limited to the range $1 < p,q < \infty$ (since, e.g., $B^{p,q}_{s-1}(\partial\Omega)$ fails to be a dual space if $\min\{p,q\} \leq 1$). Nonetheless, the following assertion holds.

Theorem 6. *Let Ω be a bounded Lipschitz domain in $\mathbb{R}^3$ and assume that $\frac{2}{3} < p \leq \infty$ and $2(\frac{1}{p} - 1)_+ < s < 1$, $0 < q \leq \infty$. Then one can define a concept of traction derivative such that*

$$\Big\{(\vec{u}, \vec{f}) \in B^{p,q}_{s+\frac{1}{p}}(\Omega) \oplus B^{p,q}_{s+\frac{1}{p}-2,0}(\Omega) : \mathcal{L}\vec{u} = \vec{f}|_{\Omega}\Big\} \ni (\vec{u}, \vec{f}) \mapsto \partial_\nu^{\vec{f}} \vec{u} \in B^{p,q}_{s-1}(\partial\Omega) \qquad (13.29)$$

is well defined, linear, and bounded, i.e.,

$$\|\partial_\nu^{\vec{f}} \vec{u}\|_{B^{p,q}_{s-1}(\partial\Omega)} \leq C\Big(\|\vec{f}\|_{B^{p,q}_{s+\frac{1}{p}-2,0}(\Omega)} + \|\vec{u}\|_{B^{p,q}_{s+\frac{1}{p}}(\Omega)}\Big)$$

holds, and which reduces to (13.28) when $p, q > 1$.

Finally, similar conclusions are valid in the context of Triebel–Lizorkin spaces for the map

$$\Big\{(\vec{u}, \vec{f}) \in F^{p,q}_{s+\frac{1}{p}}(\Omega) \oplus F^{p,q}_{s+\frac{1}{p}-2,0}(\Omega) : \mathcal{L}\vec{u} = \vec{f}|_{\Omega}\Big\} \ni (\vec{u}, \vec{f}) \mapsto \partial_\nu^{\vec{f}} \vec{u} \in B^{p,p}_{s-1}(\partial\Omega), \qquad (13.30)$$

assuming that, as before, $p \neq \infty$ and $\Big(\min\{\frac{1}{3p} + 1, \frac{s}{2} + 1\}\Big)^{-1} < q$ and that, in addition, $q < \Big(\frac{1}{p} - 1\Big)^{-1}$ if $p \leq 1$ and $q \leq \infty$ if $p > 1$.

Proof. We shall break down the argument into a series of steps.

Step I. First, we introduce an operator

$$\partial_\nu^{\mathcal{L}} : \mathbf{L}^p_{s+\frac{1}{p}}(\Omega) \longrightarrow B^{p,p}_{s-1}(\partial\Omega), \qquad (13.31)$$

which is well defined, linear, and bounded if $\frac{2}{3} < p < \infty$, $2(\frac{1}{p} - 1)_+ < s < 1$, and which reduces to (13.4) if $\vec{u}$ is sufficiently regular (say, $\vec{u} \in C^\infty(\overline{\Omega})$).

To this end, consider the system of elastostatics represented in the form (13.26) and suppose that $\varphi : \mathbb{R}^2 \to \mathbb{R}$ is a Lipschitz function such that $\varphi(0) = 0$ and, for some $M, R > 0$,

$$\begin{aligned} \Sigma_R &:= \{(x', \varphi(x')) : x' \in \mathbb{R}^2,\ |x'| < R\} \subset \partial\Omega, \\ D_{R,M} &:= \{x + te_3 : x \in \Sigma_R,\ 0 < t < 2M\} \subset \Omega. \end{aligned}$$

For every vector field $\vec{u} = (u^1, u^2, u^3)$ such that $\mathcal{L}\vec{u} = 0$ in Ω we set

$$u_j^\alpha(x) := -\int_0^M (\partial_j u^\alpha)(x + te_3)\, dt, \quad 1 \leq j, \alpha \leq 3, \quad x \in D_{R,M/2}, \qquad (13.32)$$

$$\vec{v}_j := (u_j^\alpha)_{1 \leq \alpha \leq 3} \ \text{ in } \ D_{R,M/2}, \quad 1 \leq j \leq 3,$$

where $e_3 = (0,0,1) \in \mathbb{R}^3$. Then the following identities can be easily checked:

$$\partial_k u_j^\alpha = \partial_j u_k^\alpha \ \text{ in } \ D_{R,M/2} \quad \text{for every} \quad \alpha, j, k = 1,2,3, \tag{13.33}$$
$$u_3^\alpha(x) = u^\alpha(x) - u^\alpha(x + Me_3) \quad \forall\, x \in D_{R,M/2}, \quad \forall\, \alpha = 1,2,3, \tag{13.34}$$
$$\mathcal{L}\vec{v}_j = 0 \ \text{ in } \ D_{R,M/2} \quad \text{for every} \quad j = 1,2,3. \tag{13.35}$$

Next, we prove an incisive claim to the effect that, generally speaking,

$$\vec{u} \in \mathbf{L}^p_\theta(\Omega), \quad 0 < p < \infty,\ \theta \geq 0 \implies \sum_{j=1}^{3} \|\vec{v}_j\|_{\mathbf{L}^p_\theta(D_{R,M/2})} \leq C\|\vec{u}\|_{\mathbf{L}^p_\theta(\Omega)}. \tag{13.36}$$

Indeed, since $\operatorname{dist}(x, \partial D_{R,M/2}) \leq \delta(x) = \operatorname{dist}(x, \partial\Omega)$, uniformly for $x \in D_{R,M/2}$, we may write

$$\begin{aligned}
&\int_{D_{R,M/2}} \Big(\operatorname{dist}(x, \partial D_{R,M/2})^{\langle\theta\rangle - \theta} |\nabla^{\langle\theta\rangle} \vec{v}_j(x)|\Big)^p\, dx \\
&\leq C \int_{D_{R,M/2}} \delta(x)^{p(\langle\theta\rangle - \theta)} \left(\int_0^M |(\nabla^{\langle\theta\rangle+1}\vec{u})(x + te_3)|\, dt\right)^p dx \\
&\leq C \int_{\substack{x' \in \mathbb{R}^2 \\ |x'| < R}} \int_0^M r^{p(\langle\theta\rangle - \theta)} \left(\int_0^M |(\nabla^{\langle\theta\rangle+1}\vec{u})((x', \varphi(x') + r + t))|\, dt\right)^p dr\, dx' \\
&\leq C \int_{\substack{x' \in \mathbb{R}^2 \\ |x'| < R}} \int_0^{2M} r^{p(\langle\theta\rangle - \theta)} \left(\int_r^{2M} |\nabla^{\langle\theta\rangle+1}\vec{u}|^*((x', \varphi(x') + \lambda))\, d\lambda\right)^p dr\, dx'
\end{aligned} \tag{13.37}$$

by changing variables, first,

$$D_{R,M/2} \ni x = (x', \varphi(x') + r) \mapsto (x', r) \in \{x' \in \mathbb{R}^2 : |x'| < R\} \times (0, M),$$

and, second, $\lambda := t + r \in (0, 2M)$. Above, $(\cdot)^*$ denotes the maximal radial operator defined (for a function w defined in $D_{R,M}$) as

$$w(x) := \sup_{t > 0:\, x + te_n \in D_{R,M}} |w(x + te_n)|, \quad x \in D_{R,M}.$$

This subtlety (i.e., the consideration of the radial maximal operator) is only needed in order to be able to use Hardy's inequality when $p \leq 1$ (when $p > 1$, we shall disregard $(\cdot)^*$, i.e., interpret it as the identity operator). This is because the ordinary Hardy's inequality continues to hold in the L^p-setting even when $0 < p \leq 1$, provided the function involved is nonincreasing (which is certainly the case for $|\nabla^{\langle\theta\rangle+1}\vec{u}|^*$ in the transversal direction); cf., e.g., the appendix in [22] for a precise statement. Invoking Hardy's inequality just alluded to, we can further dominate the last expression in (13.37) by

$$C \int_{\substack{x' \in \mathbb{R}^2 \\ |x'|<R}} \int_0^{2M} \lambda^{p(\langle\theta\rangle-\theta+1)} \Big(|\nabla^{\langle\theta\rangle+1} \vec{u}|^*((x', \varphi(x') + \lambda)) \Big)^p \, d\lambda \, dx'$$

$$\leq C \int_{D_{R,M}} \delta(x)^{p(\langle\theta\rangle-\theta+1)} \Big(|\nabla^{\langle\theta\rangle+1} \vec{u}|^*(x) \Big)^p \, dx.$$

To continue, we consider a Whitney decomposition of Ω (cf. [21]). Specifically, there exists a family of balls $B_j = B_{r_j}(z_j)$, with $z_j \in \Omega$ and $\delta(z_j) \approx r_j$, which cover Ω and such that their concentric doubles have finite overlap. Then Fatou's lemma implies

$$\int_{D_{R,M}} \delta(x)^{p(\langle\theta\rangle-\theta+1)} \Big(|\nabla^{\langle\theta\rangle+1} \vec{u}|^*(x) \Big)^p \, dx$$

$$\leq \sum_j \int_{D_{R,M}} \delta(x)^{p(\langle\theta\rangle-\theta+1)} \Big(\chi_{B_j \cap D_{R,M}} |\nabla^{\langle\theta\rangle+1} \vec{u}|^p \Big)^* (x) \, dx. \qquad (13.38)$$

Note that $(\chi_{B_j \cap D_{R,M}} |\nabla^{\langle\theta\rangle+1} \vec{u}|^p)^*$ is supported in a set of measure less than or equal to Cr_j^3. Also, on that set, $\delta(x)^{p(\langle\theta\rangle-\theta+1)} \leq Cr_j^{p(\langle\theta\rangle-\theta+1)}$. Finally, $\|(\chi_{B_j \cap D_{R,M}} |\nabla^{\langle\theta\rangle+1} \vec{u}|^p)^*\|_{L^\infty(D_{R,M})} \leq \|\nabla^{\langle\theta\rangle+1} \vec{u}\|^p_{L^\infty(Bj)}$. All in all,

$$\int_{D_{R,M}} \delta(x)^{p(\langle\theta\rangle-\theta+1)} \Big(\chi_{B_j \cap D_{R,M}} |\nabla^{\langle\theta\rangle+1} \vec{u}|^p \Big)^* (x) \, dx$$

$$\leq C \, r_j^{3+p(\langle\theta\rangle-\theta+1)} \|\nabla^{\langle\theta\rangle+1} \vec{u}\|^p_{L^\infty(Bj)}$$

$$\leq C \, r_j^{p(\langle\theta\rangle-\theta)} \int_{B_{2r_j}(z_j)} |\nabla^{\langle\theta\rangle} \vec{u}(x)|^p \, dx, \qquad (13.39)$$

where the last inequality follows from the sub-averaging property of $\nabla^{\langle\theta\rangle} \vec{u}$ (note that this is still a null solution for the constant coefficient operator $\mathcal{L}$, so (13.24) applies). Since the family $\{B_{2r_j}(z_j)\}_j$ has finite overlap and $\delta(x) \approx r_j$ for $x \in B_{2r_j}(z_j)$, adding up inequalities of the form (13.39) in j and recalling (13.38) leads to the conclusion that

$$\int_{D_{R,M}} \delta(x)^{p(\langle\theta\rangle-\theta+1)} \Big(|\nabla^{\langle\theta\rangle+1} \vec{u}|^*(x) \Big)^p \, dx$$

$$\leq C \int_\Omega \delta(x)^{p(\langle\theta\rangle-\theta)} |\nabla^{\langle\theta\rangle} \vec{u}(x)|^p \, dx.$$

This readily implies (13.36).

After this preamble, we can introduce the operator $\partial_\nu^{\mathcal{L}}$ acting on $\vec{u} \in \mathbf{L}^p_{s+\frac{1}{p}}(\Omega)$ by

$$(\partial_\nu^{\mathcal{L}} \vec{u})^\alpha\Big|_{\Sigma_R} := \sum_{\beta,j,k} a_{jk}^{\alpha\beta}\, \partial_{\tau_{j3}} \Big[\mathrm{Tr}\, u_k^\beta\Big]\Big|_{\Sigma_R} + \sum_{\beta,j,k} a_{jk}^{\alpha\beta} \nu_j (\partial_k u^\beta)(\cdot + Me_3)\Big|_{\Sigma_R}, \tag{13.40}$$

where $1 \leq \alpha \leq 3$, the u_j^β are as in (13.32), and $\partial_{\tau_{jk}} := \nu_j \partial_k - \nu_k \partial_j$, $j, k = 1, 2, 3$, are the tangential derivatives. Note that

$$\partial_{\tau_{jk}} \circ \mathrm{Tr} : F^{p,2}_{s+\frac{1}{p}}(\Omega) \longrightarrow B^{p,p}_{s-1}(\partial\Omega), \quad 1 \leq j, k \leq 3,$$

are (by Theorem 5) well-defined and bounded operators, so the first sum in (13.40) is meaningful. The second sum above is easily handled (in fact, in L^∞) since the first-order partial derivatives of u^β are evaluated away from $\partial\Omega$. Then (13.40) can be used to define $\partial_\nu^{\mathcal{L}} \vec{u}$ globally on $\partial\Omega$, by gluing the various local definitions together using a smooth partition of unity.

By (13.20), it follows that the operator (13.31) is bounded. There remains to show that its action is compatible with (13.4) in the case when $\vec{u} \in C^\infty(\overline{\Omega})$, i.e., that, in this situation, the formula (13.40) coincides with the definition of conormal (traction) derivative (13.27). Indeed, for $\alpha = 1, 2, 3$, we may write, based on (13.33)–(13.35), that

$$\begin{aligned}
\sum_{\beta,j,k} a_{jk}^{\alpha\beta} \nu_j \partial_k u^\beta &= \sum_{\beta,j,k} a_{jk}^{\alpha\beta} \nu_j \Big(\partial_3 u_k^\beta + \partial_k u^\beta(\cdot + Me_3)\Big) \\
&= \sum_{\beta,j,k} a_{jk}^{\alpha\beta} \Big(\nu_j \partial_3 - \nu_3 \partial_j\Big) u_k^\beta + \sum_{\beta,j,k} a_{jk}^{\alpha\beta} \nu_j \partial_k u^\beta(\cdot + Me_3) \\
&= \sum_{\beta,j,k} a_{jk}^{\alpha\beta} \partial_{\tau_{j3}} u_k^\beta + \sum_{\beta,j,k} a_{jk}^{\alpha\beta} \nu_j \partial_k u^\beta(\cdot + Me_3),
\end{aligned}$$

where in the second step we have used the fact that

$$\sum_\beta \sum_{j,k} a_{jk}^{\alpha\beta} \partial_j u_k^\beta(x) = -\int_0^M \Big(\sum_\beta \sum_{j,k} a_{jk}^{\alpha\beta} \partial_j \partial_k u^\beta\Big)(x + te_n)\, dt = 0.$$

This finishes the proof of the claim made in Step I.

Step II. We next claim that

$$\mathbf{L}^p_\alpha(\Omega) = \{\vec{u} \in F^{p,2}_\alpha(\Omega) : \mathcal{L}\vec{u} = 0 \text{ in } \Omega\} \text{ if } 0 < p < \infty \text{ and } \alpha > 3(\tfrac{1}{p} - 1)_+. \tag{13.41}$$

This, however, is proved using similar arguments to those employed in the proof of (13.20); see [14] for details in the case of harmonic functions. In particular, (13.41) shows that

$$\partial_\nu^{\mathcal{L}} : F^{p,q}_{s+\frac{1}{p}}(\Omega) \longrightarrow B^{p,p}_{s-1}(\partial\Omega) \tag{13.42}$$

is a bounded operator provided

$$\frac{2}{3} < p < \infty, \quad 2\Big(\frac{1}{p} - 1\Big)_+ < s < 1, \quad q = 2.$$

By embeddings, the same is true for $0 < q \leq 2$ and, by the discussion just prior to the statement of Theorem 6, for $1 < q < \infty$ granted that $1 < p < \infty$. Interpolating between these various partial results allows us to conclude that the operator (13.42) is bounded for the range of indices specified in the statement of the theorem in connection with (13.30).

Step III. Finally, assume that $\vec{u} \in F^{p,q}_{s+\frac{1}{p}}(\Omega)$ and $\vec{f} \in F^{p,q}_{s+\frac{1}{p}-2,0}(\Omega)$ are such that $\mathcal{L}\vec{u} = \vec{f}|_{\Omega}$. We then set

$$\partial_\nu^{\vec{f}}\vec{u} := \partial_\nu \Pi \vec{f} + \partial_\nu^{\mathcal{L}}\big(\vec{u} - (\Pi\vec{f})|_{\Omega}\big), \tag{13.43}$$

where Π stands for the elastic Newtonian potential (13.17) and $\partial_\nu^{\mathcal{L}}$ is the operator (13.31), (13.42). Now the properties of $(\vec{u}, \vec{f}) \mapsto \partial_\nu^{\vec{f}}\vec{u}$ which have been advertised in the statement of the theorem can be checked directly from (13.43) keeping in mind the mapping properties of the operators involved. This finishes the proof of the claims made about (13.30). The case of (13.29) follows from this and real interpolation.

13.4 Boundary Integral Operators and Proofs of the Main Results

Recall the linear space of vector-valued functions Ψ introduced in Section 13.1 and convention (13.9).

Theorem 7. *For each bounded Lipschitz domain Ω in $\mathbb{R}^3$ there exists $\varepsilon = \varepsilon(\partial\Omega) \in (0,1]$ with the following significance. Let*

$$\tfrac{2}{3} < p \leq \infty, \quad 2(\tfrac{1}{p} - 1)_+ < s < 1$$

be such that one of the conditions in (13.6) *is satisfied. Then the operators*

$$\tfrac{1}{2}I + K : B^{p,p}_s(\partial\Omega) \longrightarrow B^{p,p}_s(\partial\Omega),$$
$$-\tfrac{1}{2}I + K^* : \Psi^\perp \cap B^{p,p}_{s-1}(\partial\Omega) \longrightarrow \Psi^\perp \cap B^{p,p}_{s-1}(\partial\Omega)$$

are invertible.

Proof. When $1 \leq p, q \leq \infty$ this is proved in [17]. An inspection of the argument there shows that it is possible to further extend this range, by means of the interpolation techniques recently developed in [14], in the fashion indicated in the statement of the theorem.

We are now ready to prove Theorems 1 and 2.

Proof of Theorems 1 and 2. The solution of the Poisson problem with Dirichlet boundary conditions admits the integral representation formula

$$\vec{u} = (R_\Omega \circ \Pi \circ E_\Omega)\vec{f} + \mathcal{D} \circ \Big(\tfrac{1}{2}I + K\Big)^{-1}\Big(\vec{g} - \mathrm{Tr}\,(R_\Omega \circ \Pi \circ E_\Omega \vec{f})\Big) \quad \text{in } \Omega, \quad (13.44)$$

where R_Ω and E_Ω are, respectively, the restriction from $\mathbb{R}^3$ to Ω and the extension from Ω to $\mathbb{R}^3$, with preservation of class (cf. [23] for a discussion and references). Then the relevant properties of $\vec{u}$ can be read off (13.44), given that all operators involved are well understood. The uniqueness part is a consequence of embeddings and the results for $1 \le p, q \le \infty$ from [17].

As for Theorem 2, we proceed similarly, starting with the integral representation formula

$$\vec{u} = (R_\Omega \circ \Pi)\vec{f} + \mathcal{S} \circ \Big(-\tfrac{1}{2}I + K^*\Big)^{-1}\Big(\vec{g} - \partial_\nu \Pi \vec{f}\Big) \quad \text{in } \Omega.$$

Condition (13.11) ensures that $\vec{g} - \partial_\nu \Pi \vec{f} \in \Psi^\perp \cap B^{p,p}_{s-1}(\partial\Omega)$, so Theorem 7 applies.

13.5 Regularity of Green Potentials in Lipschitz Domains

Given $\Omega \subset \mathbb{R}^3$ bounded, Lipschitz domain, we let $\mathbf{G}_D$, $\mathbf{G}_N$ denote the solution operators for the Poisson problem for the Lamé operator with homogeneous Dirichlet and traction boundary conditions, respectively. In the case of the traction problem, it is understood that the given datum $\vec{f}$ is first normalized so that $\langle \vec{f}, \vec{\psi} \rangle = 0$ for each $\vec{\psi} \in \Psi$.

Theorem 8. *For each bounded, connected Lipschitz domain Ω in $\mathbb{R}^3$ there is $\varepsilon = \varepsilon(\Omega) \in (0, 1]$ with the following significance. Suppose that the numbers p and s, $\frac{2}{3} < p \le \infty$, $2(\frac{1}{p} - 1)_+ < s < 1$, are such that one of the three conditions in* (13.6) *is satisfied. Then, with $\alpha := s + \frac{1}{p} - 2$, the operators*

$$\mathbf{G}_D : B^{p,q}_\alpha(\Omega) \longrightarrow B^{p,q}_{\alpha+2}(\Omega) \quad \textit{if } 0 < q \le \infty,$$

$$\mathbf{G}_D : F^{p,q}_\alpha(\Omega) \longrightarrow F^{p,q}_{\alpha+2}(\Omega) \quad \textit{if (13.8) holds and } p \ne \infty,$$

are well defined and bounded. Similar results hold for traction boundary conditions, i.e., for

$$\mathbf{G}_N : B^{p,q}_{\alpha,0}(\Omega) \longrightarrow B^{p,q}_{\alpha+2}(\Omega) \quad \textit{if } 0 < q \le \infty,$$

$$\mathbf{G}_N : F^{p,q}_{\alpha,0}(\Omega) \longrightarrow F^{p,q}_{\alpha+2}(\Omega) \quad \textit{if (13.8) holds and } p \ne \infty.$$

These results are sharp in the class of Lipschitz domains. For $\partial\Omega \in C^1$ one can take $\varepsilon = 1$.

The two-dimensional pentagonal region consisting of all points with coordinates $(\alpha, 1/p)$ such that α, p are as in the statement of this theorem has the shape shown in Fig. 2.

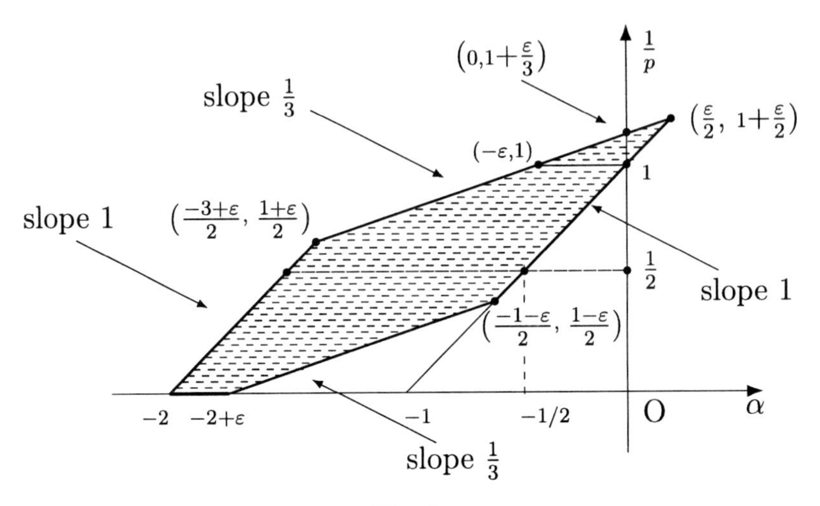

Fig. 2.

Theorem 8 is a direct consequence of Theorems 1 and 2, given the results discussed in Sections 13.2–13.4.

Recall that the Triebel–Lizorkin space $F^{p,q}_{\alpha}(\mathbb{R}^3)$ reduces to the *local* Hardy class $h^p(\mathbb{R}^3)$ if $0 < p < \infty$, $q = 2$, $\alpha = 0$ (as is well known, the latter coincides with the Lebesgue class $L^p(\mathbb{R}^3)$ if $1 < p < \infty$). Consequently, for $0 < p < \infty$ and $\Omega \subset \mathbb{R}^3$ Lipschitz domain, we set $h^p(\Omega) := F^{p,2}_0(\Omega)$ and $h^p_0(\Omega) := F^{p,2}_{0,0}(\Omega)$ (so that $h^p(\Omega) = h^p_0(\Omega) = L^p(\Omega)$ if $1 < p < \infty$).

Corollary 1. *Retain the notation and hypotheses made in Theorem 8 on the indices involved and the domain Ω. Then*

$$\nabla^2 \mathbf{G}_D : B^{p,q}_{\alpha}(\Omega) \longrightarrow B^{p,q}_{\alpha}(\Omega), \quad \nabla^2 \mathbf{G}_N : B^{p,q}_{\alpha,0}(\Omega) \longrightarrow B^{p,q}_{\alpha}(\Omega)$$

if $0 < q \leq \infty$, and

$$\nabla^2 \mathbf{G}_D : F^{p,q}_{\alpha}(\Omega) \longrightarrow F^{p,q}_{\alpha}(\Omega), \quad \nabla^2 \mathbf{G}_N : F^{p,q}_{\alpha,0}(\Omega) \longrightarrow F^{p,q}_{\alpha}(\Omega)$$

if (13.8) *holds and $p \neq \infty$, are well-defined, bounded operators.*

In particular, there exists $\varepsilon = \varepsilon(\Omega) \in (0, 1]$ such that the operators

$$\nabla^2 \mathbf{G}_D : h^p(\Omega) \longrightarrow h^p(\Omega), \quad \text{whenever } \tfrac{3}{3+\varepsilon} < p < 1, \tag{13.45}$$

$$\nabla^2 \mathbf{G}_N : h^p_0(\Omega) \longrightarrow h^p(\Omega), \quad \text{whenever } \tfrac{3}{3+\varepsilon} < p < 1, \tag{13.46}$$

are well defined and bounded. Once again, one can take $\varepsilon = 1$ whenever Ω has a C^1 boundary, in which case the range of p in (13.45)–(13.46) *becomes $\frac{3}{4} < p < 1$.*

The second part of the corollary is related to a result proved by D.-C. Chang, S.G. Krantz, and E.M. Stein in [3] regarding the Green potentials associated with the Laplacian on C^∞ domains. In [13] we have successfully dealt with the Laplacian on arbitrary Lipschitz domains, thus answering in the affirmative a conjecture made by Stein et al. in [3]. In this paper, we take the first steps in the direction of further extending this circle of ideas to elliptic systems.

To state our next result, let $W^p_k(\Omega)$ denote the usual L^p-based Sobolev space of order (of smoothness) k in Ω. Also, by $\mathring{W}^p_k(\Omega)$ we denote the closure of the space of C^∞, compactly supported functions in $W^p_k(\Omega)$, and by $L^{p,\infty}(\Omega)$ the usual weak-L^p space over Ω. Finally, recall that $\delta(x)$ denotes the distance from x to $\partial\Omega$.

Corollary 2. *Assume that Ω is a bounded Lipschitz domain in $\mathbb{R}^3$, and denote by $G_D(x,y)$ the integral kernel of the elastic Dirichlet Green potential operator $\mathbf{G}_D$ associated with Ω. Then the following are true.*

(i) *$\nabla G_D(x,\cdot)\in L^{3/2,\infty}(\Omega)$, uniformly with respect to $x\in\Omega$. In particular, $G_D(x,\cdot)\in \mathring{W}^p_1(\Omega)$, uniformly with respect to $x\in\Omega$, for any $1<p<\frac{3}{2}$.*

(ii) *$\delta(\cdot)^{-\alpha}G_D(x,\cdot)\in L^{3/(1+\alpha),\infty}(\Omega)$ for each $\alpha\in[0,1]$, uniformly with respect to $x\in\Omega$. In particular, $\delta(\cdot)^{-1}G_D(x,\cdot)\in L^{3/2,\infty}(\Omega)$ and $G_D(x,\cdot)\in L^{3,\infty}(\Omega)$, uniformly with respect to $x\in\Omega$.*

Proof. For each $k=0,1,2,...$ and $1<p<\infty$, introduce the weak Sobolev space

$$W^{p,\infty}_k(\Omega):=\{f\in L^{p,\infty}(\Omega):\,\partial^\alpha f\in L^{p,\infty}(\Omega),\ |\alpha|\leq k\},$$

plus a similar definition for $W^{p,\infty}_k(\mathbb{R}^3)$ with Ω replaced by $\mathbb{R}^3$. We shall proceed in several steps.

Step I. We claim that, with $(\cdot,\cdot)_{\theta,q}$ denoting the usual real-interpolation bracket,

$$\Big(W^{p_0}_k(\mathbb{R}^3),W^{p_1}_k(\mathbb{R}^3)\Big)_{\theta,\infty}=W^{p,\infty}_k(\mathbb{R}^3),$$

for each $k=0,1,2,...$, $1<p_0,p_1<\infty$, $0<\theta<1$, $1/p=(1-\theta)/p_0+\theta/p_1$.

To see this, bring in the Bessel potential of complex order z, acting on a tempered distribution f according to

$$\widehat{J_zf}(\xi):=(1+4\pi^2|\xi|^2)^{-z/2}\hat{f}(\xi),$$

where "hat" denotes the Fourier transform. As is well known [2], J_k maps $L^p(\mathbb{R}^3)$ isomorphically onto $W^p_k(\mathbb{R}^3)$ for $1<p<\infty$ (with J_{-k} as inverse). On the one hand, this and real interpolation immediately yield that $J_k:L^{p,\infty}(\mathbb{R}^3)\longrightarrow\Big(W^{p_0}_k(\mathbb{R}^3),W^{p_1}_k(\mathbb{R}^3)\Big)_{\theta,\infty}=W^{p,\infty}_k(\mathbb{R}^3)$ is an isomorphism whenever $1<p_0,p_1<\infty$, $0<\theta<1$, and $1/p=(1-\theta)/p_0+\theta/p_1$. On the other hand, proceeding much as in the proof of Theorem 7 in [2] we deduce that, under the same assumptions on the indices, the operator

$J_k : L^{p,\infty}(\mathbb{R}^3) \longrightarrow W_k^{p,\infty}(\mathbb{R}^3)$ is an isomorphism as well. All in all, the claim made at the beginning of Step I follows.

Step II. We shall now show that

$$\big(W_k^{p_0}(\Omega), W_k^{p_1}(\Omega)\big)_{\theta,\infty} = W_k^{p,\infty}(\Omega),$$

for each $k = 0, 1, 2, \ldots$, $1 < p_0, p_1 < \infty$, $0 < \theta < 1$, $1/p = (1-\theta)/p_0 + \theta/p_1$.

The left-to-right inclusion is trivial and, granted the result in Step I, the key step in proving the opposite one is the claim that Stein's extension operator $E : W_k^p(\Omega) \to W_k^p(\mathbb{R}^3)$ (cf. [21]) also extends as a bounded operator from $W_k^{p,\infty}(\Omega)$ into $W_k^{p,\infty}(\mathbb{R}^3)$.

To see this, we note the commutator identity $\partial^\alpha(Ef) = E(\partial^\alpha f) + \sum_{|\beta|\leq|\alpha|} T_{\alpha\beta}(\partial^\beta f)$ (cf. (26) on p. 185 of [21]) where the linear operators $T_{\alpha\beta}$ map $L^p(\Omega)$ boundedly into $L^p(\mathbb{R}^3)$ for $1 < p < \infty$. This latter claim is seen as in p. 174 of [11] and p. 187 of [21]. By real interpolation we may therefore conclude that $E : W_k^{p,\infty}(\Omega) \to W_k^{p,\infty}(\mathbb{R}^3)$ boundedly, as desired. This concludes the discussion in Step II.

Step III. Consider now the linear operator

$$T\vec{u}(x) := \vec{u}(x) - \mathcal{D}\big(\big(\tfrac{1}{2}I + K\big)^{-1}(\mathrm{Tr}\,\vec{u})\big)(x), \quad x \in \Omega,$$

which, we claim, maps $W_1^{p,\infty}(\Omega)$ boundedly into itself for any $\frac{3}{2} - \varepsilon < p < 3 + \varepsilon$.

By Step II and real interpolation, it suffices to show that $T : W_1^p(\Omega) \to W_1^p(\Omega)$ boundedly for $\frac{3}{2} - \varepsilon < p < 3 + \varepsilon$. This, in turn, follows from the mapping properties of the operators involved. Indeed, Theorem 5 ensures that $\mathrm{Tr}\,\vec{u} \in B^{p,p}_{1-\frac{1}{p}}(\partial\Omega)$ if $\vec{u} \in W_1^p(\Omega)$, Theorem 7 guarantees that the operator $\frac{1}{2}I + K$ is invertible on $B^{p,p}_{1-\frac{1}{p}}(\partial\Omega)$ for any $\frac{3}{2} - \varepsilon < p < 3 + \varepsilon$, and Theorem 4 implies that $\mathcal{D}$ maps this latter space into $F_1^{p,2}(\Omega) = W_1^p(\Omega)$. This finishes the proof of the claim made in Step III.

Step IV. Recall the matrix $\Gamma = (\Gamma_{\alpha\beta})_{\alpha,\beta}$ with entries as in Section 13.2. To prove the first claim made in the statement of the corollary, we make the observation that, in the y variable, the functions $|x-y|^{-1}$, $\frac{(x_\alpha - y_\alpha)(x_\beta - y_\beta)}{|x-y|^3}$, $1 \leq \alpha, \beta \leq 3$, belong to $W_1^{\frac{3}{2},\infty}(\Omega)$ uniformly in $x \in \Omega$. Thus, $\Gamma(x - \cdot) \in W_1^{\frac{3}{2},\infty}(\Omega)$, uniformly in $x \in \Omega$ and since $G_D(x,y) = T(\Gamma(x-\cdot))(y)$ we may conclude with the aid of Step III that $\nabla G_D(x,\cdot) \in L^{3/2,\infty}(\Omega)$, uniformly in $x \in \Omega$.

Step V. The proof of the fact that $\delta(\cdot)^{-\alpha} G_D(x,\cdot) \in L^{3/(1+\alpha),\infty}(\Omega)$ for each $\alpha \in [0,1]$, uniformly in $x \in \Omega$, follows a similar pattern. The departure point is the observation that $T : W_1^p(\Omega) \to \mathring{W}_1^p(\Omega)$ is well defined and bounded for $\frac{3}{2} - \varepsilon < p < 3 + \varepsilon$. The second observation is that $\delta^{-\alpha}$, viewed as a multiplication operator, maps $\mathring{W}_1^p(\Omega)$ boundedly into $L^{p_\alpha}(\Omega)$ for $1 < p < 3$, $\alpha \in [0,1]$, and $1/p_\alpha := 1/p - (1-\alpha)/3$. Indeed, this follows from interpolating between $\alpha = 0$ (Sobolev's embedding theorem) and $\alpha = 1$ (Hardy's inequality), via E. Stein's interpolation theorem for analytic families of operators.

Combining the two observations just made above and then relying on real interpolation eventually yields that $\delta^{-\alpha}T$ maps $W_1^{p,\infty}(\Omega)$ boundedly into $L^{p_\alpha,\infty}(\Omega)$ for $\frac{3}{2}-\varepsilon < p < 3$, $\alpha \in [0,1]$ and, as before, $1/p_\alpha = 1/p-(1-\alpha)/3$. For $p = 3/2$, in which case p_α becomes $3/(1+\alpha)$, this and the identity $G_D(x,y) = T(\Gamma(x-\cdot))(y)$ justifies the conclusion in (ii). This finishes the proof of the corollary.

Of course, there is a similar result for the (matrix-valued) function $G_N(x,y)$, the integral kernel of the operator $\mathbf{G}_N$.

More can be said when extra geometrical information about the domain Ω is available. For example, according to Theorems 5.4 and 5.5 in [16], [10], and Theorem 6.1 in [19],

$$\nabla^2 \mathbf{G}_D : L^2(\Omega) \longrightarrow L^2(\Omega), \tag{13.47}$$

whenever Ω is a *convex polyhedron* in $\mathbb{R}^3$. By interpolation with (13.45) we see that for a convex polyhedron $\Omega \subset \mathbb{R}^3$,

$$\nabla^2 \mathbf{G}_D : h^p(\Omega) \longrightarrow h^p(\Omega), \quad \tfrac{3}{3+\varepsilon} < p \leq 2. \tag{13.48}$$

In particular, $\nabla^2 \mathbf{G}_D$ maps $L^p(\Omega)$ boundedly to itself for $1 < p \leq 2$. A similar set of conclusions applies to the operator $\nabla^2 \mathbf{G}_N$. More specifically, with Ω as before,

$$\nabla^2 \mathbf{G}_N : h_0^p(\Omega) \longrightarrow h^p(\Omega), \quad \tfrac{3}{3+\varepsilon} < p < \tfrac{3}{2}, \tag{13.49}$$

where the endpoint $3/2$ is as in [16].

If we denote by $L^{1,\infty}(\Omega)$ the standard weak-L^1 space on Ω, then interpolating between (13.47)–(13.49) and the results in Corollary 1 further yields the following.

Theorem 9. *Let $\Omega \subset \mathbb{R}^3$ be a convex polyhedron. Then the following operators are bounded:*

$$\nabla^2 \mathbf{G}_D : h^1(\Omega) \longrightarrow h^1(\Omega), \qquad \nabla^2 \mathbf{G}_N : h_0^1(\Omega) \longrightarrow h^1(\Omega),$$
$$\nabla^2 \mathbf{G}_D : L^1(\Omega) \longrightarrow L^{1,\infty}(\Omega), \quad \nabla^2 \mathbf{G}_N : L^1(\Omega) \longrightarrow L^{1,\infty}(\Omega).$$

We conclude this section with a brief discussion of the mapping properties of the *gradient* of the elastic Green potentials. Recall that the classical $L^p - L^q$ estimates for the gradient of the Green potential associated with the Laplacian in the half-space have been extended to Lipschitz domains by B. Dahlberg in [4]. In [13] we gave new proofs to Dahlberg's estimates and extended his main results to other types of function spaces. The following assertion, pertaining to the mapping properties of the Green potentials associated with the system of elastostatics, can be proved in a similar manner, starting with Theorems 1 and 2.

Theorem 10. *Let $\Omega \subset \mathbb{R}^3$ be a bounded Lipschitz domain. There exists $\varepsilon = \varepsilon(\Omega) \in (0,1]$ such that for every $\frac{3}{3+\varepsilon} < p < \frac{3}{2-\varepsilon}$ and $\frac{1}{q} = \frac{1}{p} - \frac{1}{3}$, the operators*

$$\nabla \mathbf{G}_D : h^p(\Omega) \to L^q(\Omega), \qquad \nabla \mathbf{G}_N : h^p_0(\Omega) \to L^q(\Omega),$$
$$\nabla \mathbf{G}_D : L^1(\Omega) \to L^{3/2,\infty}(\Omega), \quad \nabla \mathbf{G}_N : L^1(\Omega) \to L^{3/2,\infty}(\Omega)$$

are well defined and bounded. In particular,

$$\nabla \mathbf{G}_D : h^1(\Omega) \to L^{3/2}(\Omega), \quad \nabla \mathbf{G}_N : h^1_0(\Omega) \to L^{3/2}(\Omega)$$

are bounded operators. The ranges of indices are optimal in the class of Lipschitz domains. For $\partial\Omega \in C^1$ one may take $\varepsilon = 1$.

13.6 The Two-dimensional Setting

All our main results discussed so far continue to hold in the two-dimensional setting, albeit with possibly different conditions imposed on the intervening indices (p, q, s), when Ω is a Lipschitz domain in the *two-dimensional* Euclidean space $\mathbb{R}^2$.

Somewhat more specifically, the conditions (13.6) should be replaced with the demand that the indices $\frac{1}{2} < p \leq \infty$ and $(\frac{1}{p} - 1)_+ < s < 1$ satisfy any of the following three conditions:

$$\begin{aligned} (I') : &\quad \tfrac{2}{1+\varepsilon} \leq p \leq \tfrac{2}{1-\varepsilon} \quad \text{and} \quad 0 < s < 1; \\ (II') : &\quad \tfrac{2}{3+\varepsilon} < p < \tfrac{2}{1+\varepsilon} \quad \text{and} \quad \tfrac{1}{p} - \tfrac{1+\varepsilon}{2} < s < 1; \\ (III') : &\quad \tfrac{2}{1-\varepsilon} < p \leq \infty \quad \text{and} \quad 0 < s < \tfrac{1}{p} + \tfrac{1+\varepsilon}{2}. \end{aligned}$$

Also, the two-dimensional analog of (13.8) takes the form

$$\left(\min\left\{\tfrac{1}{2p} + 1,\, s + 1\right\}\right)^{-1} < q \leq \infty.$$

With the aforementioned alterations, Theorems 1, 2, and 8 and Corollary 1 remain valid for bounded Lipschitz domains in $\mathbb{R}^2$.

In closing, we want to point out that $\nabla^2 \mathbf{G}_D : L^2(\Omega) \longrightarrow L^2(\Omega)$ is a bounded operator if either $\Omega \subset \mathbb{R}^2$ is a convex polygon (see p. 149 in [10], and [1]), or the Lamé moduli satisfy $|\lambda| < \sqrt{3}\mu$ and $\Omega \subset \mathbb{R}^2$ is a bounded convex domain ([9]; see also [15] for related matters). In particular, we may state the following assertion.

Theorem 11. *Suppose that the Lamé moduli satisfy $|\lambda| < \sqrt{3}\mu$ and that $\Omega \subset \mathbb{R}^2$ is a bounded convex domain. Then*

$$\nabla^2 \mathbf{G}_D : L^p(\Omega) \longrightarrow L^p(\Omega)$$

is a bounded operator for each $1 < p \leq 2$. *In this setting, the operator* $\nabla^2 \mathbf{G}_D$ *also maps* $h^1(\Omega)$ *boundedly to itself. Thus, by embeddings,* $\mathbf{G}_D$ *maps* $h^1(\Omega)$ *boundedly into* $C(\bar{\Omega})$, *the space of continuous functions on* $\bar{\Omega}$.

References

1. C. Bacuta and J. Bramble, Regularity estimates for solutions of the equations of linear elasticity in convex plane polygonal domains, *Z. Angew. Math. Phys.* **54** (2003), 874–878.
2. A.P. Calderón, Lebesgue spaces of differentiable functions and distributions, in *Proc. Symposium Pure Math.* **IV**, Amer. Math. Soc., Providence, RI, 1961, 33–49.
3. D.-C. Chang, S.G. Krantz, and E.M. Stein, H^p theory on a smooth domain in R^N and elliptic boundary value problems, *J. Functional Anal.* **114** (1993), 286–347.
4. B. Dahlberg, L^q-estimates for Green potentials in Lipschitz domains, *Math. Scand.* **44** (1979), 149–170.
5. B. Dahlberg and C. Kenig, Hardy spaces and the Neumann problem in L^p for Laplace's equation in Lipschitz domains, *Ann. Math.* **125** (1987), 437–465.
6. B. Dahlberg and C. Kenig, L^p-estimates for the three-dimensional system of elastostatics on Lipschitz domains, in *Lect. Notes Pure Appl. Math.* **122**, C. Sadosky, ed., 1990, 621–634.
7. B. Dahlberg, C. Kenig, and G. Verchota, Boundary value problems for the system of elastostatics on Lipschitz domains, *Duke Math. J.* **57** (1988), 795–818.
8. E. Fabes, O. Mendez, and M. Mitrea, Boundary layers on Sobolev-Besov spaces and Poisson's equation for the Laplacian in Lipschitz domains, *J. Functional Anal.* **159** (1998), 323–368.
9. P. Grisvard, Le problème de Dirichlet pour les équations de Lamé, *C.R. Acad. Sci. Paris Sér. I Math.* **304** (1987), 71–73.
10. P. Grisvard, *Singularities in Boundary Value Problems*, Research Appl. Math. **22**, Masson, Paris and Springer-Verlag, Berlin, 1992.
11. D. Jerison and C. Kenig, The inhomogeneous problem in Lipschitz domains, *J. Functional Anal.* **130** (1995), 161–219.
12. V.D. Kupradze, T.G. Gegelia, M.O. Basheleishvili, and T.V. Burchuladze, *Three-dimensional Problems of the Mathematical Theory of Elasticity and Thermoelasticity*, North-Holland, Amsterdam, 1979.
13. S. Mayboroda and M. Mitrea, Sharp estimates for Green potentials on non-smooth domains, *Math. Res. Lett.* **11** (2004), 481–492.
14. S. Mayboroda and M. Mitrea, Green potential estimates and the Poisson problem on Lipschitz domains, preprint (2004).

15. V.G. Maz'ya and J. Rossmann, Weighted L_p estimates of solutions to boundary value problems for second order elliptic systems in polyhedral domains, *Z. Angew. Math. Mech.* **83** (2003), 435–467.

16. V.G. Maz'ya, Personal communication.

17. O. Mendez and M. Mitrea, The Banach envelopes of Besov and Triebel-Lizorkin spaces and applications to partial differential equations, *J. Fourier Anal. Appl.* **6** (2000), 503–531.

18. A. Miyachi, On the extension properties of Triebel-Lizorkin spaces, *Hokkaido Math. J.* **27** (1998), 273–301.

19. S. Nicaise, Regularity of the solutions of elliptic systems in polyhedral domains, *Bull. Belg. Math. Soc. Simon Stevin* **4** (1997), 411–429.

20. Z. Shen, The L^p Dirichlet problem for elliptic systems on Lipschitz domains, preprint (2004).

21. E.M. Stein, *Singular Integrals and Differentiability Properties of Functions,* Princeton Univ. Press, Princeton, NJ, 1970.

22. E. Straube, Interpolation between Sobolev and between Lipschitz spaces of analytic functions on star-shaped domains, *Trans. Amer. Math. Soc.* **316** (1989), 653–671.

23. H. Triebel, Function spaces on Lipschitz domains and on Lipschitz manifolds. Characteristic functions as pointwise multipliers, *Rev. Mat. Complut.* **15** (2002), 475–524.

14 Analysis of Boundary-domain Integral and Integro-differential Equations for a Dirichlet Problem with a Variable Coefficient

Sergey E. Mikhailov

14.1 Introduction

The Dirichlet boundary value problem for the "Laplace" linear differential equation with a variable coefficient is reduced to boundary-domain integral or integro-differential equations (BDIEs or BDIDEs) based on a specially constructed parametrix. The BDI(D)Es contain potential-type integral operators defined on the domain under consideration and acting on the unknown solution as well as integral operators defined on the boundary and acting on the trace and/or conormal derivative of the unknown solution or on an auxiliary function. Some of the considered BDIDEs are to be supplemented by the original boundary conditions, thus constituting boundary-domain integro-differential problems (BDIDPs). Solvability, solution uniqueness, and equivalence of the BDIEs/BDIDEs/BDIDPs to the original boundary value problem (BVP) are investigated in appropriate Sobolev spaces.

Reduction of BVPs with arbitrarily variable coefficients to boundary integral equations is usually not effective for numerical implementations, since the fundamental solution necessary for such reduction is generally not available in an analytical form (except some special dependence of the coefficients on coordinates, see e.g. [1]). Using a parametrix (Levi function) as a substitute of a fundamental solution, it is possible however to reduce such a BVP to a BDIE (see, e.g., [2], [3], [4, Sect. 18], [5], and [6], where the Dirichlet, Neumann, and Robin problems for some partial differential equations (PDEs) were reduced to indirect BDIEs).

In [7], [8], and [10], the 3D mixed (Dirichlet–Neumann) BVP for the variable-coefficient "Laplace" equation was considered. Such equations appear, e.g., in electrostatics, stationary heat transfer and other diffusion problems for inhomogeneous media. The BVP has been reduced to either segregated or united direct BDI(D)Es or BDIDPs. Some of the BDI(D)Es/BDIDPs are associated with the BDIDE and BDIE formulated in [9]. Although several of the integral and integro-differential formulations for the mixed problem in [7], [8], and [10] look like equations of the second kind, the spaces for the out-of-integral terms are different from the spaces for the right-hand sides of the equations, thus the equations are of "almost" second kind.

Analysis of the four different (systems of) BDI(D)Es/BDIDP, to which the Dirichlet problem for the same PDE is reduced in the present paper, requires special consideration. Equivalence of the considered BDI(D)Es/BDIDP to the original BVP is proved along with their solvability, solution uniqueness, and the operator invertibility in corresponding Sobolev–Slobodetski spaces. In particular, it is shown that the Dirichlet problem can be reduced to a genuine second-kind integral or integro-differential equation.

14.2 Formulation of the Boundary Value Problem

Let Ω be a bounded open three-dimensional region of $\mathbb{R}^3$. For simplicity, we assume that the boundary $S := \partial\Omega$ is a simply connected, closed, infinitely smooth surface. Let $a \in C^\infty(\bar{\Omega})$, $a(x) > 0$ for $x \in \bar{\Omega}$. Let also $\partial_{x_j} := \partial/\partial x_j$ $(j = 1, 2, 3)$, $\partial_x = (\partial_{x_1}, \partial_{x_2}, \partial_{x_3})$.

We consider the scalar elliptic differential equation

$$Lu(x) := L(x, \partial_x)\, u(x) := \sum_{i=1}^{3} \frac{\partial}{\partial x_i}\left(a(x)\, \frac{\partial u(x)}{\partial x_i} \right) = f(x), \quad x \in \Omega, \tag{14.1}$$

where u is an unknown function and f is a given function in Ω.

In what follows, $H^s(\Omega) = H_2^s(\Omega)$ and $H^s(S) = H_2^s(S)$ are the Bessel potential spaces, where $s \in \mathbb{R}$ is an arbitrary real number (see, e.g., [11] and [12]). We recall that H^s coincide with the Sobolev–Slobodetski spaces W_2^s for any nonnegative or integer s.

For a linear operator L_*, we introduce the subspace of $H^1(\Omega)$ [13]

$$H^{1,0}(\Omega; L_*) := \{g :\ g \in H^1(\Omega),\ L_* g \in L_2(\Omega)\},$$

endowed with the norm

$$\|g\|_{H^{1,0}(\Omega; L_*)} := \|g\|_{H^1(\Omega)} + \|L_* g\|_{L_2(\Omega)}.$$

In this paper, we will particularly use the space $H^{1,0}(\Omega; L_*)$ for L_* being either the operator L from (14.1) or the Laplace operator Δ. Since

$$Lu - \Delta u = \sum_{i=1}^{3} \frac{\partial a}{\partial x_i} \frac{\partial u}{\partial x_i} \in L_2(\Omega)$$

for $u \in H^1(\Omega)$, we have $H^{1,0}(\Omega; L) = H^{1,0}(\Omega; \Delta)$.

From the trace theorem (see, e.g., [11] and [13]–[15]) for $u \in H^1(\Omega)$, it follows that $u^+ := \tau_S^+ u \in H^{1/2}(S)$, where τ_S^+ is the trace operator on S from Ω.

For $u \in H^2(\Omega)$ we can denote by T^+ the corresponding conormal differentiation operator on S in the sense of traces,

$$T^+(x, n^+(x), \partial_x)\, u(x) := \sum_{i=1}^{3} a(x)\, n_i^+(x) \left(\frac{\partial u(x)}{\partial x_i}\right)^+ = a(x) \left(\frac{\partial u(x)}{\partial n^+(x)}\right)^+,$$

where $n^+(x)$ is the exterior (to Ω) unit normal vectors at the point $x \in S$.

Let $u \in H^{1,0}(\Omega; \Delta)$. We can correctly define the generalized conormal derivative $T^+u \in H^{-1/2}(S)$ with the help of Green's formula (see, for example, [13] and [15], Lemma 4.3),

$$\left\langle T^+u\,,\, v^+\right\rangle_S := \int_\Omega v(x) L u(x)\, dx + \int_\Omega \sum_{i=1}^{3} a(x)\, \frac{\partial u(x)}{\partial x_i}\, \frac{\partial v(x)}{\partial x_i} dx \quad \forall\, v \in H^1(\Omega), \tag{14.2}$$

where $\langle\,\cdot\,,\,\cdot\,\rangle_S$ denotes the duality brackets between the spaces $H^{-1/2}(S)$ and $H^{1/2}(S)$, extending the usual L_2 scalar product.

We will investigate the following *Dirichlet boundary value problem.*

Find a function $u \in H^1(\Omega)$ satisfying the conditions

$$L\, u = f \quad \text{in } \Omega, \tag{14.3}$$

$$u^+ = \varphi_0 \quad \text{on } S, \tag{14.4}$$

where $\varphi_0 \in H^{1/2}(S)$ and $f \in L_2(\Omega)$.

Equation (14.3) is understood in the distributional sense and condition (14.4) in the trace sense.

We have the following uniqueness theorem.

Theorem 1. *BVP* (14.3)–(14.4) *with* $\varphi_0 \in H^{1/2}(S)$ *and* $f \in L_2(\Omega)$ *has at most one solution in* $H^1(\Omega)$.

Proof. The assertion follows immediately from Green's formula (14.2) with $v = u$ as a solution of the homogeneous Dirichlet problem, i.e., with $f = 0$ and $\varphi_0 = 0$.

14.3 Parametrix and Potential-type Operators

We say that a function $P(x,y)$ of two variables $x, y \in \Omega$ is a parametrix (the Levi function) for the operator $L(x, \partial_x)$ in $\mathbb{R}^3$ if (see, e.g., [2]–[6] and [9])

$$L(x, \partial_x)\, P(x,y) = \delta(x-y) + R(x,y), \tag{14.5}$$

where $\delta(\cdot)$ is the Dirac distribution and $R(x,y)$ possesses a weak (integrable) singularity at $x = y$, i.e.,

$$R(x,y) = \mathcal{O}\left(|x-y|^{-\kappa}\right) \quad \text{with } \kappa < 3. \tag{14.6}$$

It is easy to see that for the operator $L(x, \partial_x)$ given by the right-hand side in (14.1), the function

$$P(x,y) = \frac{-1}{4\pi\, a(y)\, |x-y|}, \quad x, y \in \mathbb{R}^3, \tag{14.7}$$

is a parametrix; the corresponding remainder function is

$$R(x,y) = \sum_{i=1}^{3} \frac{x_i - y_i}{4\pi\, a(y)\, |x-y|^3} \frac{\partial\, a(x)}{\partial x_i}, \quad x, y \in \mathbb{R}^3, \tag{14.8}$$

and satisfies estimate (14.6) with $\kappa = 2$, due to the smoothness of the function $a(x)$.

Evidently, the parametrix $P(x,y)$ given by (14.7) is a fundamental solution to the operator $L(y, \partial_x) := a(y)\Delta(\partial_x)$ with "frozen" coefficient $a(x) = a(y)$, i.e.,

$$L(y, \partial_x)\, P(x,y) = \delta(x-y).$$

Note that remainder (14.8) is not smooth enough for the parametrix (14.7) and the corresponding potential operators to be treated as in [15].

For some scalar function g, let

$$Vg(y) := -\int_S P(x,y)\, g(x)\, dS_x, \quad y \notin S, \tag{14.9}$$

$$Wg(y) := -\int_S \left[T(x, n(x), \partial_x)\, P(x,y)\right]\, g(x)\, dS_x, \quad y \notin S, \tag{14.10}$$

be the single-layer and the double-layer surface potential operators, where the integrals are understood in the distributional sense if g is not integrable.

The corresponding boundary integral (pseudodifferential) operators of direct surface values of the single-layer potential $\mathcal{V}$ and of the double-layer potential $\mathcal{W}$, and the conormal derivatives of the single-layer potential, $\mathcal{W}'$, and of the double-layer potential, $\mathcal{L}^+$, are, respectively,

$$\mathcal{V}\, g(y) := -\int_S P(x,y)\, g(x)\, dS_x, \tag{14.11}$$

$$\mathcal{W}\, g(y) := -\int_S \left[T(x, n(x), \partial_x)\, P(x,y)\right]\, g(x)\, dS_x, \tag{14.12}$$

$$\mathcal{W}'\, g(y) := -\int\limits_S \big[\, T(y, n(y), \partial_y)\, P(x,y)\,\big]\, g(x)\, dS_x, \tag{14.13}$$

$$\mathcal{L}^+ g(y) := [T(y, n(y), \partial_y)\, W g(y)]^+, \tag{14.14}$$

where $y \in S$.

The parametrix-based volume potential operator and the remainder potential operator, corresponding to parametrix (14.7) and to remainder (14.8) are

$$\mathcal{P}g(y) := \int\limits_\Omega P(x,y)\, g(x)\, dx, \tag{14.15}$$

$$\mathcal{R}g(y) := \int\limits_\Omega R(x,y)\, g(x)\, dx. \tag{14.16}$$

For $g_1 \in H^{-1/2}(S)$ and $g_2 \in H^{1/2}(S)$, there hold the jump relations on S

$$\begin{aligned} [V g_1(y)]^+ &= \mathcal{V} g_1(y) \\ [W g_2(y)]^+ &= -\tfrac{1}{2}\, g_2(y) + \mathcal{W} g_2(y), \\ [T(y, n(y), \partial_y) V g_1(y)]^+ &= \tfrac{1}{2}\, g_1(y) + \mathcal{W}' g_1(y), \end{aligned}$$

where $y \in S$.

The jump relations as well as mapping properties of potentials and operators (14.9)–(14.16) are well known for the case $a = \text{const}$. They were extended to the case of variable coefficient $a(x)$ in [7] and [8].

14.4 Green Identities and Integral Relations

Let $u \in H^{1,0}(\Omega; \Delta)$, $v \in H^{1,0}(\Omega; \Delta)$ be some real functions. Then, subtracting (14.2) from its counterpart with exchanged roles of u and v, we obtain the so-called second Green identity for the operator $L(x, \partial_x)$,

$$\int\limits_\Omega \big[\, v\, L(x, \partial_x) u - u\, L(x, \partial_x) v\,\big]\, dx = \big\langle T^+ u\,,\, v^+ \big\rangle_S - \big\langle u^+\,,\, T^+ v\,\big\rangle_S. \tag{14.17}$$

For $u \in H^{1,0}(\Omega; \Delta)$ and $v(x) = P(x,y)$, where the parametrix $P(x,y)$ is given by (14.7), we obtain from (14.17), (14.5) by the standard limiting procedures (cf. [4]) the third Green identity,

$$u(y) + \mathcal{R}u(y) - V T^+ u(y) + W u^+(y) = \mathcal{P} L u(y), \quad y \in \Omega. \tag{14.18}$$

If $u \in H^{1,0}(\Omega; \Delta)$ is a solution of equation (14.1), then (14.18) gives

$$G u := u + \mathcal{R} u - V T^+ u + W u^+ = \mathcal{P} f \quad \text{in } \Omega, \tag{14.19}$$

$$\mathcal{G}u := \tfrac{1}{2}u^+ + [\mathcal{R}u]^+ - \mathcal{V}T^+u + \mathcal{W}u^+ = [\mathcal{P}f]^+ \quad \text{on } S, \qquad (14.20)$$

$$\mathcal{T}u := \tfrac{1}{2}T^+u + T^+\mathcal{R}u - \mathcal{W}'T^+u + \mathcal{L}^+u^+ = T^+\mathcal{P}f \quad \text{on } S. \quad (14.21)$$

For some functions f, Ψ, Φ, let us consider a more general "indirect" integral relation, associated with (14.19), namely,

$$u(y) + \mathcal{R}u(y) - V\Psi(y) + W\Phi(y) = \mathcal{P}f(y), \quad y \in \Omega. \qquad (14.22)$$

Lemma 1. *Let $\Psi \in H^{-1/2}(S)$, $\Phi \in H^{1/2}(S)$, and $f \in L_2(\Omega)$. Suppose a function $u \in H^1(\Omega)$ satisfies* (14.22). *Then $u \in H^{1,0}(\Omega;\Delta)$, it is a solution of PDE* (14.3) *in Ω, and*

$$V(\Psi - T^+u)(y) - W(\Phi - u^+)(y) = 0, \quad y \in \Omega.$$

Proof. First of all, equation (14.22) and mapping properties of the operators $\mathcal{R}$, $\mathcal{P}$, V and W imply $u \in H^{1,0}(\Omega;\Delta)$. The rest of the lemma claims follow from its counterpart proved in [7], Lemma 4.1.

The following statement is well known (see, for example, [7], Lemma 4.2).

Lemma 2. (i) *Let $\Psi^* \in H^{-1/2}(S)$. If $V\Psi^*(y) = 0$, $y \in \Omega$, then $\Psi^* = 0$.*
(ii) *Let $\Phi^* \in H^{1/2}(S)$. If $W\Phi^*(y) = 0$, $y \in \Omega$, then $\Phi^* = 0$.*

Theorem 2. *Let $f \in L_2(\Omega)$. A function $u \in H^{1,0}(\Omega;\Delta)$ is a solution of PDE* (14.3) *in Ω if and only if it is a solution of BDIDE* (14.19).

Proof. If $u \in H^{1,0}(\Omega;\Delta)$ solves PDE (14.3) in Ω, then it satisfies (14.19). On the other hand, if $u \in H^{1,0}(\Omega;\Delta)$ solves BDIDE (14.19), then Lemma 1 with $\Psi = T^+u$ and $\Phi = u^+$ completes the proof.

14.5 Segregated Boundary-domain Integral Equations

Let us consider a segregated purely *integral* boundary-domain formulation for the Dirichlet problem, similar to the formulations introduced and analyzed in [7], [8], and [10] for the mixed problem with $u \in H^1(\Omega)$ and $u \in H^{1,0}(\Omega;\Delta)$.

14.5.1 Integral Equation System $(G\mathcal{G})$

To reduce BVP (14.3)–(14.4) to a BDIE system in this section, we will use equation (14.19) in Ω and equation (14.20) on S, where the known function φ_0 is substituted for u^+ and an auxiliary unknown function $\psi \in H^{-1/2}(S)$ for T^+u. Then we arrive at the system

$$u(y) + \mathcal{R}u(y) - V\psi(y) = F_0(y), \quad y \in \Omega, \qquad (14.23)$$

$$\mathcal{R}^+u(y) - \mathcal{V}\psi(y) = F_0^+(y) - \varphi_0(y), \quad y \in S, \qquad (14.24)$$

where

$$F_0(y) := \mathcal{P}f(y) - W\varphi_0(y), \quad y \in \Omega. \tag{14.25}$$

Note that for $f \in L_2(\Omega)$ and $\varphi_0 \in H^{1/2}(S)$, we have the inclusion $F_0 \in H^{1,0}(\Omega; \Delta)$ due to the mapping properties of the Newtonian (volume) and layer potentials.

Remark 1. $F_0 = 0$ if and only if $(f, \varphi_0) = 0$. Indeed, the latter equality evidently implies the former. Inversely, let $F_0 = 0$. Keeping in mind equation (14.25), Lemma 1 with $F_0 = 0$ for u implies $f = 0$ and $W\varphi_0 = 0$ in Ω. Lemma 2(ii) then gives $\varphi_0 = 0$ on S.

Let us prove that BVP (14.3)–(14.4) in Ω is equivalent to the system of BDIEs (14.23)–(14.24).

Theorem 3. *Let $f \in L_2(\Omega)$ and $\varphi_0 \in H^{1/2}(S)$.*

(i) *If some $u \in H^1(\Omega)$ solves BVP* (14.3)–(14.4) *in Ω, then the solution is unique and the couple $(u, \psi) \in H^1(\Omega) \times H^{-1/2}(S)$, where*

$$\psi = T^+ u \quad \text{on } S, \tag{14.26}$$

solves BDIE system (14.23)–(14.24)*;*

(ii) *If a couple $(u, \psi) \in H^1(\Omega) \times H^{-1/2}(S)$ solves BDIE system* (14.23)–(14.24)*, then the solution is unique, u solves BVP* (14.3)–(14.4)*, and ψ satisfies* (14.26).

Proof. Remark that if $u \in H^1(\Omega)$ is a solution of the BVP (14.3)–(14.4) with $f \in L_2(\Omega)$, then $u \in H^{1,0}(\Omega; \Delta)$. On the other hand, if $\mathcal{U} = (u, \psi) \in H^1(\Omega) \times H^{-1/2}(S)$ is a solution of system (14.23)–(14.24) with $f \in L_2(\Omega)$ and $\varphi_0 \in H^{1/2}(S)$, then $u \in H^{1,0}(\Omega; \Delta)$ due to mapping properties of operators $\mathcal{R}$, V, and W (see [10] and [8]).

Let $u \in H^1(\Omega)$ be a solution to BVP (14.3)–(14.4). It is unique due to Theorem 1. Set ψ by (14.26), evidently, $\psi \in H^{-1/2}(S)$. Then it immediately follows from relations (14.19)–(14.20) that the couple (u, ψ) solves system (14.23)–(14.24), which completes the proof of item (i).

Let now a couple $(u, \psi) \in H^1(\Omega) \times H^{-1/2}(S)$ solve BDIE system (14.23)–(14.24). Taking trace of equation (14.23) on S and subtracting equation (14.24) from it, we obtain

$$u^+(y) = \varphi_0(y), \quad y \in S, \tag{14.27}$$

i.e., u satisfies the Dirichlet condition (14.4).

Equation (14.23) and Lemma 1 with $\Psi = \psi$ and $\Phi = \varphi_0$ imply that u is a solution of PDE (14.3) and

$$V\Psi^*(y) - W\Phi^*(y) = 0, \quad y \in \Omega,$$

where $\Psi^* = \psi - T^+ u$ and $\Phi^* = \varphi_0 - u^+$. Due to equation (14.27), $\Phi^* = 0$. Lemma 2(i) implies that $\Psi^* = 0$, which completes the proof of conditions (14.26).

Uniqueness of the solution to BDIE system (14.23)–(14.24) follows from (14.26) along with Remark 1 and Theorem 1.

System (14.23)–(14.24) can be rewritten in the form

$$\mathcal{A}^{G\mathcal{G}}\mathcal{U} = \mathcal{F}^{G\mathcal{G}},$$

where $\mathcal{U}^\top := (u, \psi) \in H^{1,0}(\Omega; \Delta) \times H^{-1/2}(S)$,

$$\mathcal{A}^{G\mathcal{G}} := \begin{bmatrix} I - \mathcal{R} & -V \\ \mathcal{R}^+ & -\mathcal{V} \end{bmatrix}, \quad \mathcal{F}^{G\mathcal{G}} := \begin{bmatrix} F_0 \\ F_0^+ - \varphi_0 \end{bmatrix}.$$

Due to the mapping properties of operators V, $\mathcal{V}$, W, $\mathcal{W}$, $\mathcal{P}$, $\mathcal{R}$, and $\mathcal{R}^+$ (see [10] and [8]), we have $\mathcal{F}^{G\mathcal{G}} \in H^{1,0}(\Omega; \Delta) \times H^{1/2}(S)$, and the operators

$$\mathcal{A}^{G\mathcal{G}} : H^{1,0}(\Omega; \Delta) \times H^{-1/2}(S) \to H^{1,0}(\Omega; \Delta) \times H^{1/2}(S) \qquad (14.28)$$

$$: H^1(\Omega) \times H^{-1/2}(S) \to H^1(\Omega) \times H^{1/2}(S) \qquad (14.29)$$

are continuous. By Theorem 3 and the uniqueness Theorem 1, both operators (14.28) and (14.29) are injective.

Theorem 4. *Operators* (14.28) *and* (14.29) *are continuous and invertible.*

Proof. Let us consider the proof for the operator $\mathcal{A}^{G\mathcal{G}}$ given by (14.29) first. The continuity and injectivity are proved above. To prove the invertibility, let us consider the operator

$$\mathcal{A}_0^{G\mathcal{G}} := \begin{bmatrix} I & -V \\ 0 & -\mathcal{V} \end{bmatrix}.$$

As a result of compactness properties of the operators $\mathcal{R}$ and $\mathcal{R}^+$, the operator $\mathcal{A}_0^{G\mathcal{G}}$ is a compact perturbation of the operator $\mathcal{A}^{G\mathcal{G}}$.

The operator $\mathcal{A}_0^{G\mathcal{G}}$ is an upper triangular matrix operator with the scalar diagonal invertible operators

$$I: \; H^1(\Omega) \to H^1(\Omega),$$
$$\mathcal{V}: \; H^{-1/2}(S) \to H^{1/2}(S)$$

(see [14], Ch. XI, Part B, Sect. 2, Theorem 3 for $\mathcal{V}$). This implies that

$$\mathcal{A}_0^{G\mathcal{G}} \; : \; H^1(\Omega) \times H^{-1/2}(S) \to H^1(\Omega) \times H^{1/2}(S)$$

is an invertible operator. Hence, the operator $\mathcal{A}^{G\mathcal{G}}$ possesses the Fredholm property and its index is zero. The injectivity of the operator $\mathcal{A}^{G\mathcal{G}}$, already proved, completes the theorem proof for this operator.

Let us now construct an inverse to operator (14.28). Let $(\mathcal{A}^{G\mathcal{G}})^{-1}$: $H^1(\Omega) \times H^{1/2}(S) \to H^1(\Omega) \times H^{-1/2}(S)$ be the operator inverse to (14.29).

Thus, for any $\mathcal{F}^{G\mathcal{G}} \in H^{1,0}(\Omega;\Delta) \times H^{1/2}(S)$, the solution of the system $\mathcal{A}^{G\mathcal{G}}\mathcal{U} = \mathcal{F}^{G\mathcal{G}}$ in $H^1(\Omega) \times H^{-1/2}(S)$ is $\mathcal{U} = (\mathcal{A}^{G\mathcal{G}})^{-1}\mathcal{F}^{G\mathcal{G}}$. Taking into account that the operators $V : H^{-1/2}(S) \to H^{1,0}(\Omega;\Delta)$ and $\mathcal{R} : H^1(\Omega) \to H^{2,0}(\Omega;\Delta)$ are continuous [9: Th. A.1, Remark B.2], the first equation of this system then implies $u = \mathcal{U}_1 \in H^{1,0}(\Omega;\Delta)$ and the operator $(\mathcal{A}^{G\mathcal{G}})^{-1}$ is continuous also from $H^{1,0}(\Omega;\Delta) \times H^{1/2}(S)$ to $H^{1,0}(\Omega;\Delta) \times H^{-1/2}(S)$.

The original BVP (14.3)–(14.4) can be written in the form

$$A^D u = F^D,$$

where

$$A^D := \begin{bmatrix} L \\ \tau^+ \end{bmatrix}, \quad F^D = \begin{bmatrix} f \\ \varphi_0 \end{bmatrix}.$$

The operator $A^D : H^{1,0}(\Omega;L) \to L_2(\Omega) \times H^{1/2}(S)$ is evidently continuous and, due to the uniqueness theorem for the BVP, it is also injective.

The invertibility of the operator (14.28) and equivalence Theorem 3 lead to the following assertion.

Corollary. *The operator $A^D : H^{1,0}(\Omega;L) \to L_2(\Omega) \times H^{1/2}(S)$ is continuous and continuously invertible.*

Note that the above statement and the uniqueness Theorem 1 evidently imply the following existence theorem in $H^1(\Omega)$.

Theorem 5. *Let $\varphi_0 \in H^{1/2}(S)$ and $f \in L_2(\Omega)$. Then BVP (14.3)–(14.4) is uniquely solvable in $H^1(\Omega)$.*

14.5.2 Integral Equation System (GT)

To obtain a segregated BDIE system of *the second kind*, we will use equation (14.19) in Ω and equation (14.21) on S, where again the known function φ_0 is substituted for u^+ and an auxiliary unknown function $\psi \in H^{-1/2}(S)$ for T^+u. Then we arrive at the following system (GT):

$$u + \mathcal{R}u - V\psi = F_0 \quad \text{in } \Omega, \tag{14.30}$$

$$\tfrac{1}{2}\psi + T^+\mathcal{R}u - \mathcal{W}'\psi = T^+F_0 \quad \text{on } S, \tag{14.31}$$

where F_0 is given by (14.25).

Let us prove that BVP (14.3)–(14.4) is equivalent to system (14.30)–(14.31).

Theorem 6. *Let $f \in L_2(\Omega)$ and $\varphi_0 \in H^{1/2}(S)$.*

(i) *If some $u \in H^1(\Omega)$ solves BVP (14.3)–(14.4) in Ω, then the couple $(u,\psi)^\top \in H^1(\Omega) \times H^{-1/2}(S)$, where*

$$\psi = T^+u \quad \text{on } S, \tag{14.32}$$

solves BDIE system (14.30)–(14.31).

(ii) *If a couple* $(u,\psi)^\top \in H^1(\Omega)\times H^{-1/2}(S)$ *solves BDIE system* (14.30)–(14.31), *then the solution is unique,* u *solves BVP* (14.3)–(14.4), *and* ψ *satisfies* (14.32).

Proof. As in the proof of Theorem 3, $u \in H^{1,0}(\Omega;\Delta)$ if either hypothesis (i) or (ii) is satisfied.

Let $u \in H^1(\Omega)$ be a solution to BVP (14.3)–(14.4). Set $\psi = T^+u$. Evidently, $\psi \in H^{-1/2}(S)$. Then it immediately follows from relations (14.19) and (14.21) that the couple (u,ψ) solves system (14.30)–(14.31), which completes the proof of item (i).

Let now a couple $(u,\psi) \in H^1(\Omega)\times H^{-1/2}(S)$ solve BDIE system (14.30)–(14.31).

Take the conormal derivative of equation (14.30) on S and subtract it from equation (14.31) to obtain

$$\Psi^* := \psi - T^+u = 0 \quad \text{on } S,$$

that is, equation (14.32) is proved.

Equation (14.30) and Lemma 1 with $\Psi = \psi$ and $\Phi = \varphi_0$ imply that u is a solution of equation (14.1) and

$$V\Psi^*(y) - W\Phi^*(y) = 0, \quad y \in \Omega, \tag{14.33}$$

where $\Phi^* = \varphi_0 - u^+$. Since $\Psi^* = 0$ on S, (14.33) reduces to

$$W\Phi^*(y) = 0, \quad y \in \Omega,$$

and Lemma 2(ii) implies that $\Phi^* = 0$ on S. This means that u satisfies the Dirichlet condition (14.4).

By Remark 1, the unique solvability of BDIE system (14.30)–(14.31) then follows from (14.32) along with the unique solvability of (14.3)–(14.4).

System (14.30)–(14.31) can be rewritten in the form

$$\mathcal{A}^{GT}\mathcal{U} = \mathcal{F}^{GT},$$

where $\mathcal{U}^\top := (u,\psi)^\top \in H^1(\Omega) \times H^{-1/2}(S)$,

$$\mathcal{A}^{GT} := \begin{bmatrix} I+\mathcal{R} & -V \\ T^+\mathcal{R} & \frac{1}{2}I - \mathcal{W}' \end{bmatrix}, \quad \mathcal{F}^{GT} := \begin{bmatrix} F_0 \\ T^+F_0 \end{bmatrix}. \tag{14.34}$$

Due to the mapping properties of the operators involved in (14.34) we have $\mathcal{F}^{GT} \in H^{s,0}(\Omega^+;\Delta) \times H^{-1/2}(S)$, and the operators

$$\mathcal{A}^{GT} : H^{1,0}(\Omega;\Delta) \times H^{-1/2}(S) \to H^{1,0}(\Omega;\Delta) \times H^{-1/2}(S) \tag{14.35}$$

$$: H^1(\Omega) \times H^{-1/2}(S) \to H^1(\Omega) \times H^{-1/2}(S) \tag{14.36}$$

are continuous. By Theorem 6 and the uniqueness Theorem 1, both operators (14.35) and (14.36) are injective.

Theorem 7. *Operators* (14.35) *and* (14.36) *are continuous and invertible.*

Proof. The operator

$$\mathcal{A}_0^{GT} := \begin{bmatrix} I & -V \\ 0 & \frac{1}{2}I \end{bmatrix}$$

is a compact perturbation of both operators (14.35) and (14.36), due to compactness properties of the operators $\mathcal{R}$ and $\mathcal{W}$ (see [7], [8], and [10]). The invertibility of operators (14.35) and (14.36) then follows by arguments similar to those in the proof of Theorem 4.

14.6 United Boundary-domain Integro-differential Equations and Problems

Instead of introducing an auxiliary function ψ, we will work in this section with the original form of the equations containing the conormal derivative operator T^+ of the internal field.

14.6.1 Integro-differential Problem (GD)

The equivalence of the differential and boundary-domain integro-differential equations proved in Theorem 2 allows us to supplement BDIDE (14.19) with the original Dirichlet boundary conditions and arrive at BDIDP (GD) constituted by (14.19), (14.4). The BDIDP is equivalent to the Dirichlet boundary value problem (14.3)–(14.4) in Ω, in the following sense.

Theorem 8. *Let* $f \in L_2(\Omega)$, $\varphi_0 \in H^{1/2}(S)$. *A function* $u \in H^{1,0}(\Omega;\Delta)$ *solves BVP* (14.3)–(14.4) *in* Ω *if and only if* u *solves BDIDP* (14.19), (14.4). *Such a solution does exist and is unique.*

Proof. A solution of BVP (14.3)–(14.4) does exist and is unique, by Theorem 5, and provides a solution to BDIDP (14.19), (14.4), by Theorem 2. On the other hand, any solution of BDIDP (14.19), (14.4) also satisfies (14.3), due to the same Theorem 2.

BDIDP (14.19), (14.4) can be written in the form

$$\mathcal{A}^{GD} u = \mathcal{F}^{GD}, \tag{14.37}$$

where

$$\mathcal{A}^{GD} := \begin{bmatrix} I + \mathcal{R} - VT^+ + W\tau^+ \\ \tau^+ \end{bmatrix}, \quad \mathcal{F}^{GD} = \begin{bmatrix} \mathcal{P}f \\ \varphi_0 \end{bmatrix}.$$

Owing to the mapping properties of operators V, W, $\mathcal{P}$, and $\mathcal{R}$, we have $\mathcal{F}^{GD} \in H^{1,0}(\Omega;\Delta) \times H^{1/2}(S)$, and the operator $\mathcal{A}^{GD} : H^{1,0}(\Omega;\Delta) \to H^{1,0}(\Omega;\Delta) \times H^{1/2}(S)$ is continuous. By Theorem 8, it is also injective. Let us now characterize the range of the operator $\mathcal{A}^{GD}$ in the whole space $H^{1,0}(\Omega;\Delta) \times H^{1/2}(S)$.

Theorem 9. *Let*

$$\mathcal{F}^{GD} = (\mathcal{F}_1^{GD}, \mathcal{F}_2^{GD}) \in H^{1,0}(\Omega; \Delta) \times H^{1/2}(S).$$

System (14.37) *has a solution in* $H^{1,0}(\Omega; \Delta)$ *if and only if there exists* $f_* \in L_2(\Omega)$ *such that*

$$\mathcal{F}_1^{GD} = \mathcal{P} f_* \quad \text{in } \Omega. \tag{14.38}$$

When the solution does exist, it is unique.

Proof. If condition (14.38) is satisfied, then, according to Theorem 7, there exists a unique solution of system (14.37).

On the other hand, if $u \in H^{1,0}(\Omega; \Delta)$ is a solution of system (14.37), then it satisfies the third Green identity (14.18). Comparing it with the first equation of system (14.37) implies representation (14.38) with $f_* = Lu$.

Let T_Δ^+, V_Δ, and W_Δ denote the operators of the conormal derivative, single-layer potential, and double-layer potential associated with the Laplace operator, that is, for the coefficient $a = 1$.

Remark 2. Condition (14.38) for an $\mathcal{F}_1^{GD} \in H^{1,0}(\Omega; \Delta)$ is equivalent to the condition

$$V_\Delta T_\Delta^+(a\mathcal{F}_1^{GD}) - W_\Delta(a\mathcal{F}_1^{GD})^+ = 0 \quad \text{in } \Omega, \tag{14.39}$$

or

$$V\left[T^+\mathcal{F}_1^{GD} + \mathcal{F}_1^{GD+}\frac{\partial a}{\partial n^+}\right] - W(\mathcal{F}_1^{GD})^+ = 0 \quad \text{in } \Omega. \tag{14.40}$$

Indeed, condition (14.38) can be rewritten as

$$a\mathcal{F}_1^{GD} = \mathcal{P}_\Delta f_* \quad \text{in } \Omega. \tag{14.41}$$

The third Green identity (14.18) for $u = a\mathcal{F}_1^{GD}$ and for the potentials associated with the operator Δ gives

$$a\mathcal{F}_1^{GD} - V_\Delta T_\Delta^+(a\mathcal{F}_1^{GD}) + W(a\mathcal{F}_1^{GD})^+ = \mathcal{P}_\Delta \Delta(a\mathcal{F}_1^{GD}) \quad \text{in } \Omega. \tag{14.42}$$

Thus (14.39) implies (14.41) with $f_* = \Delta(a\mathcal{F}_1^{GD})$.

On the other hand, if (14.41) is satisfied, then application of the Laplace operator to it gives $\Delta(a\mathcal{F}_1^{GD}) = f_*$. Substituting this into (14.42) and comparing with (14.41) implies (14.39).

Condition (14.40) follows from (14.39) and the definitions of V and W.

To realize how restrictive condition (14.38), or, equivalently, conditions (14.39) and (14.40), are, we prove the following assertion.

Lemma 3. *For any function $\mathcal{F}_1 \in H^{1,0}(\Omega;\Delta)$, there exists a unique couple $(f_*, \Phi_*) = \mathcal{C}_{\Phi}\mathcal{F}_1 \in L_2(\Omega) \times H^{1/2}(S)$ such that*

$$\mathcal{F}_1(y) = \mathcal{P}f_*(y) - W\Phi_*(y), \quad y \in \Omega, \tag{14.43}$$

and $\mathcal{C}_{\Phi} : H^{1,0}(\Omega;\Delta) \to L_2(\Omega) \times H^{1/2}(S)$ is a linear bounded operator.

Proof. We adapt here the proof scheme from [8], Lemma 5.2.

Suppose first that there exist some functions $f_*(y)$, $\Phi_*(y)$ satisfying (14.43) and find their expressions in terms of $\mathcal{F}_1(y)$. Taking into account definitions (14.7) and (14.9) for the volume and double layer potentials, ansatz (14.43) can be rewritten as

$$a(y)\mathcal{F}_1(y) = \mathcal{P}_\Delta f_*(y) - W_\Delta[a\Phi_*](y), \quad y \in \Omega, \tag{14.44}$$

where $\mathcal{P}_\Delta = \mathcal{P}|_{a=1}$, $W_\Delta = W|_{a=1}$ are the volume and double layer potential operators associated with the Laplace operator Δ.

Applying the Laplace operator to (14.44) we obtain that

$$f_* = \Delta(a\mathcal{F}_1) \quad \text{in } \Omega. \tag{14.45}$$

Then (14.44) can be rewritten as

$$W_\Delta[a\Phi_*](y) = Q(y), \quad y \in \Omega, \tag{14.46}$$

where

$$Q(y) := \mathcal{P}_\Delta[\Delta(a\mathcal{F}_1)](y) - a(y)\mathcal{F}_1(y), \quad y \in \Omega. \tag{14.47}$$

The trace of (14.46) on the boundary gives

$$\left[-\tfrac{1}{2}I + \mathcal{W}_\Delta\right][a\Phi_*](y) = Q^+(y), \quad y \in S, \tag{14.48}$$

where $\mathcal{W}_\Delta := \mathcal{W}|_{a=1}$ is the direct value on S of the double layer operator associated with the Laplace operator.

Since $\left[-(1/2)I + \mathcal{W}_\Delta\right]$ is an isomorphism (see, e.g., [14], Ch. XI, Part B, Sect. 2, Remark 8) and $a(y) \neq 0$, we obtain the following expression for Φ_*:

$$\Phi_*(y) = \frac{1}{a(y)}\left[-\tfrac{1}{2}I + \mathcal{W}_\Delta\right]^{-1}Q^+(y), \quad y \in S, \tag{14.49}$$

Now we have to prove that $f_*(y)$, $\Phi_*(y)$ given by (14.45) and (14.49) do satisfy (14.43). Indeed, the potential $W_\Delta[a\Phi_*](y)$ with $\Phi_*(y)$ given by (14.49) is a harmonic function, and one can check that Q given by (14.47) is also harmonic. Since (14.48) implies that they coincide on the boundary, the two harmonic functions should also coincide in the domain, i.e., (14.46) holds true, which implies (14.43).

Thus we constructed a bounded operator $\mathcal{C}_{\Phi} : H^{1,0}(\Omega;\Delta) \to L_2(\Omega) \times H^{1/2}(S)$ given by (14.45), (14.49), (14.47).

Lemma 3 implies that ansatz (14.38) does not cover the whole space $H^{1,0}(\Omega;\Delta)$.

14.6.2 Integro-differential Equation (G)

In this section, we eliminate the Dirichlet boundary condition to deal with only one integro-differential equation. Substituting the Dirichlet boundary condition (14.4) into (14.19) leads to the following BDIE (G) for $u \in H^{1,0}(\Omega;\Delta)$:

$$\mathcal{A}^G u := u + \mathcal{R}u - VT^+u = \mathcal{F}^G \quad \text{in } \Omega, \tag{14.50}$$

where

$$\mathcal{F}^G = F_0 = \mathcal{P}f - W\varphi_0. \tag{14.51}$$

Let us prove the equivalence of the BDIDE to the BVP (14.3)–(14.4).

Theorem 10. *Let $f \in L_2(\Omega)$, $\varphi_0 \in H^{1/2}(S)$. A function $u \in H^{1,0}(\Omega;\Delta)$ solves the mixed BVP* (14.3)–(14.4) *in Ω if and only if u solves BDIDE* (14.50) *with right-hand side* (14.51). *Such a solution exists and is unique.*

Proof. Any solution of BVP (14.3)–(14.4) solves BDIDE (14.50) due to the third Green formula (14.19).

On the other hand, if u is a solution of BDIDE (14.50), then Lemma 1 implies that u satisfies equation (14.3) and $W(\varphi_0 - u^+) = 0$ in Ω. Lemma 2(ii) then implies that the Dirichlet boundary condition (14.4) is satisfied. Thus any solution of BDIDE (14.50) satisfies BVP (14.3)–(14.4). The unique solvability of the latter is implied by Theorem 5.

The mapping properties of operators V, W, $\mathcal{P}$, and $\mathcal{R}$ imply the membership $\mathcal{F}^G \in H^{1,0}(\Omega;\Delta)$ and continuity of the operator $\mathcal{A}^G$ in $H^{1,0}(\Omega;\Delta)$, while Theorem 10 implies its injectivity.

Theorem 11. *The operator $\mathcal{A}^G$ is continuous and continuously invertible in $H^{1,0}(\Omega;\Delta)$.*

Proof. The continuity of $\mathcal{A}^G$ is already proved, and we have to prove the existence of a bounded inverse operator $(\mathcal{A}^G)^{-1}$.

Let us consider equation (14.50) with an arbitrary function $\mathcal{F}^G$ from $H^{1,0}(\Omega;\Delta)$. By Lemma 3, $\mathcal{F}^G$ can be presented as

$$\mathcal{F}^G(y) = \mathcal{P}f_*(y) - W\Phi_*(y) \quad y \in \Omega,$$

where $(f_*, \Phi_*) = \mathcal{C}_\Phi \mathcal{F}^G$ and $\mathcal{C}_\Phi$ is a bounded operator from $H^{1,0}(\Omega;\Delta)$ to $L_2(\Omega) \times H^{1/2}(S)$. Then Theorem 10 and the Corollary imply that equation (14.50) has a unique solution $u = (A^D)^{-1}(f_*, \Phi_*)^\top$, where $(A^D)^{-1}$ is a bounded operator.

14.7 Concluding Remarks

The Dirichlet problem for a variable-coefficient PDE with a right-hand side function from $L_2(\Omega)$, and with the Dirichlet data from the spaces $H^{1/2}(S)$, has been considered in the paper. It was shown that the BVP can be equivalently reduced to two direct segregated boundary-domain integral equation systems, one of them of the second kind. On the other hand, the

BVP can be equivalently reduced to a united boundary-domain integro-differential problem, or to a united boundary-domain integro-differential equation of the second kind. It was shown that the operators associated with the left-hand sides of all the four systems/problems/equations are continuous and three of them continuously invertible in the corresponding Sobolev–Slobodetski spaces.

A further analysis of spectral properties of the two second kind equations obtained in the paper is needed to decide whether the resolvent theory and the Neumann series method (see [16], [17], and references therein) are efficient for solving the equations.

By the same approach, the corresponding BDIDEs/BDIDPs for unbounded domains can be analyzed as well. The approach can also be extended to more general PDEs and to systems of PDEs, while smoothness of the variable coefficients and the boundary can be essentially relaxed.

This study can serve as a starting point for approaching BDIDEs/BDIDPs based on the *localized* parametrices, leading after discretization to sparsely populated systems of linear algebraic equations, attractive for computations [9]. This can then be extended to analysis of localized BDIDEs/BDIDPs of nonlinear problems [18].

References

1. P.A. Martin, J.D. Richardson, L.J. Gray, and J.R. Berger, On Green's function for a three-dimensional exponentially graded elastic solid, *Proc. Roy. Soc. London* **A458** (2002), 1931–1947.
2. E.E. Levi, I problemi dei valori al contorno per le equazioni lineari totalmente ellittiche alle derivate parziali, *Mem. Soc. Ital. dei Sci. XL* **16** (1909), 1–112.
3. D. Hilbert, *Grundzüge einer allgemeinen Theorie der linearen Integralgleichungen,* Teubner, Leipzig, 1912.
4. C. Miranda, *Partial Differential Equations of Elliptic Type,* 2nd ed., Springer-Verlag, Berlin-Heidelberg-New York, 1970.
5. A. Pomp, Levi functions for linear elliptic systems with variable coefficients including shell equations, *Computational Mech.* **22** (1998), 93–99.
6. A. Pomp, *The Boundary-domain Integral Method for Elliptic Systems with Applications in Shells,* Lect. Notes in Math. **1683**, Springer-Verlag, Berlin-Heidelberg, 1998.
7. O. Chkadua, S.E. Mikhailov, and D. Natroshvili, Analysis of direct boundary-domain integral equations for a mixed BVP with variable coefficient. I (submitted for publication), preprint available from http://www.gcal.ac.uk/cms/contact/staff/sergey/CMS-MAT-PP-2004-1.pdf.
8. O. Chkadua, S.E. Mikhailov, and D. Natroshvili, Analysis of direct boundary-domain integral equations for a mixed BVP with variable coefficient. II (submitted for publication), preprint available from http://www.gcal.ac.uk/cms/contact/staff/sergey/CMS-MAT–2004-12.pdf.

9. S.E. Mikhailov, Localized boundary-domain integral formulation for problems with variable coefficients, *Engng. Analysis Boundary Elements* **26** (2002), 681–690.

10. S.E. Mikhailov, Analysis of united boundary-domain integro-differential and integral equations for a mixed BVP with variable coefficient (submitted for publication), preprint available from http://www.gcal.ac.uk/ cms/contact/staff/sergey/CMS-MAT-2004-11.pdf.

11. J.-L. Lions and E. Magenes, *Non-homogeneous Boundary Value Problems and Applications, vol. 1,* Springer-Verlag, Berlin-Heidelberg-New York, 1972.

12. H. Triebel, *Interpolation Theory, Function Spaces, Differential Operators*, North-Holland, Amsterdam, 1978.

13. M. Costabel, Boundary integral operators on Lipschitz domains: elementary results, *SIAM J. Math. Anal.* **19** (1988), 613–626.

14. R. Dautray and J.-L. Lions, *Mathematical Analysis and Numerical Methods for Science and Technology, vol. 4: Integral Equations and Numerical Methods,* Springer-Verlag, Berlin-Heidelberg, 1990.

15. W. McLean, *Strongly Elliptic Systems and Boundary Integral Equations*, Cambridge Univ. Press, Cambridge, 2000.

16. S.E. Mikhailov, On an integral equation of some boundary value problems for harmonic functions in plane multiply connected domains with nonregular boundary, *Mat. Sb.* **121** (1983), 533–544. (English translation: *USSR Math.-Sb.* **49** (1984), 525–536.)

17. O. Steinbach and W.L. Wendland, On C. Neumann's method for second-order elliptic systems in domains with non-smooth boundaries, *J. Math. Anal. Appl.* **262** (2001), 733–748.

18. S.E. Mikhailov, Localized direct boundary-domain integro-differential formulations for scalar nonlinear problems with variable coefficients, *J. Engng. Math.* **51** (2005), 283–302.

15 On the Regularity of the Harmonic Green Potential in Nonsmooth Domains

Dorina Mitrea

15.1 Introduction

As is well known, the free-space Green's function for the Laplace operator $\Delta = \sum_{j=1}^{n} \partial_j^2$ in $\mathbb{R}^n$ is given by

$$\Gamma(x) := \begin{cases} \dfrac{1}{2\pi} \ln |x| & \text{if } n = 2, \\ \dfrac{c_n}{|x|^{n-2}} & \text{if } n \geq 3, \end{cases} \tag{15.1}$$

where $c_n = [(2-n)\omega_n]^{-1}$ and ω_n denotes the area of the unit sphere in $\mathbb{R}^n$. This allows one to solve the Poisson problem for the Laplacian in the whole space in integral form. More specifically, the Newtonian potential operator

$$\Pi f(x) := \int_{\mathbb{R}^n} \Gamma(x-y) f(y)\, dy, \quad x \in \mathbb{R}^n, \tag{15.2}$$

satisfies

$$u := \Pi f \quad \Longrightarrow \quad \Delta u = f \quad \text{in } \mathbb{R}^n, \tag{15.3}$$

at least if f is well behaved. The mapping properties of Π are well understood. Essentially, this is a smoothing operator of order two on the scale of L^p-based Sobolev spaces $H^{s,p}$ and this readily translates into mapping properties for the solution operator $f \mapsto u = \Pi f$ of the partial differential equation (PDE) in question.

Next, take the case when the Poisson problem is considered in a bounded domain $\Omega \subset \mathbb{R}^n$. This time, boundary conditions must be imposed and we consider the homogeneous Dirichlet boundary condition in which scenario the PDE reads

$$\begin{aligned} \Delta u &= f \quad \text{in } \Omega, \\ \operatorname{Tr} u &= 0 \quad \text{on } \partial\Omega, \end{aligned} \tag{15.4}$$

where Tr stands for the operator of trace on $\partial\Omega$. Paralleling the setup (15.2)–(15.3) corresponding to the entire Euclidean space, the solution operator for (15.4) may again be expressed in integral form, namely,

$$u(x) = \mathbf{G} f(x) := \int_{\Omega} G(x,y) f(y)\, dy, \quad x \in \Omega,$$

where $G(x,y)$ is the Green's function on Ω, i.e., a version of (15.1) suitably adapted to the domain Ω. For each fixed (pole) $x \in \Omega$, it satisfies

$$\Delta_y G(x,y) = \delta_x(y), \quad y \in \Omega,$$
$$\operatorname{Tr} G(x,\cdot) = 0 \quad \text{on } \partial\Omega,$$

where δ_x is the Dirac distribution with mass at x. The Poisson problem (15.4) is intimately connected with the Dirichlet problem for the Laplacian

$$\begin{aligned} \Delta u &= 0 \quad \text{in } \Omega, \\ \operatorname{Tr} u &= g \quad \text{on } \partial\Omega. \end{aligned} \tag{15.5}$$

In terms of the Green's function, the solution operator of (15.5) is

$$u(x) =: \mathrm{PI}(g) = \int_{\partial\Omega} \partial_{\nu_y} G(x,y) g(y)\, d\sigma_y, \quad x \in \Omega, \tag{15.6}$$

the so-called *Poisson integral*, where ν and $d\sigma$ are, respectively, the unit normal and the surface measure on $\partial\Omega$. In particular,

$$G(x,y) = \Gamma(x-y) - \mathrm{PI}\,[\operatorname{Tr}\Gamma(x-\cdot)](y), \quad x,y \in \Omega.$$

Let us also record an alternative expression for the Poisson integral operator, based on the so-called (harmonic) double-layer potential

$$\mathcal{D}\psi(x) := \int_{\partial\Omega} \partial_{\nu_y}[\Gamma(x-y)]\psi(y)\, d\sigma_y, \quad x \in \Omega,$$

and its boundary trace

$$\operatorname{Tr} \circ \mathcal{D} = \tfrac{1}{2} I + K,$$

where I is the identity, and K is the principal-value operator

$$K\psi(x) := p.v. \int_{\partial\Omega} \partial_{\nu_y}[\Gamma(x-y)]\psi(y)\, d\sigma_y, \quad \text{for a.a. } x \in \partial\Omega.$$

The identity referred to above then reads

$$\mathrm{PI} = \mathcal{D} \circ \left(\tfrac{1}{2} I + K\right)^{-1}, \tag{15.7}$$

assuming that the inverse operator exists. Returning to the mainstream discussion, recall that the Green's function satisfies

$$G(x,y) \geq 0 \quad \text{and} \quad G(x,y) = G(y,x), \quad x,y \in \Omega,\ x \neq y.$$

The mapping properties one can expect of the operator $\mathbf{G}$ are inexorably linked to the nature of the singularity in $G(x,y)$ (located on the diagonal). In turn, the nature of this singularity is, to a large extent, dictated by the smoothness of the boundary of the domain Ω. It is folklore that (if, say, $n \geq 3$) $G(x,y)$ satisfies

$$G(x,y) \leq C|x-y|^{2-n}, \quad x,y \in \Omega,$$

and

$$\partial\Omega \in C^\infty \Longrightarrow |\nabla_x G(x,y)| \leq C|x-y|^{1-n}, \quad x,y \in \Omega, \tag{15.8}$$

though (15.8) may *fail* if $\partial\Omega$ contains irregularities.

Estimates such as (15.8) are important for establishing mapping properties for the first-order derivatives of the Green operator and its adjoint, i.e.,

$$\nabla \mathbf{G} f(x) = \int_\Omega \nabla_x G(x,y) f(y)\, dy, \quad x \in \Omega, \tag{15.9}$$

$$(\nabla \mathbf{G})^* f(x) = \int_\Omega \nabla_y G(x,y) f(y)\, dy, \quad x \in \Omega. \tag{15.10}$$

One way to achieve this is by invoking the celebrated Hardy–Littlewood–Sobolev fractional integration theorem (see, e.g., [1]). Recall that if

$$I_\alpha f(x) = \int_{\mathbb{R}^n} \frac{f(y)}{|x-y|^{n-\alpha}}\, dy, \quad x \in \mathbb{R}^n, \tag{15.11}$$

stands for the usual Riesz potential operator then, for $0 < \alpha < n$,

$$I_\alpha : L^p(\mathbb{R}^n) \longrightarrow L^q(\mathbb{R}^n), \quad 1 < p < \frac{n}{\alpha}, \quad \frac{1}{q} = \frac{1}{p} - \frac{\alpha}{n}. \tag{15.12}$$

It follows from (15.9)–(15.12) that if $\partial\Omega$ is sufficiently smooth (for example, $\partial\Omega \in C^{1+\gamma}$ for some $\gamma > 0$ will do), then

$$\nabla \mathbf{G} : L^p(\Omega) \longrightarrow L^q(\Omega), \quad 1 < p < n, \quad \frac{1}{q} = \frac{1}{p} - \frac{1}{n}, \tag{15.13}$$

Observe that, by duality, (15.13) also yields

$$(\nabla \mathbf{G})^* : L^p(\Omega) \longrightarrow L^q(\Omega), \quad 1 < p < n, \quad \frac{1}{q} = \frac{1}{p} - \frac{1}{n}. \tag{15.14}$$

It is perhaps worth noting that (15.13) translates into the estimate

$$\|\nabla u\|_{L^q(\Omega)} \leq C \|f\|_{L^p(\Omega)} \qquad \text{whenever } u \text{ solves (15.4)},$$

whereas (15.14) corresponds to

$$\|u\|_{L^q(\Omega)} \le C\|\vec{f}\|_{L^p(\Omega)}$$

whenever u solves

$$\Delta u = \operatorname{div} \vec{f} \quad \text{in } \Omega,$$
$$\operatorname{Tr} u = 0 \quad \text{on } \partial\Omega.$$

The extent to which the classical estimate (15.13) continues to hold for domains with *Lipschitz* boundaries has been studied by Dahlberg in [2]. Recall that a Lipschitz domain is a domain whose boundary is locally given by graphs of Lipschitz functions. Dahlberg proved that for any bounded Lipschitz domain $\Omega \subset \mathbb{R}^n$, there exists $\varepsilon = \varepsilon(\Omega) > 0$ such that (15.13) continues to hold if

$$1 < p < p_n + \varepsilon, \quad 1/q = 1/p - 1/n, \quad p_n = \begin{cases} 4/3, & n = 2, \\ 3n(n+3)^{-1}, & n \ge 3. \end{cases}$$

Dahlberg also proved that the choice of p_n is sharp in the class of Lipschitz domains. The moral is that $\nabla\mathbf{G}$ continues to behave like a fractional integration operator of order 1 but only for a more restricted range of indices than what the classical theory would warrant.

In this paper we provide a new proof of Dahlberg's result as well as an extension. Our approach is conceptually different from the one in [2] and it has the advantage of being adaptable to Neumann boundary conditions and even systems of PDEs. To state the more general problem we study here we need one more piece of notation. Throughout the paper, $H^{s,p}(\Omega)$ will stand for the usual scale of Sobolev spaces on Ω of smoothness $s \in \mathbb{R}$ and integrability $p \in (1, \infty)$. The issue we have in mind is the boundedness of the Green potential operator in the context of Sobolev spaces

$$\mathbf{G} : H^{-s,p}(\Omega) \to H^{1-s,q}(\Omega), \quad 1 < p < n, \quad \frac{1}{q} = \frac{1}{p} - \frac{1}{n}, \tag{15.15}$$

for a bounded Lipschitz domain Ω in $\mathbb{R}^n$. The question is to determine the range of indices p, s for which the operator (15.15) is bounded. In this new formulation, the problem solved by Dahlberg corresponds to taking $s = 0$ in (15.15).

Let us describe in greater detail the main idea behind our approach. The departure point is an identity relating the operators $\mathbf{G}$ and PI to the effect that

$$\mathbf{G}f = \Pi(\tilde{f})\big|_{\Omega} - \mathrm{PI}\big[\operatorname{Tr}\Pi(\tilde{f})\big], \tag{15.16}$$

where $\tilde{f}$ is a compactly supported function in $\mathbb{R}^n$ such that $\tilde{f}\,|_{\Omega} = f$. As discussed in the first part of this section, the mapping properties of the assignment $f \mapsto \Pi(\tilde{f})$ are well understood. Regarding the second operator appearing in (15.16), we intend to employ the factorization implicit in the diagram

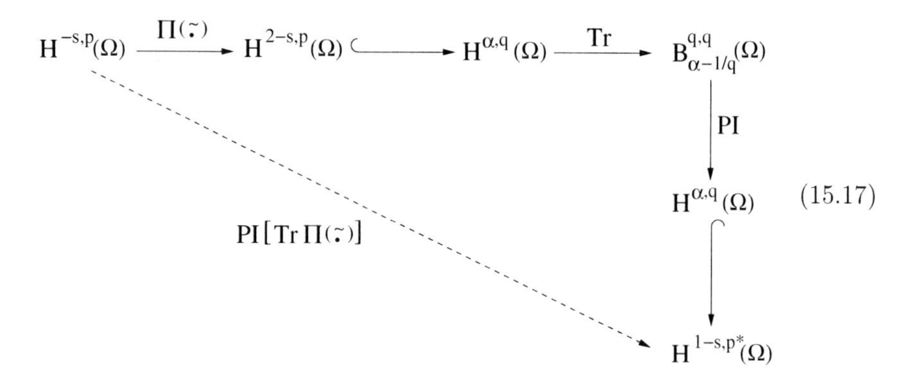

(15.17)

where the hook-arrows denote embeddings and $\frac{1}{p^*} = \frac{1}{p} - \frac{1}{n}$. In order to be able to successfully carry out this program, we must identify all pairs of indices s, p for which there exist α, q such that all operators in (15.17) are well defined and bounded. The embedding results are, of course, well understood. The trace operator (for Lipschitz domains) has been treated in [3]. Finally, the mapping properties of the Poisson integral are handled based on the identity (15.7). For this segment in our analysis we rely on the results proved in [4] and [5] for layer potentials acting on Sobolev–Besov spaces.

The plan of the paper is as follows. In Section 15.2 we state the main result pertaining to the boundedness of (15.15) and in Section 15.3 we review a number of preliminary results which are useful for us. Section 15.4 is devoted to the proof of our main result.

15.2 Statement of the Main Result

As mentioned before, we denote by $H^{s,p}$ the usual scale of Sobolev spaces of smoothness $s \in \mathbb{R}$ and integrability $p \in (1, \infty)$. These are defined in $\mathbb{R}^n$ with the help of the Fourier transform and on Ω via restriction. Recall that the Sobolev space with a negative smoothness index can also be defined by duality. In particular, $H^{-s,p}(\Omega) = (H_0^{s,p'}(\Omega))^*$, where $\frac{1}{p'} = 1 - \frac{1}{p}$ and $H_0^{s,p}(\Omega)$ is the space of distributions in $H^{s,p}(\mathbb{R}^n)$ with support contained in $\overline{\Omega}$ (see [6]). In order to state the main result of our paper we need to define a family of regions $\mathcal{Q}(n, \varepsilon)$ in $\mathbb{R}^2$, depending on $n \in \mathbb{N}$, $n \geq 2$, and a parameter $\varepsilon > 0$. These are as follows.

The region $\mathcal{Q}(2, \varepsilon)$ consists of points of the form $(s, \frac{1}{p})$ in $\mathbb{R}^2$ whose coordinates satisfy the inequalities

$$\begin{gathered}
-\tfrac{1}{2} < s \leq -\tfrac{1}{2} + \varepsilon, \quad \tfrac{1}{2} - s < \tfrac{1}{p} < 1, \\
-\tfrac{1}{2} + \varepsilon < s \leq \tfrac{1}{2} - \varepsilon, \quad -\tfrac{s}{2} + \tfrac{3}{4} - \tfrac{\varepsilon}{2} < \tfrac{1}{p} < 1, \\
\tfrac{1}{2} - \varepsilon < s \leq \tfrac{1}{2} + \varepsilon, \quad \tfrac{1}{2} < \tfrac{1}{p} < 1, \\
\tfrac{1}{2} + \varepsilon < s < 1, \quad \tfrac{1}{2} < \tfrac{1}{p} < -\tfrac{s}{2} + \tfrac{5}{4} + \tfrac{\varepsilon}{2}.
\end{gathered}$$

The region $\mathcal{Q}(2,\varepsilon)$ is the shaded region in Fig. 1. The quadrilateral with vertices $(-\frac{1}{2},1)$, $(0,\frac{1}{2})$, $(1,\frac{1}{2})$, and $(1,1)$ corresponds to the case $\varepsilon = 1/2$.

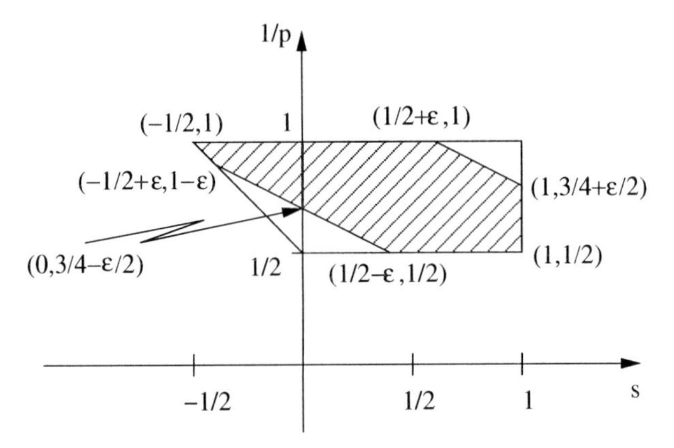

Fig. 1. The region $Q(2,\varepsilon)$.

When $n \geq 3$, the region $\mathcal{Q}(n,\varepsilon)$ consists of points $(s,\frac{1}{p})$ in $\mathbb{R}^2$ whose coordinates satisfy the inequalities

$$\begin{cases} -1+\frac{1}{n} < s \leq -\frac{1}{2}+\frac{\varepsilon}{2}, & -s+\frac{1}{n} < \frac{1}{p} < 1, \\ -\frac{1}{2}+\frac{\varepsilon}{2} < s \leq 0, & -\frac{s}{3}+\frac{1}{n}+\frac{1}{3}-\frac{\varepsilon}{3} < \frac{1}{p} < 1, \\ 0 < s < \varepsilon, & \max\left\{\frac{1}{n}, -\frac{s}{n}+\frac{1}{n}+\frac{1}{3}-\frac{\varepsilon}{3}\right\} < \frac{1}{p} < 1, \\ \varepsilon < s < 1, & \max\left\{\frac{1}{n}, -\frac{s}{n}+\frac{1}{n}+\frac{1}{3}-\frac{\varepsilon}{3}\right\} < \frac{1}{p} < -\frac{s}{3}+1+\frac{\varepsilon}{3}. \end{cases}$$

The region $\mathcal{Q}(n,\varepsilon)$ is the shaded region in Fig. 2. The quadrilateral with vertices $(-1+\frac{1}{n},1)$, $(0,\frac{1}{n})$, $(1,\frac{1}{n})$, and $(1,1)$ corresponds to the case when $\varepsilon = 1$.

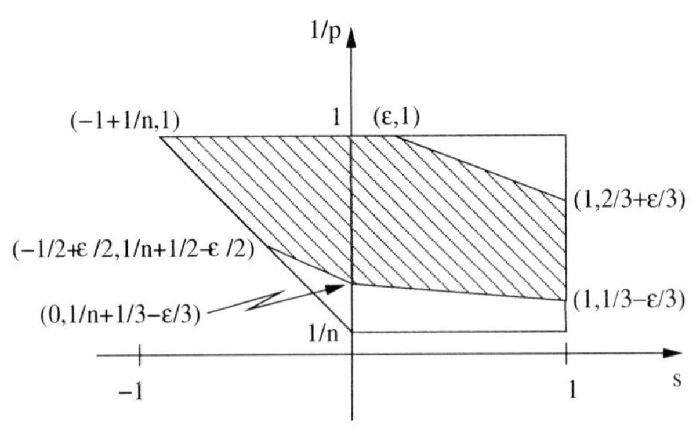

Fig. 2. The region $Q(n,\varepsilon)$.

We are now ready to state our main theorem.

Theorem 1. *For each Ω bounded Lipschitz domain in $\mathbb{R}^n$, $n \geq 2$, with connected boundary, there exists $\varepsilon = \varepsilon(\Omega) > 0$, $\varepsilon \in (0, \frac{1}{2}]$ if $n = 2$ and $\varepsilon \in (0, 1]$ if $n = 3$, such that the operator*

$$\mathbf{G} : H^{-s,p}(\Omega) \to H^{1-s,p^*}(\Omega) \tag{15.18}$$

is well defined, linear, and bounded whenever

$$\left(s, \frac{1}{p}\right) \in \mathcal{Q}(n, \varepsilon) \quad \text{and} \quad \frac{1}{p^*} = \frac{1}{p} - \frac{1}{n}.$$

Moreover, when $\partial\Omega \in C^1$ one can take $\varepsilon = \frac{1}{2}$ if $n = 2$ and $\varepsilon = 1$ if $n \geq 3$.

Remark 1. If we set $s = 0$ in (15.18) we recover the $L^p - L^q$ estimates proved by Dahlberg in [2] (see the discussion in the Introduction).

Remark 2. A result similar in concept is valid for the case of (homogeneous) Neumann boundary conditions.

The proof of Theorem 1 is carried out in Section 15.4.

15.3 Prerequisites

In this section we recall some basic definitions and summarize a number of technical prerequisites which will be used in what follows. We shall retain the notation and the conventions introduced so far.

Fix a bounded Lipschitz domain $\Omega \subset \mathbb{R}^n$ and let $L^p(\partial\Omega)$ stand for the Lebesgue space of measurable, pth power integrable functions on $\partial\Omega$. Also, let $L^p_1(\partial\Omega)$ denote the Sobolev space of functions from $L^p(\partial\Omega)$ whose tangential gradient has L^p components.

Next, for $0 < s < 1$, $1 < p, q < \infty$, denote by $B^{p,q}_s(\partial\Omega)$ the usual scale of Besov spaces over $\partial\Omega$ with smoothness s. By real interpolation,

$$(L^p(\partial\Omega), L^p_1(\partial\Omega))_{\theta,q} = B^{p,q}_\theta(\partial\Omega), \quad 0 < \theta < 1, \quad 1 < p, q < \infty.$$

Besov spaces on $\partial\Omega$ with a negative smoothness index can be introduced via duality. Specifically, if $0 < s < 1$, $1 < p, q < \infty$, $1/p + 1/p' = 1$, $1/q + 1/q' = 1$, then we set

$$B^{p,q}_{-s}(\partial\Omega) := \left(B^{p',q'}_s(\partial\Omega)\right)^*.$$

It is well known that the trace operator

$$\text{Tr} : H^{s,p}(\Omega) \longrightarrow B^{p,p}_{s-\frac{1}{p}}(\partial\Omega), \quad 1 < p < \infty, \ \ \tfrac{1}{p} < s < \tfrac{1}{p} + 1,$$

is well defined and bounded. See, e.g., [6] and [7] for references and more details.

For $\varepsilon > 0$, we define the following two regions in $\mathbb{R}^2$:

$$\mathcal{R}^1_\varepsilon := \text{the set of points inside the hexagon with vertices at}$$
$$(0,0),\ (\tfrac{1}{2}+\varepsilon,0),\ (1,\tfrac{1}{2}-\varepsilon),\ (1,1),\ (\tfrac{1}{2}-\varepsilon,1),\ (0,\tfrac{1}{2}+\varepsilon),$$
$$\mathcal{R}^2_\varepsilon := \text{the set of points inside the hexagon with vertices at}$$
$$(0,0),\ (\varepsilon,0),\ (1,\tfrac{1}{2}-\tfrac{1}{2}\varepsilon),\ (1,1),\ (1-\varepsilon,1),\ (0,\tfrac{1}{2}+\tfrac{1}{2}\varepsilon).$$

Recall the Newtonian potential from (15.2) and the Poisson integral operator (15.6).

Theorem 2. *For each bounded Lipschitz domain Ω in $\mathbb{R}^n$, $n \geq 2$, whose boundary is connected, the following hold.*

(1) The operator

$$\Pi : H_0^{-s,p}(\Omega) \to H^{2-s,p}(\Omega)$$

is bounded for each $1 < p < \infty$, $0 \leq s \leq 2$.

(2) There exists $\varepsilon = \varepsilon(\partial\Omega) \in (0, \frac{1}{2}]$ for $n = 2$ and $\varepsilon = \varepsilon(\partial\Omega) \in (0,1]$ for $n \geq 3$ such that the operator

$$\mathrm{PI} : B^{p,p}_s(\partial\Omega) \longrightarrow H^{s+\frac{1}{p},p}(\Omega)$$

is well defined and bounded whenever $(s, \frac{1}{p}) \in \mathcal{R}^1_\varepsilon$ for $n = 2$, and whenever $(s, \frac{1}{p}) \in \mathcal{R}^2_\varepsilon$ for $n \geq 3$.

Proof. The fact that the Newtonian potential is a bounded operator in the context of (15.1) follows from classical Calderón–Zygmund theory, duality and interpolation. The fact that the operator in (15.2) is well defined and bounded has been proved in [5] for $n = 2$ and in [4] for $n \geq 3$, based on the integral representation (15.7).

15.4 Proof of Theorem 1

We start with $f \in H^{-s,p}(\Omega)$, for some $-1 \leq s \leq 1$, and $1 < p < n$. Then, the solution to (15.4) can be formally written as

$$u = \Pi(\tilde{f})\,|_\Omega - \mathrm{PI}\left[\mathrm{Tr}\,\Pi(\tilde{f})\right], \tag{15.19}$$

where $\tilde{f} \in H^{-s,p}(\mathbb{R}^n)$, has compact support, and satisfies $\tilde{f}|_\Omega = f$.

The goal is to show that there exists $\varepsilon > 0$ such that the right-hand side of (15.19) is well defined and contained in $H^{1-s,p^*}(\Omega)$ for all $(s, \frac{1}{p}) \in \mathcal{Q}(n,\varepsilon)$, where $\frac{1}{p^*} = \frac{1}{p} - \frac{1}{n}$. From Theorem 2 it follows that $\Pi(\tilde{f})\,|_\Omega \in H^{2-s,p}(\Omega)$ for all indices s and p as above.

The task at hand is to inspect (15.17) and analyze each operator in this diagram. The first horizontal arrow (essentially the Newtonian potential) has already been discussed. The second horizontal arrow in the diagram (an inclusion) is well defined provided

$$2 - s \geq \alpha \quad \text{and} \quad \tfrac{1}{q} - \tfrac{\alpha}{n} = \tfrac{1}{p} - \tfrac{2-s}{n}. \tag{15.20}$$

The third arrow (a trace operator) is well defined whenever

$$\tfrac{1}{q} < \alpha < 1 + \tfrac{1}{q}. \tag{15.21}$$

The first vertical arrow is the operator PI. By Theorem 2, there is $\varepsilon > 0$ such that this is well defined and bounded between the given spaces if

$$\left(\alpha - \tfrac{1}{q}, \tfrac{1}{q}\right) \in \mathcal{R}_\varepsilon^1 \quad \text{for } n = 2, \tag{15.22}$$

$$\left(\alpha - \tfrac{1}{q}, \tfrac{1}{q}\right) \in \mathcal{R}_\varepsilon^2 \quad \text{for } n \geq 3. \tag{15.23}$$

Finally, the last vertical arrow (an inclusion) is well defined provided

$$\alpha \geq 1 - s \quad \text{and} \quad \tfrac{1}{p^*} - \tfrac{1-s}{n} = \tfrac{1}{q} - \tfrac{\alpha}{n}. \tag{15.24}$$

Let us combine (15.20)–(15.24) above, for $\varepsilon > 0$ as in Theorem 2. For each pair $(s, \frac{1}{p})$ with $-1 \leq s \leq 1$, $1 < p < n$, we set $t := 2 - s > 0$ (this is done in order to simplify the exposition that follows). Then we should have

$$1 \leq t \leq 3, \quad t - 1 \leq \alpha \leq t, \tag{15.25}$$

where the second set of inequalities in (15.25) is equivalent to the first conditions in (15.20) and (15.24). The last condition in (15.24) is a consequence of (15.20) and the definition of p^*.

The condition (15.21) is equivalent with $(\alpha, \frac{1}{q})$ belonging to the region between the lines $y = x$ and $y = x - 1$, while based on Theorem 2, the conditions (15.22)–(15.23) hold if and only if $(\alpha, \frac{1}{q})$ belongs to the hexagon $\mathcal{H}_\varepsilon^1$ for $n = 2$ and to the hexagon $\mathcal{H}_\varepsilon^2$ for $n \geq 3$ (see Fig. 3 and 4).

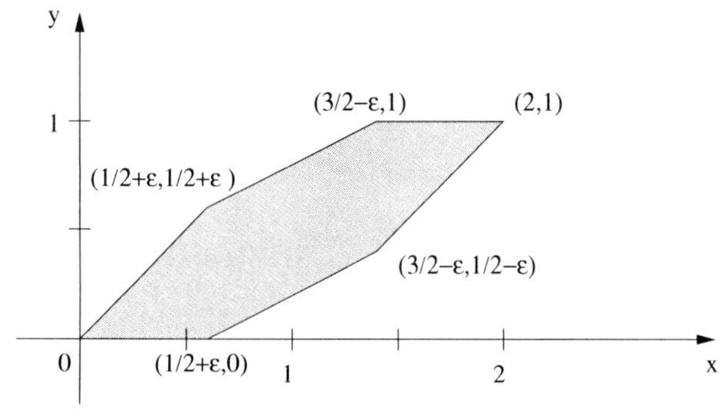

Fig. 3. The region $\mathcal{H}_\varepsilon^1$.

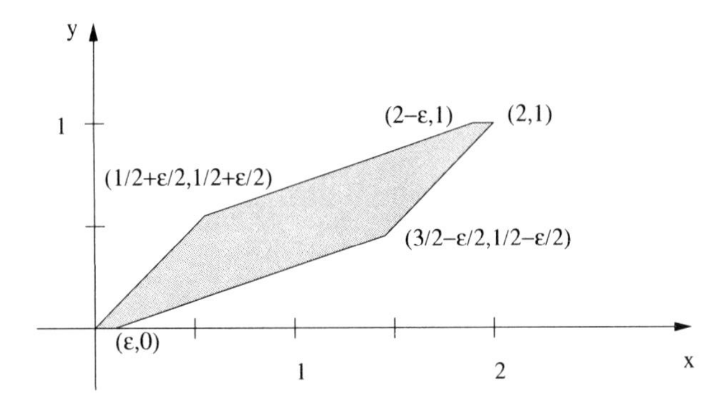

Fig. 4. The region $\mathcal{H}^2_\varepsilon$.

The last condition in (15.20) is equivalent to the requirement that

$$\left(\alpha, \tfrac{1}{q}\right) \text{ belongs to the line through the point } \left(t, \tfrac{1}{p}\right) \text{ and of slope } \tfrac{1}{n}.$$

In particular, if for a fixed $(t, \frac{1}{p})$, we select a point $(\alpha, \frac{1}{q})$ on the line of slope $\frac{1}{n}$ and passing through the point $(t, \frac{1}{p})$, which lies to the left of $(t, \frac{1}{p})$, then the conditions in (15.20) are satisfied.

Case 1: $n = 2$.

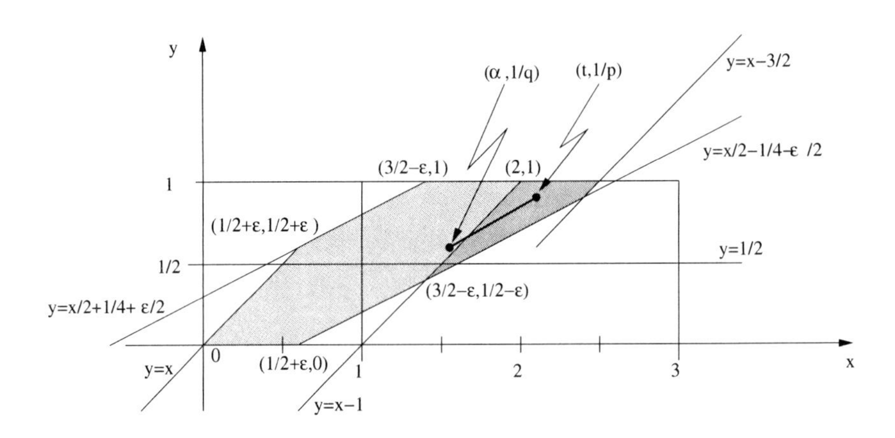

Fig. 5.

If we take a point $(t, \frac{1}{p})$ inside the region bounded by the lines $y = 1$, $y = x - 1$, $y = \frac{1}{2}$, $y = \frac{1}{2}x - \frac{1}{4} - \frac{\varepsilon}{2}$, and $y = x - \frac{3}{2}$ (see Fig. 5), then clearly $1 < p < 2$, $1 \leq t \leq 3$. Also, there exists a point $(\alpha, \frac{1}{q})$ on the line of slope $\frac{1}{2}$ through $(t, \frac{1}{p})$ such that $t-1 \leq \alpha \leq t$ and $(\alpha, \frac{1}{q}) \in \mathcal{H}^1_\varepsilon$. The same is true for the points $(t, \frac{1}{p})$ on the line $y = x-1$ having $\frac{1}{2} < \frac{1}{p} < 1$. Hence, for all these points $(t, \frac{1}{p})$ we can find $(\alpha, \frac{1}{q})$ satisfying (15.20)–(15.24) with $s = 2 - t$. On the other hand, if $(t, \frac{1}{p})$ is taken inside the region bounded by the lines $y = 1$, $y = \frac{1}{2}x + \frac{1}{4} + \frac{\varepsilon}{2}$, $x = 1$, $y = \frac{1}{2}$, and $y = x - 1$, we can take $\alpha = t$ and $q = p$ and again (15.20)–(15.24) are verified when $s = 2 - t$, and hence the solution (15.19) is in $H^{1-s,p^*}(\Omega)$. Now if we apply the transformation $(x, y) \mapsto (2 - x, y)$ to the region bounded by the lines $y = 1$, $y = \frac{1}{2}x + \frac{1}{4} + \frac{\varepsilon}{2}$, $x = 1$, $y = \frac{1}{2}$, $y = \frac{1}{2}x - \frac{1}{4} - \frac{\varepsilon}{2}$, and $y = x - \frac{3}{2}$, the resulting region is $\mathcal{Q}(2, \varepsilon)$. What we have proved is that for $(s, \frac{1}{p}) \in \mathcal{Q}(2, \varepsilon)$, the operator (15.18) is bounded.

Case 2: $n \geq 3$.

The main ideas of the proof are similar to those in the case $n = 2$. This time we have to analyze the region shown in Fig. 6.

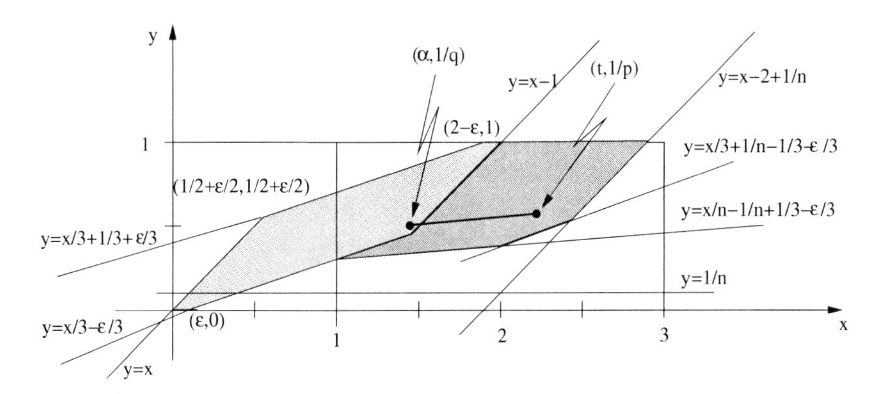

Fig. 6.

Whereas the lines $y = \frac{1}{n}$ and $y = \frac{1}{n}x - \frac{1}{n} + \frac{1}{3} - \frac{\varepsilon}{3}$ do not intersect in our Fig. 6, that might change depending on the values of n and ε. If we take a point $(t, \frac{1}{p})$ inside the region bounded by the lines $y = 1$, $y = x - 1$, $y = \frac{1}{3}x - \frac{\varepsilon}{3}$, $y = \frac{1}{n}x - \frac{1}{n} + \frac{1}{3} - \frac{\varepsilon}{3}$, $y = \frac{1}{3}x + \frac{1}{n} - \frac{1}{3} - \frac{\varepsilon}{3}$, and $y = x - 2 + \frac{1}{n}$, with $1 < p < n$ then clearly $1 \leq t \leq 3$. Also, there exists a point $(\alpha, \frac{1}{q})$ on the line of slope $\frac{1}{n}$ passing through $(t, \frac{1}{p})$ such that $t - 1 \leq \alpha \leq t$ and $(\alpha, \frac{1}{q}) \in \mathcal{H}^2_\varepsilon$. The same is true for the points $(t, \frac{1}{p})$ on the line $y = x - 1$ having $\frac{1}{2} - \frac{\varepsilon}{2} < \frac{1}{p} < 1$ and for the points $(t, \frac{1}{p})$ on the line $y = \frac{1}{3}x - \frac{\varepsilon}{3}$ having $\frac{1}{3} - \frac{\varepsilon}{3} < \frac{1}{p} \leq \frac{1}{2} - \frac{\varepsilon}{2}$. Hence, for all these points $(t, \frac{1}{p})$ we can find $(\alpha, \frac{1}{q})$ satisfying (15.20)–(15.24) with $s = 2 - t$. On the other hand, if $(t, \frac{1}{p})$ is taken inside the region bounded by the lines $y = 1$, $y = \frac{1}{3}x + \frac{1}{3} + \frac{\varepsilon}{3}$, $x = 1$, $y = \frac{1}{3}x - \frac{\varepsilon}{3}$, and $y = x - 1$, we can take $\alpha = t$ and $q = p$ and again (15.20)–(15.24) are verified when $s = 2 - t$ and, hence, the solution (15.19) is in $H^{1-s,p^*}(\Omega)$. Applying the transformation $(x, y) \mapsto (2 - x, y)$ to the region bounded by the lines $y = 1$, $y = \frac{1}{3}x + \frac{1}{3} + \frac{\varepsilon}{3}$, $x = 1$, $y = \frac{1}{n}x - \frac{1}{n} + \frac{1}{3} - \frac{\varepsilon}{3}$,

$y = \frac{1}{3}x + \frac{1}{n} - \frac{1}{3} - \frac{\varepsilon}{3}$, and $y = x - 2 + \frac{1}{n}$, the resulting region is $\mathcal{Q}(n, \varepsilon)$. What we have proved is that for $(s, \frac{1}{p}) \in \mathcal{Q}(n, \varepsilon)$, the operator (15.18) is bounded.

References

1. E. Stein, *Singular Integrals and Differentiability Properties of Functions,* Princeton Math. Ser. **30**, Princeton Univ. Press, Princeton, NJ, 1970.
2. B. Dahlberg, L^q-estimates for Green potentials in Lipschitz domains, *Math. Scand.* **44** (1979), 149–170.
3. A. Jonsson and H. Wallin, Function spaces on subsets of R^n, *Math. Reports* **2** (1984), no. 1.
4. E. Fabes, O. Mendez, and M. Mitrea, Boundary layers on Sobolev-Besov spaces and Poisson's equation for the Laplacian in Lipschitz domains, *J. Functional Anal.* **159** (1998), 323–368.
5. D. Mitrea, Layer potentials and Hodge decompositions in two-dimensional Lipschitz domains, *Math. Ann.* **322** (2002), 75–101.
6. D. Jerison and C.E. Kenig, The inhomogeneous Dirichlet problem in Lipschitz domains, *J. Functional Anal.* **130** (1995), 161–219.
7. T. Runst and W. Sickel, *Sobolev Spaces of Fractional Order, Nemytskij Operators, and Nonlinear Partial Differential Operators,* de Gruyter, Berlin-New York, 1996.

16 Applications of Wavelets and Kernel Methods in Inverse Problems

Zuhair Nashed

16.1 Introduction and Perspectives

In many areas of applied sciences, engineering, and technology there are three problems dealing with data and signals: (i) data compression, (ii) signal representations, and (iii) recovery of signals from partial or indirect information about the signals, often contaminated by noise. Major advances in these problems have been achieved in recent years where wavelets, multiresolution analysis, and kernel methods have played key roles. We consider problem (iii) and provide glimpses of specific contributions to inverse and ill-posed problems.

Inverse problems deal with determining, for a given input-output system, an input that produces an observed output, or determining an input that produces a desired output. In terms of an operator T acting between, say, two normed spaces X and Y, the problem of solving the equation $T(x) = y$ for given data $y \in Y$ is a canonical example of an inverse problem. Typically, inverse problems are *ill posed.* Important examples of ill-posed inverse problems include integral equations of the first kind, tomography, and inverse scattering. *Signal analysis/processing* deals with digital representations of signals and their analog reconstructions from digital representations. Its principal tools are sampling expansions, filters, transforms, and, more recently, wavelets. *Image analysis* deals with problems such as image recovery, enhancement, feature extraction, and motion detection. *Medical imaging* is an important branch of *image science* and deals with image analysis in medical applications.

Two or three decades ago, the common thread between inverse problems, signal processing, and imaging problems was rather tenuous; these fields appeared dissimilar in their point of view, scope, and intent. Researchers in one of these fields were often unfamiliar with the other two areas. This situation has changed drastically in recent years, and there is now significant interaction among these fields. This is partly due to some of the unifying common concepts and tools that are used in these fields (such as expansion theorems, wavelets, certain function spaces, etc.) and the impact of functional analysis, harmonic analysis, and computational mathematics on recent developments. Some perspectives on these interactions can be found, for example, in [12], [18], and related references cited therein, which also contain some papers on applications of wavelets in these areas.

The common thread among the areas of inverse problems, signal analysis, and image analysis is a canonical problem: recovery of an object (function, signal, picture) from partial or indirect information about the object (often contaminated by noise). Both inverse problems and imaging science have emerged in recent years as interdisciplinary research fields with profound applications in many areas of science, engineering, technology, and medicine. Research in inverse problems and image processing has rich interactions with several areas of mathematics, and strong links to signal processing, variational problems, applied harmonic analysis, and computational mathematics.

Series expansions and integral representations of functions and operators play a fundamental role in the analysis of *direct problems* of applied mathematics—witness the role of Fourier series, power series, and eigenfunction expansions associated with self-adjoint linear operators, and the role of integral representations in potential theory, boundary value problems, complex analysis, and other areas.

Expansion methods also play a useful role in certain *inverse problems* (see, for example, [17], on which this exposition is partly based). Two important problems are (i) the recovery of a function from inner products (moments), and (ii) the recovery of a function from its values on a subset of its domain. We consider as examples wavelet sampling solutions of ill-posed problems, and wavelet solutions of moment problems. For perspectives and applications on (i), see [15], [30], [33], [35], and [36]. For perspectives and various results on (ii), see [4], [5], [10], [13], [22], [28], [29], and [34].

To motivate our presentation, we first consider two simple recovery problems for a function f belonging to a function space V of dimension n. Let $\{u_1, u_2, \ldots, u_n\}$ be a basis for V. Then f has the unique representation $f(x) = \sum_{i=1}^{n} c_i u_i(x)$. The form of the coefficients c_i and the ease of their computation depend on the type of *recovery problem* considered, and the choice of a basis that would be most convenient for computation.

(i) Recovery of a function from inner products (moments). If we know $\alpha_j := \langle f, u_j \rangle$, then the c_i are the solutions of the system

$$\sum_{i=1}^{n} c_i \langle u_i, u_j \rangle = \alpha_j, \quad j = 1, \ldots, n.$$

In particular, if $\{u_1, \ldots, u_n\}$ is an orthonormal basis, then $c_j = \langle f, u_j \rangle$, and so we have the Fourier expansion

$$f(x) = \sum_{j=1}^{n} \langle f, u_j \rangle \, u_j(x). \tag{16.1}$$

(ii) Recovery of a function from its values on a subset of its domain. If we know $f(t_j), j = 1, \ldots, n$, then the c_i are uniquely determined from the system

$$\sum_{i=1}^{n} c_i u_i(t_j) = f(t_j), \quad j = 1, \ldots, n.$$

Then we obtain the expansion

$$f(t) = \sum_{i=1}^{n} f(t_i) u_i(t), \tag{16.2}$$

which may be viewed as a Fourier-type expansion in terms of the *discrete* orthonormal sequence $\{u_i(t_j)\}_{i,j=1}^n$. This is a finite-dimensional version of a *sampling expansion* in signal processing.

Both forms, (16.1) and (16.2), are special cases of the expansion

$$f(t) = \sum_{i=1}^{n} l_i(f) u_i(t),$$

where the l_i are (continuous) linear functionals. The extensions to infinite-dimensional problems lead to Fourier series and Shannon's sampling theorem, as examples.

The problem of reconstruction (recovery) of a function in an infinite-dimensional space (such as a finite-energy signal) from *partial* or *indirect* information about the function arises in many areas of applied mathematics, and can be viewed as an *inverse* problem. Examples of recovery problems from partial information include

(a) the *sampling problem:* given $f(t_n)$, $n \in N$, construct f;
(b) the *extrapolation problem:* given $f(t)$ for $t \in T$, extrapolate f to a larger domain;
(c) given $f(t_n)$ and $f'(t_n)$ for a set of points $\{t_n\}$, construct f.

Clearly, these problems do not have a solution in general. They are classical problems in signal analysis and approximation theory.

We consider the role of a *reproducing kernel Hilbert space* (RKHS) in such problems. We recall that a Hilbert space H of functions f on an interval T is said to be an RKHS if all the evaluation functionals $E_t(f) := f(t)$, $f \in H$, for each fixed $t \in T$, are continuous. Then, by the Riesz representation theorem, for each $t \in T$ there exists a unique element $k_t \in H$ such that

$$f(t) = \langle f, k_t \rangle, \quad f \in H. \tag{16.3}$$

Let $k(t, s) := \langle k_s, k_t \rangle$ for $s, t \in T$. Then $k_s(t) = k(t, s)$, which is called the *reproducing kernel* of H. It is clear that every finite-dimensional function space V is an RKHS with reproducing kernel $k(s,t) = \sum_{i=1}^n u_i(s)\overline{u_i(t)}$, where $\{u_i\}_{i=1}^n$ is any orthonormal basis for V.

If H is a separable Hilbert space with reproducing kernel $k(t, s)$ and orthonormal basis $\{u_n\}_{n=1}^\infty$, then

$$k(t, s) = \sum_{i=1}^{n} u_i(t)\overline{u_i}(s),$$

the series converging absolutely. Note that $L_2(T)$ is not an RKHS, while the subspace of all functions in $L_2(T)$ which are absolutely continuous and whose derivatives are in $L_2(T)$ is an RKHS under the appropriate inner product. $L_2(T)$ contains many interesting subspaces that are RKHS. Basic properties of RKHS related to applications in this presentation can be found in [1], [11], [19], and [24].

16.2 Sampling Solutions of Integral Equations of the First Kind

Fredholm integral equations of the first kind

$$(\mathbf{K}x)(t) = \int_a^b K(t,s)x(s)ds = y(t) \tag{16.4}$$

are notoriously difficult to solve. The reason is that in most cases, solving the equation is an *ill-posed* problem. It may not have a solution, and even if it does, the solution does not depend continuously on the data.

The difficulty is compounded if the data $y(t)$ are not known exactly but only approximately for a discrete set of t, and, moreover, are contaminated by noise. That is, instead of (16.4) we have to solve

$$y_i = \int_a^b K(t_i,s)x(s)ds + \epsilon_i, \quad i = 1,2,\ldots,n, \tag{16.5}$$

where ϵ_i is a random variable of noise.

A number of approaches are available (see, for example, [3], [8], [9], [14], [16], and [19]), the most popular of which is regularization (see [8], [9], [20], and [27]). However, most of them seem to suffer from either slow convergence or a tedious asymptotic analysis to determine an optimal value for a regularization (or related) parameter. There is a technique, due principally to Stenger (see [25] and [26]), which is used to solve certain integral equations numerically and whose rate of convergence is exponential. This technique uses cardinal series and has been applied to a number of different integral operators, most of which, however, are not compact.

Fortunately, a number of new results have appeared recently (such as [2], [22], [30], and [34]) which will enable us to use signal analysis techniques for a wider class of problems. These may be coupled with an older approach involving problems such as (16.4) and (16.5) in a reproducing kernel Hilbert space setting (see [19], [20], [21], and [31]).

In this section we describe an approach to the solutions of inverse and ill-posed problems considered by G.G. Walter and the author. The approach uses general sampling theorems in RKHS developed in [22].

Rather than attempting to solve (16.5), we use the standard engineering approach and replace the discrete values $\{y_i\}$ by the impulse train

$$y^*(t) = \sum_{i=1}^{n} y_i \delta(t-i), \tag{16.6}$$

where $\delta(t-i)$ is the Dirac delta "function" (impulse) at $t=i$. Both the domain and range of the integral operator in (16.4) are assumed to be RKHSs with reproducing kernels P and Q, respectively, which we denote by H_P and H_Q. (This assumption is satisfied in many problems of interest.) We also suppose that H_Q is a closed subspace of the Sobolev space H^{-1} (so we can relate y to y^* as in [22]).

Our proposed procedure is as follows:

(i) project the data y^* into the range space H_Q;
(ii) approximate the projected function $\mathbf{Q}y^*$ by a *sampling* series;
(iii) solve the problem for the *partial* sums of the sampling series with a sampling series in H_P;
(iv) find the resulting error and establish its order.

That (i) and (ii) are possible is shown in [22]. Step (iii) is feasible if the reproducing kernel has isolated zeros. Then, as shown in [22], $\{Q(t_n,t)\}$ and $\{P(s_n,s)\}$ are orthogonal sequences.

The integral operator $\mathbf{K}$ in (16.4) maps $P(s_n,s)$ into $K(t,s_n)$. If we expand x with respect to $\{P(s_n,s)\}$, we obtain

$$x(s)=\sum_n x(s_n)\frac{P(s_n,s)}{P(s_n,s_n)}$$

and, similarly,

$$y(t)=\sum_n y(t_n)\frac{Q(t_n,t)}{Q(t_n,t_n)}.$$

Hence, we must solve

$$\sum_n y(t_n)\frac{Q(t_n,t)}{Q(t_n,t_n)}=(\mathbf{K}x)(t)=\sum_n x(s_n)\frac{K(t,s_n)}{K(s_n,s_n)}.$$

But at $t=t_m$ we have

$$y(t_m)=\sum_n x(s_n)\frac{K(t_m,s_n)}{P(s_n,s_n)},$$

so $x(s_n)$ may be obtained by solving a linear system whose matrix is

$$\left[\frac{K(t_m,s_n)}{P(s_n,s_n)}\right].$$

The error calculations may be quite complex. However, some types of errors have been considered in [22]. The procedure can also be extended to treat (more abstract) operator equations acting between two RKHSs, and to determine numerically least-squares solutions when y is not in the range of the operator, using the general framework of [14], [19], [20], and [21].

In summary, the proposed procedure is to attach a semi-discrete (or moment discretization) version (16.5) of (16.4) by using the associated impulse train (16.6). This is projected onto the RKHS H_Q, and is then approximated by the partial sums of a sampling series. The problem is finally solved for these partial sums and the error found and compared to other methods.

16.3 Wavelet Sampling Solutions of Integral Equations of the First Kind

We conclude our presentation by considering wavelet sampling solutions of (16.4)–(16.5). Associated with each RKHS $V_0 \subset L^2(\mathbb{R})$ there is a sequence $\{V_n\}$ of *dilation spaces*

$$V_{i+1} = \{f \in L^2(\mathbb{R}) : f(x/2) \in V_i\}, \quad i = 0, 1, \ldots.$$

Clearly, each V_i is an RKHS. If $V_{i+1} \supset V_i$, we obtain a *multiresolution ladder*

$$V_0 \subset V_1 \subset \ldots \subset V_i \subset V_{i+1} \subset \ldots \subset L^2(\mathbb{R}).$$

This is the setting for *wavelet analysis* [6]. The initial space V_0 is chosen as the closure of the span of translates of the *scaling function* $\phi(x)$. The reproducing kernel for this space is

$$k(x, y) = \sum_{n=-\infty}^{\infty} \phi(x - n)\phi(y - n),$$

where $\{\phi(\cdot - n)\}$ is an orthonormal system. These scaling functions satisfy a scaling equation

$$\phi(x) = \sum c_k \phi(2x - k), \ \{c_k\} \in \ell^2.$$

The *"mother wavelet"* associated with ϕ is given by

$$\psi(x) = \sum (-1)^{k-1} c_k \phi(2x + k - 1).$$

Its translates form an orthonormal basis of the orthogonal complement W_1 of V_0 in V_1; that is, $V_1 = V_0 \oplus W_1$ and

$$\psi_{jk}(x) = 2^{j/2} \psi(2^j x - k)$$

form an orthogonal basis of $L^2(\mathbb{R})$. For the classical Paley–Wiener case, the space V_0 is the space of π-bandlimited functions and $\phi(x) = \operatorname{sinc} \pi x$. Then V_1 is the space of 2π-bandlimited functions and W_1 is the set of functions whose Fourier transforms have support in $[-2\pi, -\pi) \cup [\pi, 2\pi)$.

We now apply the sampling solution procedure to (16.4) or (16.5) but restrict consideration to H_Q, which are wavelet subspaces. These subspaces are usually associated with a multiresolution analysis, a nested sequence $\{V_m\}$ of RKHSs contained in $L^2(\mathbb{R})$ such that

$$\ldots \subset V_{-1} \subset V_0 \subset V_1 \subset \ldots \subset V_m \subset \ldots \subset L^2(\mathbb{R}).$$

Each of these subspaces has an associated sampling theorem under very broad hypotheses (see [30] and [22]) and fits the theory developed in our previous work. A similar approach in a wavelet context has been considered by Donoho and Johnstone, who in a series of papers (see, for example, [7]) have attacked the problem (16.5) by first projecting (16.6) onto V_m, but then using a "shrinking" procedure to change the wavelet coefficients. Their method is based on a statistical optimization technique and works best if $\{\epsilon_i\}$ is white noise. It uses the sample values $\{y_i\}$ as an approximation to the wavelet coefficients. Our method, because it uses the sampling series rather than the wavelet series, is exact when the noise is zero and $y \in V_m$. It also works when the $\{t_i\}$ are not uniformly distributed since nonuniform sampling exists in some wavelet subspaces (see [2], [30], and [32]).

RKHS methods have been effectively developed for the simultaneous regularization and approximation of ill-posed integral and operator equations (see, for example, [20]). RKHS methods can also be coupled with other approaches to improperly posed problems, as described in the classification given by Payne [23]. The proposed approach should increase the effectiveness of the RKHS approach since *sampling expansions* (with derived error bounds) in wavelet and reproducing kernel subspaces will replace the integral representations. The interlocking in the areas of inverse/ill-posed problems, signal analysis, and moment problems will advance all these areas.

References

1. N. Aronszajn, Theory of reproducing kernels, *Trans. Amer. Math. Soc.* **68** (1950), 337–404.

2. J.J. Benedetto and M.W. Frazier (eds.), *Wavelets: Mathematics and Applications*, CRC Press, Boca Raton, FL, 1993.

3. M. Bertero, C. De Mol, and E.R. Pike, Linear inverse problems with discrete data, I, II, *Inverse Problems*, **1** (1985), 301–330; **4** (1988), 573–594.

4. P. Butzer, A survey of the Whittaker-Shannon sampling theorem and some of its extensions, *J. Math. Res. Exposition* **3** (1983), 185–212.

5. P. Butzer, W. SplettstöSer, and R. Stens, The sampling theorem and linear predictions in signal analysis, *Jahresber, Deutsch. Math.-Verein* **90** (1988), 1–60.

6. I. Daubechies, *Ten Lectures on Wavelets*, SIAM, Philadelphia, 1992.

7. D.L. Donoho and I.M. Johnstone, Minimax estimation via wavelet shrinkage, *Ann. Statistics* **26** (1998), 879–921.

8. H.W. Engl, M. Hanke, and A. Neubauer, *Regularization of Inverse Problems*, Kluwer, Dordrecht, 1996.
9. C.W. Groetsch, *The Theory of Tikhonov Regularization for Fredholm Equations of the First Kind,* Pitman, London-Boston, 1984.
10. J. Higgins, Five short stories about the cardinal series, *Bull. Amer. Math. Soc.* **12** (1985), 45–89.
11. E. Hille, Introduction to the general theory of reproducing kernels, *Rocky Mountain J. Math.* **2** (1972), 321–368.
12. M.E.H. Ismail, M.Z. Nashed, A.I. Zayed, and A.F. Ghaleb (eds.), *Mathematical Analysis, Wavelets, and Signal Processing,* Contemporary Mathematics, vol. 190, Amer. Math. Soc., Providence, RI, 1995.
13. A.J. Jerri, The Shannon sampling theorem—its various extensions and applications: a tutorial review, *Proc. IEEE* **65** (1977), 1565–1596.
14. M.Z. Nashed, Approximate regularized solutions to improperly posed linear integral and operator equations, in *Constructive and Computational Methods for Differential and Integral Equations,* D. Colton and R.P. Gilbert (eds.), Springer-Verlag, Berlin-Heidelberg-New York, 1974, 289–332.
15. M.Z. Nashed, On moment-discretization and least-squares solutions of linear integral equations of the first kind, *J. Math. Anal. Appl.* **53** (1976), 359–366.
16. M.Z. Nashed, Operator-theoretic and computational approaches to ill-posed problems with applications to antenna theory, *IEEE Trans. Antennas and Propagation,* **AP-29** (1981), 220–231.
17. M.Z. Nashed, On expansion methods for inverse and recovery problems from partial information, in *Nonlinear Problems in Applied Mathematics,* T.S. Angell et al. (eds.), SIAM, Philadelphia, 1996, 177–189.
18. M.Z. Nashed and O. Scherzer (eds.), *Inverse Problems, Image Analysis, and Medical Imaging,* Contemporary Mathematics, vol. 313, Amer. Math. Soc., Providence, RI, 2002.
19. M.Z. Nashed and G. Wahba, Convergence rates of approximate least-squares solutions of linear integral and operator equations of the first kind, *Math. Comp.* **28** (1974), 69–80.
20. M.Z. Nashed and G. Wahba, Regularization and approximation of linear operator equations in reproducing kernel spaces, *Bull. Amer. Math. Soc.* **80** (1974), 1213–1218.
21. M.Z. Nashed and G. Wahba, Generalized inverses in reproducing kernel spaces: an approach to regularization of linear operator equations, *SIAM J. Math. Anal.* **5** (1974), 974–987.
22. M.Z. Nashed and G.G. Walter, General sampling theorems for functions in reproducing kernel Hilbert spaces, *Math. Control, Signals & Systems* **4** (1991), 363–390.

23. L.E. Payne, *Improperly Posed Problems in Partial Differential Equations,* Regional Conference Series in Applied Mathematics, vol. **22**, SIAM, Philadelphia, 1975.

24. H.S. Shapiro, *Topics in Approximation Theory,* Springer-Verlag, Berlin-Heidelberg-New York, 1981.

25. F. Stenger, Numerical methods based on Whittaker cardinal, or sinc functions, *SIAM Rev.* **23** (1981), 165–224.

26. F. Stenger, *Numerical Methods Based on Sinc and Analytic Functions,* Springer-Verlag, New York, 1993.

27. A.N. Tikhonov and V.Y. Arsenin, *Solutions of Ill-Posed Problems,* Winston, Wiley, New York, 1977.

28. V.K. Tuan and M.Z. Nashed, Stable recovery of analytic functions using basic hypergeometric series, *J. Comput. Anal. Appl.* **3** (2001), 33–51.

29. C.V. van der Mee, M.Z. Nashed, and S. Seatzu, Sampling expansions and interpolation in unitarily translation invariant reproducing kernel Hilbert spaces, *Adv. Computat. Math.* 19 (2003), 355–372.

30. G.G. Walker and X. Shen, *Wavelets and Other Orthogonal Systems with Applications,* 2nd ed., CRC Press, Boca Raton, FL, 2001.

31. X.-G. Xia and M.Z. Nashed, The Backus-Gilgert method for signals in reproducing kernel Hilbert spaces and wavelet subspaces, *Inverse Problems* **10** (1994), 785–804.

32. X.-G. Xia and M.Z. Nashed, A modified minimum norm solution method for band-limited signal extrapolation with inaccurate data, *Inverse Problems* **13** (1997), 1641–1661.

33. R.M. Young, *An Introduction to Nonharmonic Fourier Series,* Academic Press, New York, 1980.

34. A.I. Zayed, *Advances in Shannon's Sampling Theory,* CRC Press, Boca Raton, FL, 1993.

35. M. Zwaan, Approximation of the solution to the moment problem in a Hilbert space, *Numer. Functional Anal. Optimization* **11** (1990), 601–612.

36. A. Zwaan, MRI reconstruction as a moment problem, *Math. Methods Appl. Sci.* **15** (1992), 661–675.

17 Zonal, Spectral Solutions for the Navier–Stokes Layer and Their Aerodynamical Applications

Adriana Nastase

17.1 Introduction

The starting point for this paper is the partial differential equations (PDEs) of the three-dimensional stationary, compressible Navier–Stokes layer (NSL), without any simplifications, as given in [1] and [2]. The new zonal, spectral solutions proposed here for the NSL's PDEs are useful for the computation of the flow over flattened, flying configurations (FCs). Let us introduce a new spectral coordinate, namely:

$$\eta = (x_3 - Z(x_1, x_2))/\delta(x_1, x_2).$$

Here $Z(x_1, x_2)$ is the equation operator of the FC, $\delta(x_1, x_2)$ is the thickness of the NSL, and the coordinate x_3 is measured perpendicular to the planform of the FC. The spectral forms of the axial, lateral, and vertical components u_δ, v_δ, and w_δ of velocity, the density function $R = \ln \rho$, and the absolute temperature T are, as in [3]–[5],

$$u_\delta = u_e \sum_{i=1}^{N} u_i \eta^i, \quad v_\delta = v_e \sum_{i=1}^{N} v_i \eta^i, \quad w_\delta = w_e \sum_{i=1}^{N} w_i \eta^i, \qquad (17.1\text{a–c})$$

$$R = R_w + (R_e - R_w) \sum_{i=1}^{N} r_i \eta^i, \quad T = T_w + (T_e - T_w) \sum_{i=1}^{N} t_i \eta^i. \qquad (17.2\text{a,b})$$

The exponential law of the viscosity μ versus the absolute temperature T and the physical equation of an ideal gas for the pressure p used here are

$$\mu = \mu_\infty \left[\frac{T}{T_\infty}\right]^{n_1}, \quad p \equiv R_g \rho T = R_g e^R T. \qquad (17.3\text{a,b})$$

Here, R_g is the universal gas constant, T_∞ and μ_∞ are the values of T and μ of the undisturbed flow, n_1 is the viscosity exponent, R_w and T_w are the given values of R and T at the wall, u_e, v_e, w_e, R_e, and T_e are the values of u, v, w, R, and T at the NSL's edge, and u_i, v_i, w_i, r_i, and t_i are free spectral coefficients, which are determined by satisfying the NSL's PDE and the boundary conditions at the NSL's edge. The inviscid potential flow over the FC, obtained after the solidification of the NSL, is used here as outer flow (instead of the parallel undisturbed flow used by Prandtl in his boundary layer theory). The boundary conditions on the FC (at $\eta = 0$) are automatically satisfied. The boundary conditions for the velocity components at the NSL's edge (at $\eta = 1$) are written in a special explicit form, as in [3]–[5], and those of the absolute temperature T and of the density function R are written in implicit form:

$$
\begin{gathered}
u_{N-2} = \alpha_{0,N-2} + \sum_{i=1}^{N-3} \alpha_{i,N-2}\, u_i, \quad v_{N-2} = \alpha_{0,N-2} + \sum_{i=1}^{N-3} \alpha_{i,N-2}\, v_i, \\
u_{N-1} = \alpha_{0,N-1} + \sum_{i=1}^{N-3} \alpha_{i,N-1}\, u_i, \quad v_{N-1} = \alpha_{0,N-1} + \sum_{i=1}^{N-3} \alpha_{i,N-1}\, v_i, \\
u_{N} = \alpha_{0,N} + \sum_{i=1}^{N-3} \alpha_{i,N}\, u_i, \quad v_{N} = \alpha_{0,N} + \sum_{i=1}^{N-3} \alpha_{i,N}\, v_i, \\
w_N = \gamma_{0,N} + \sum_{i=1}^{N-1} \gamma_{i,N}\, w_i, \\
\sum_{i=1}^{N} r_i = 1, \quad \sum_{i=1}^{N} t_i = 1.
\end{gathered}
\tag{17.4a–i}
$$

If the spectral forms of the velocity's components are introduced in the NSL's PDE of impulse, the boundary conditions at the NSL's edge are eliminated and the collocation method is used; the spectral coefficients are obtained by iteration from the equivalent quadratic algebraic system

$$
\begin{aligned}
&\sum_{i=1}^{N-3} u_i \left[\sum_{j=1}^{N-3} \left(\bar{A}^{(1)}_{ijk}\, u_j + \bar{B}^{(1)}_{ijk}\, v_j \right) + \sum_{j=1}^{N-1} \bar{C}^{(1)}_{ijk}\, w_j \right] \\
&\qquad = \bar{D}^{(1)}_{k} + \sum_{i=1}^{N-3} \left(\bar{A}^{(1)}_{ik}\, u_i + \bar{B}^{(1)}_{ik}\, v_i \right) + \sum_{i=1}^{N-1} \bar{C}^{(1)}_{ik}\, w_i, \\
&\sum_{i=1}^{N-3} v_i \left[\sum_{j=1}^{N-3} \left(\bar{A}^{(2)}_{ijk}\, u_j + \bar{B}^{(2)}_{ijk}\, v_j \right) + \sum_{j=1}^{N-1} \bar{C}^{(2)}_{ijk}\, w_j \right] \\
&\qquad = \bar{D}^{(2)}_{k} + \sum_{i=1}^{N-3} \left(\bar{A}^{(2)}_{ik}\, u_i + \bar{B}^{(2)}_{ik}\, v_i \right) + \sum_{i=1}^{N-1} \bar{C}^{(2)}_{ik}\, w_i, \\
&\sum_{i=1}^{N-3} w_i \left[\sum_{j=1}^{N-3} \left(\bar{A}^{(3)}_{ijk_1}\, u_j + \bar{B}^{(3)}_{ijk_1}\, v_j \right) + \sum_{j=1}^{N-1} \bar{C}^{(3)}_{ijk_1}\, w_j \right] \\
&\qquad = \bar{D}^{(3)}_{k_1} + \sum_{i=1}^{N-3} \left(\bar{A}^{(3)}_{ik_1}\, u_i + \bar{B}^{(3)}_{ik_1}\, v_i \right) + \sum_{i=1}^{N-1} \bar{C}^{(3)}_{ik_1}\, w_i,
\end{aligned}
$$

with slightly variable coefficients. These NSL solutions are reinforced because they have correct asymptotic behavior along singular lines such as subsonic leading edges, wing/fuselage junction lines, and wing/leading edge flaps junction lines, a correct last behavior at infinity, and, for the supersonic flow, they also satisfy the boundary condition on a characteristic surface. These zonal, spectral solutions are useful for the qualitative analysis of the asymptotic behavior of the NSL's PDE in the vicinity of their singular points, for the determination of the friction drag coefficient of FC, and for the global optimal design of the FC shape.

17.2 Qualitative Analysis of the Asymptotic Behavior of the NSL's PDE

The NSL's impulse equations are quadratic algebraic equations (QAEs) with variable coefficients, namely:

$$\sum_{i=1}^{M}\left[\sum_{j=1}^{M} a_{ij}^{(k)} X_i X_j + 2a_{i,M+1}^{(k)} X_i\right] + a_{M+1,M+1}^{(k)} = 0. \tag{17.5a}$$

In each QAE, only the free term $a_k = a_{M+1,M+1}^{(k)}$ depends on the pressure at the edge of the NSL and, therefore, has a greater variation. In addition, the free term of a kth QAE is systematically varied and all the other coefficients are maintained constant. The canonical form of the kth QAE is

$$\sum_{i=1}^{M} \lambda_i^{(k)} X_{ik}''^{2} + a_k'' = 0. \tag{17.5b}$$

In the QAE (17.5), the free term is $a_k'' = \Delta_k/\delta_k$, where δ_k is the discriminant and Δ_k the great determinant of this QAE, given by

$$\delta_k = \begin{vmatrix} a_{11}^{(k)} & a_{12}^{(k)} & \cdots & a_{1M}^{(k)} \\ a_{21}^{(k)} & a_{22}^{(k)} & \cdots & a_{2M}^{(k)} \\ \vdots & \vdots & & \vdots \\ a_{M,1}^{(k)} & a_{M,2}^{(k)} & \cdots & a_{M,M}^{(k)} \end{vmatrix},$$

$$\Delta_k = \begin{vmatrix} a_{11}^{(k)} & a_{12}^{(k)} & \cdots & a_{1,M}^{(k)} & a_{1,M+1}^{(k)} \\ a_{21}^{(k)} & a_{22}^{(k)} & \cdots & a_{2,M}^{(k)} & a_{2,M+1}^{(k)} \\ \vdots & \vdots & & \vdots & \vdots \\ a_{M,1}^{(k)} & a_{M,2}^{(k)} & \cdots & a_{M,M}^{(k)} & a_{M,M+1}^{(k)} \\ a_{M+1,1}^{(k)} & a_{M+1,2}^{(k)} & \cdots & a_{M+1,M}^{(k)} & a_{M+1,M+1}^{(k)} \end{vmatrix},$$

$\lambda_i^{(k)}$ are the eigenvalues of the QAE, obtained as solutions of its characteristic equation

$$\Delta_c \equiv \begin{vmatrix} a_{11}^{(k)} - \lambda & a_{12}^{(k)} & \dots & a_{1,M}^{(k)} \\ a_{21}^{(k)} & a_{22}^{(k)} - \lambda & \dots & a_{2,M}^{(k)} \\ \vdots & \vdots & & \vdots \\ a_{M,1}^{(k)} & a_{M,2}^{(k)} & \dots & a_{M,M}^{(k)} - \lambda \end{vmatrix} = 0,$$

and X_i'' is the canonical system of coordinates of this QAE, obtained from the initial system X_i of coordinates after successive translation and rotation. The canonical form of the kth QAE (17.5) is used for the qualitative analysis and the visualization of the asymptotic behavior of the initial kth QAE, in the vicinity of its critical point. The geometric visualization in the M-dimensional space of the independent spectral coefficients u_i, v_i, w_i is presented in the form of hyperellipsoids if all the eigenvalues λ_i are of the same sign, and hyperhyperboloids if one or more eigenvalues are of different sign. The character of each QAE remains unchanged during the variation of its free term (because they do not enter in the discriminant δ) and the elliptic and hyperbolic equations reach their critical points, located at their centers. The critical value $a_k = a_c$ is obtained from the linear equation $\Delta_k \equiv \delta a_c + d = 0$ which, for elliptic and hyperbolic QAEs, has a unique solution (because $\delta \neq 0$). The asymptotic behavior of the elliptic and hyperbolic QAEs in the vicinity of their singular points is very different and, therefore, the qualitative analysis and the visualization of this behavior in the vicinity of singular points will be treated separately in what follows.

If the free term of the elliptic QAE is systematically varied, the QAE is visualized in the form of coaxial hyperellipsoids of decreasing sizes if the free term $a = a_k$ increases; for the critical value $a = a_c$, they degenerate into one point. For $a > a_c$, the coaxial hyperellipsoids collapse. The visualization is made in sections for a space of dimension $M = 2$. The elliptic QAE chosen as an example for $M = 2$ is

$$F_1 \equiv 3x^2 + 5y^2 + 4xy - 6x - 3y + a = 0. \tag{17.6}$$

The canonical form of the equation (17.6), after translation and rotation, is

$$F_1 \equiv \lambda_1 x''^2 + \lambda_2 y''^2 + a'' = 0, \quad a'' = \Delta_1/\delta_1. \tag{17.7}$$

The determinants δ_1 and Δ_1 of this QAE are

$$\delta_1 = 11, \quad \Delta_1 = \tfrac{1}{4}(44a - 135),$$

and the eigenvalues λ_i are of the same sign:

$$\lambda_1 = 1.764, \quad \lambda_2 = 6.236;$$

consequently, equation (17.6) is elliptic. The critical value of the free term of equation (17.6) ($a \equiv a_c = 3.068$) is obtained by cancelling the great determinant $\Delta_1 = 0$.

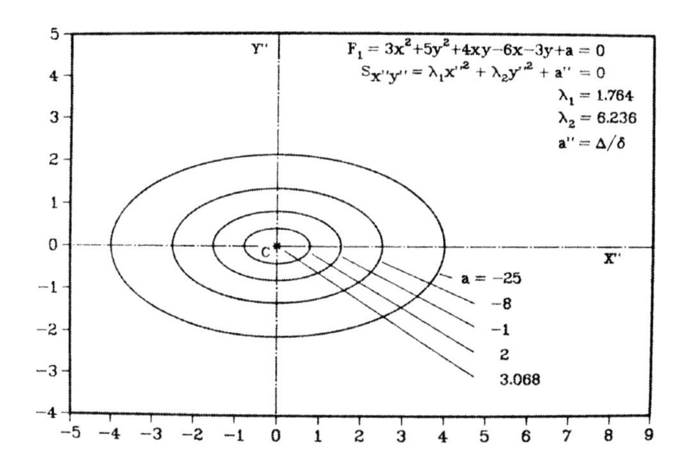

Fig. 1. The collapse of the coaxial ellipses F_1 for $M = 2$.

If the free term a in the equation (17.7) is systematically varied from $-\infty$ to $+\infty$, then for $a < a_c$, the canonical equations $F_1 = 0$ are visualized in the form of coaxial ellipses (Fig. 1) centered at $C(x^{''} = y^{''} = 0)$, which shrink to their common center C; for $a \equiv a_c = 3.068$, the corresponding ellipse degenerates into one point; and for $a > a_c$, the ellipses collapse. A black point occurs for $a = a_c$.

If the free term $b = a_k$ of the hyperbolic QAE is systematically varied, then the QAE is visualized in the form of coaxial hyperboloids. For $b = b_c$, these hyperhyperboloids have a saddle point, i.e., they degenerate into their common asymptotic hypersurface and, as b varies from $b < b_c$ to $b > b_c$, the coaxial hyperhyperboloids are jumping from one side of their asymptotic hypersurface to the other. Below, the qualitative analysis and the visualization are made in sections for a space of dimension $M = 2$.

The hyperbolic QAE chosen as an example for $M = 2$ is

$$F_2 \equiv 4x^2 + 7y^2 + 12xy - 4x - 5y + b = 0. \tag{17.8}$$

The canonical form of this equation, after translation and rotation, is

$$F_2 = \lambda_1 x''^2 + \lambda_2 y''^2 + b'' = 0, \quad b'' = \Delta_2/\delta_2).$$

The determinants δ_2 and Δ_2 of this QAE are

$$\delta_2 = -8, \quad \Delta_2 = -8b + 7,$$

and the eigenvalues λ_i are of opposite sign: $\lambda_1 = -0.685$, $\lambda_2 = 11.685$; therefore, equation (17.8) is hyperbolic. The critical value ($b \equiv b_c = 0.875$)

of the free term of this equation is obtained by cancelling the great determinant $\Delta_2 = 0$.

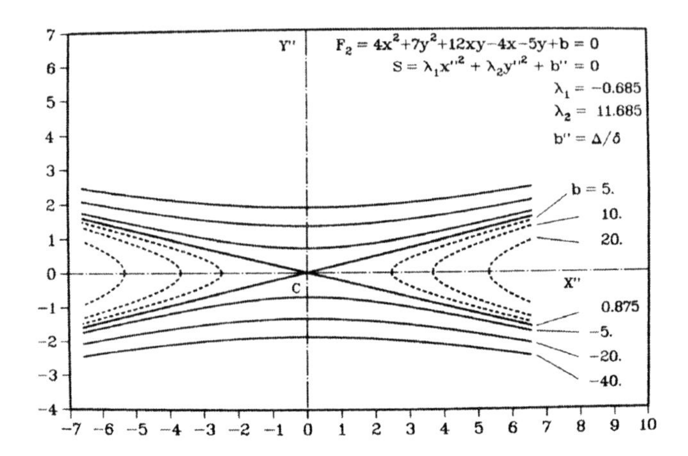

Fig. 2. The jump of the coaxial hyperbolas F_2 for $M = 2$.

If the free term b in equation (17.8) is systematically varied from $-\infty$ to $+\infty$, then for $b < b_c$, the canonical equation $F_2 = 0$ is represented in the form of coaxial hyperbolas with two sheets (Fig. 2), centered at $C(x'' = 0,\ y'' = 0)$, which approach their common, intersecting asymptotic lines; for $b \equiv b_c = 0.875$, the corresponding hyperbola degenerates into its asymptotic lines; and for $b > b_c$, the coaxial hyperbolas jump in the other pair of opposite angles of their intersecting asymptotic lines and move away from them. A saddle point occurs for $b = b_c$.

17.3 Determination of the Spectral Coefficients of the Density Function and Temperature

The continuity equation is written in a special form by using the density function $R = \ln \rho$. This equation, which is nonlinear in ρ, is linear in R. If the relations (17.1a–c) and (17.2a) are substituted in the continuity equation and (17.4h) and the collocation method are used, then the spectral coefficients r_i of R are obtained only as functions of the velocity's spectral coefficients u_i, v_i, and w_i, by solving the linear algebraic system

$$\sum_{i=1}^{N} g_{ip} r_i = \gamma_p, \quad p = 1, 2, \ldots, N.$$

The coefficients g_{ip} and γ_p depend only on the velocity's spectral coefficients. Similarly, if the relations (17.1a–c) and (17.2b) are used and the viscosity μ, computed from the exponential law (17.3a), and the pressure p, computed from the physical equation of gas (17.3b), are substituted in the temperature's PDE, and if (17.4i) and the collocation method are also used, then the spectral coefficients t_i of the absolute temperature T are expressed only as functions of the spectral coefficients of the velocity by solving the transcendental algebraic system

$$\sum_{i=1}^{N} h_{ip}\, t_i + h_{0p}\, (T^{n_1})_p = \theta_p, \quad p = 1, 2, \ldots, N.$$

The coefficients h_{ip}, h_{0p}, and θ_p depend only on the velocity's spectral coefficients.

17.4 Computation of the Friction Drag Coefficient of the Wedged Delta Wing

The shear stress τ_w at the wall and the global friction drag coefficient $C_d^{(f)}$ of the delta wing are

$$\tau_w = \mu \left.\frac{\partial u_\delta}{\partial \eta}\right|_{\eta=0} = \mu u_1 u_e, \quad C_d^{(f)} = 8\nu_f u_1 \int\limits_{\tilde{O}\tilde{A}_1\tilde{C}} u_e \tilde{x}_1\, d\tilde{x}_1\, d\tilde{y}.$$

The total drag coefficient $C_d^{(t)}$ is obtained by adding the friction coefficient $C_d^{(f)}$ to the inviscid drag $C_d^{(i)}$, given in [4]. Fig. 3 shows the wedged delta wing model of LAF (Lehr- und Forschungsgebiet Aerodynamik des Fluges). In Fig. 4 are visualized the variations of its inviscid and total drag coefficients $C_d^{(i)}$ and $C_d^{(t)}$, including the friction effect, versus the angle of attack α, for supersonic cruising Mach number $M_\infty = 2.0$. Fig. 5 illustrates the inviscid and total polars of the wedged delta wing. From Fig. 4 and 5 it can be seen that the influence of the viscosity in the total drag coefficient cannot be neglected. The computation of the total drag is possible only by using a viscous solver, like the NSL's zonal, spectral solutions proposed here.

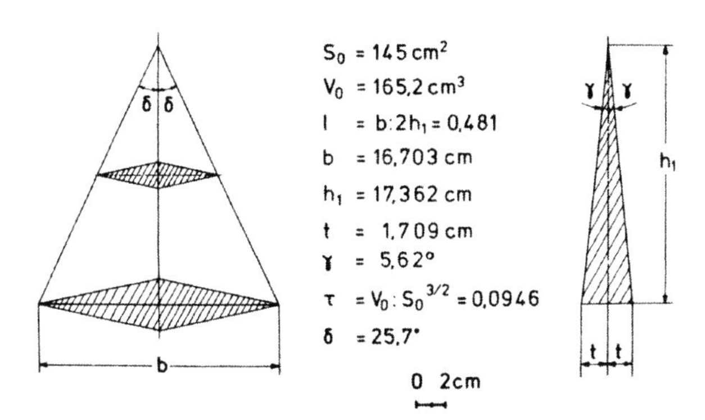

Fig. 3. The wedged delta wing model of LAF.

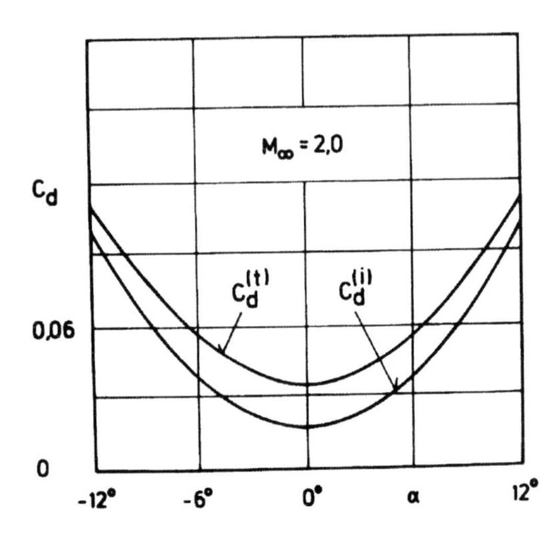

Fig. 4. The influence of the angle of attack α on the inviscid $(C_d^{(i)})$ and the total $(C_d^{(t)})$ drag coefficients of the LAF wedged delta wing model.

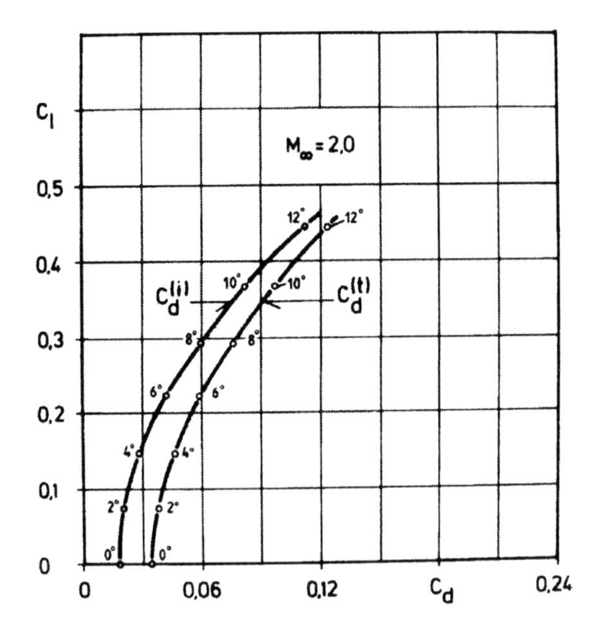

Fig. 5. The inviscid $(C_d^{(i)})$ and the total $(C_d^{(t)})$ polars of the LAF wedged delta wing model.

17.5 Conclusions

This hybrid analytic-numerical method is more accurate and needs less computer time than full-numerical methods because it needs no grid generation, the derivatives of all parameters can be easily and exactly computed, and the NSL's PDEs are satisfied exactly (at an arbitrary number N of chosen points).

References

1. H. Schlichting, *Boundary Layer Theory,* McGraw-Hill, New York, 1979.
2. A.D. Young, *Boundary Layers,* Blackwell, London, 1989.
3. A. Nastase, Aerodynamical applications of zonal, spectral solutions for the compressible boundary layer, *Z. Angew. Math. Mech.* **81** (2001), 929–930.
4. A. Nastase, Spectral solutions for the Navier-Stokes equations and shape optimal design, ECCOMAS 2000, Barcelona.
5. A. Nastase, A new spectral method and its aerodynamic applications, in *Proc. Seventh Internat. Symp. on CFD,* Beijing, China, 1997.

18 Hybrid Laplace and Poisson Solvers. Part III: Neumann BCs

Fred R. Payne

18.1 Introduction

A new hybrid DE solver, "DFI" (direct, formal integration) [1], has many analytic and numeric advantages (IMSE93 volume [2] lists 44). Most of these improve CPU numerics and permit multiple analytic forms for analysis and coding; some major ones inherent in this optimum DE solver are as follows:

1. PDEs yield $D(N+1)$ distinct, equivalent algorithms for DEs of order N and dimension D; ODEs yield $(N+1)$ equivalent algorithms for analysis and/or coding.
2. This solver eliminates all derivatives at the user's choice, avoiding large FDM or FEM matrices and their consequent computing error and time penalties.
3. Digital computers treat integrals (quadrature sums) numerically better than derivatives (divided differences).
4. No iteration of consequent Volterra IDEs/IEs is needed for any order 2 or higher system which has no $(n-1)$-order derivative.
5. There is easy extension from 2D to 3D (add a DO-loop) and to 4D (add 2 DO-loops); only one more DO-loop is needed per each new dimension.

The author's series of IMSE papers, 1985–2002, on various systems serve as inducements to apply DFI to any linear or nonlinear ODE/PDE system. The 1980 DFI discovery was a major impetus for authors founding, 1985, and chairing IMSE, 1985, 1990. Fredholm IE applications in turbulence, 1965–85, was another. The author's 40 years' experience with IEs, trapezoidal quadratures, and FORTRAN are dominant factors in this work.

18.2 Solution Techniques

The 2D Poisson equation, where $f(x,y)=0$ yields Laplace, is

$$G_{xx}+G_{yy}=f(x,y).$$

Two successive integrations on $[0,y]$ with $y\varepsilon[\Delta y,1]$, yield the pair of Volterra IDEs

$$G_y(x, y) = G_y(x, 0) - \int_0^y [G_{xx}(x, s) - f(x, s)]\, ds, \tag{18.1}$$

$$G(x, y) = G(x, 0) + yG_y(x, 0) - \int_0^y (y - s)[G_{xx}(x, s) - f(x, s)]\, ds, \tag{18.2}$$

where the "lag factor" $[y - s]$, unique to DFI-type solvers, is due to the Lovitt [3] form for repeated integrals. Either (18.1) or (18.2) can be used as the algorithm. (18.2) has the massive advantage of bypassing the usual Volterra iterations for any 2-point quadrature; cost and time savings are huge. Elliptic DFI still requires sweeping but relaxation is automatic. Full DFI (add two x-integrations here) eliminates all derivatives as the $(N-1)$-order derivative is missing in these systems; a few full DFI cases such as Euler (elliptic in subsonic flows) [4] obtained the solution in a single sweep. Both (18.1) and (18.2) are IDE and thus mixed quadratures and finite differences are required for numerics. Results for ten analytic trial $G(x, y)$ for Laplace and Poisson Neumann BC serve as accuracy checks using form (18.3) below. Neumann BCs imply the iterative and undesirable computational form (18.1); however, the simple identity

$$G(x, y) = G(x, 0) + \int_0^y G_s(x, s)\, ds$$

converts (18.1) to (18.3) below with a massive computational advantage since no iteration for any 2-point quadrature is needed or possible:

$$\begin{aligned} G_y(x, y) = G_y(x, 0) - yG_{xx}(x, 0) \\ - \int_0^y [(y - s)G_{yxx}(x, s) + f(x, s)]\, ds. \end{aligned} \tag{18.3}$$

In (18.3), the "shooter" parameter for $G_y(x, 1)$, where $y = 1$ is the upper boundary, is

$$G_{xx}(x, 0),$$

and Volterra iteration is again bypassed due to a structure almost identical to (18.2) since $f(x, y)$ is known and iteration in (18.3) is impossible for any 2-point quadrature; the Dirichlet BC shooter for $G(x, 1)$ in (18.2) is

$$G_y(x, 0),$$

as described in [5]; Robin BCs [6] allow either equation (18.2) or (18.3) to be used, depending upon BCs and user choice. $G(0, 0)$ is always set to zero, the Laplace arbitrary constant. The algorithm has four parts:

1. Establish a $[0, 1] \times [0, 1]$ grid and insert the BCs.
2. Sweep $x \in [\Delta x, 1 - \Delta x]$ over the field, using a DFI y-trajectory at each fixed x. The first and last internal points require extrapolation if 5-point CDs are used. Hence, 3-point central differences (Δx^2) and

simple trapezoidal quadratures are used. Alternatively, an x trajectory can be used with y-sweeping. Any coordinate can be chosen as the integration trajectory.

3. Compute the global DE, $\triangle G - f$, RMS errors which continually diminish roughly as $1/e$ as the Laplace operator on the computed solution aproaches $f(x, y)$; thus, the numeric solution is obtained.
4. Repeat steps 2–3 until the desired RMS accuracy is attained.

18.3 Results for Five of Each of Laplace and Poisson Neumann BC Problems

There was a twofold rationale for the ten test cases.

1. Laplace: progressively high-order solutions.
2. Poisson: increasingly complex forcing functions ($f = \Delta G$ residuals) from constants to y-linear, xy-bilinear to biquartic.

The ten cases computed are listed in Table 1.

Table 1. Chosen Laplace and Poisson test cases for DFI.

Case	G	$G_y(x, y)$	Laplace residual $f(x, y)$
0	$x + y$	1	0
1	xy	x	0
2	$x^2 - y^2$	$-2y$	0
3	$x^2y - y^3/3$	$x^2 - y^2$	0
4	$x^3y - xy^3$	$x^3 - 3xy^2$	0
5	$x^2 + y$	1	2
6	$x^2 + y^2$	$2y$	4
7	x^2y	x^2	$2y$
8	$x^3 + y^3$	$3y^2$	$6(x + y)$
9	$x^4y^2 - x^2y^4$	$2x^4y - 4x^2y^3$	$2(x^4 - y^4)$

A curious result first noted for the Laplace equation in 1985 and recurring in IMSE work (see [5]–[7]) is the sharp dependence of DFI on the grid "aspect ratio" AR, defined by

$$AR = NY \setminus NX,$$

where NY is the number of y-grid points and NX is the number of x-points. For a range of the number of y-points from 1K to 8K (binary) a good AR value is 128 (or nearby in binary: 256, 512).

Conjectured is that the "hammerhead" stencil of DFI (see below) may exhibit (at least under secant shooter) Euler-like "column instability" for small values of AR which translate to excessively large y-steps over a too-narrow x-base. This AR behavior may be special to the Laplace operator and its averaging property. This question is left to numerical analysts. ("A good piece of research leaves work for others." (FP, ca. 1970))

Consider the DFI "hammerhead" stencil; let k be the known BC values, or values already computed at previous y-steps on the expanding trajectory starting at the boundary. Let u be the unknown value to be computed at the current y-step (under Lovitt). DFI/Lovitt decouples the implicit, standard Volterra IE dependence of the solution on itself to explicit dependence on already available values. The "hammerhead" stencil for the first three y-steps is shown in Table 2.

Table 2. DFI "hammerhead" stencil (3-point central differences).

		u
	u	$k\ k\ k$
u	$k\ k\ k$	$k\ k\ k$
$k\ k\ k$	$k\ k\ k$	$k\ k\ k$
1st step off wall	2nd step off wall	3rd step off wall

Picture the 128th step, a tall, narrow stencil. This may become unstable near the top of the long trajectory for a poor choice of grid sizes (aspect ratio AR and the number of y-steps, NY).

Unused 5-point CD, $O(\Delta x^4)$ complexity [6] are the first ($x = 2\Delta x$) and last ($x = 1 - 2\Delta x$) y-trajectories. At the first DFI x-step, G_y values for $x = \Delta x$ are extrapolated from BCs at $x = 0$; at the last step, $x = 1 - 2\Delta x$, G_y values for $x = 1 - \Delta x$ are backward extrapolated from $x = 1$. DFI fills in these values on subsequent sweeps. 5-point CD accuracy gains were small in [6].

Yet another DFI serendipity simulates the unsteady problem of zeroed initial internal field (at time $t = 0$) under given BCs and develops the "solution wave" as it approaches a "steady" state of small or zero error. This facet will be valuable for unsteady problems.

Results for the ten $G(x, y)$ test cases are given in Table 3. Errors are RMS order of magnitudes over the field in powers of 10. Unless noted otherwise, $NY/AR = 1\text{K}$ (binary y-steps)/128 and 32 global sweeps were made for each case.

Unless noted otherwise, runs used 1K(1024) digital y-steps (each y-step was 0.00098) and $AR = 128$ (x-step was 0.125). No CPU timing is listed for runs on an earlier machine (Intel PRO 200); F77 timer code was used in the later stages of work on Pentium III 733 MHz, which increased RAM from 128MB for the PRO 200 to 512MB for the faster machine, used in Part IV [7] for 3D and 4D Laplace equations.

18.4 Discussion

The usual grid of 1024 y-steps and $AR = 128$ suggests that y-accuracy is about $O(10^{-6})$ and x-accuracy is $O(10^{-2})$. The accuracy of runs, generally not exceeding $O(10^{-6})$, is better than expected; this may be due to smoothing by the Laplace operator combined with similar quadrature effects. Rather large x-steps and x-finite differences may add to this. DFI has shown good success with a broad range of test functions. Global errors are dominated by the size of the y-step since 128–512 times as many y-steps as x-steps are taken.

Table 3. Numerical results and global RMS errors (missing entries are lost).

Case	PDE	G	G_y	NY/AR	Sweeps	CPUsec
0	$x+y$	-12	-16	-29	126[1]	0.14
1	xy	-13	-14	-13		
2	x^2-y^2	-6	-6	-15	2K/128	
3	$x^2y-y^3/3$	-6	-6	-10		
4	x^3y-xy^3	-6	-6	-5		
5	x^2+y	-6	-6	-9		
6	x^2+y^2	-6	-6	-9		
7	x^2y	-6	-6	-9	8K/512	
8	x^3+y^3	-6	-7	-15	8K/512	
9	$x^4y^2-x^2y^4$	-7	-7	-3	8K/256	17[2]

[1] Tests a maximum possible accuracy on this machine.
[2] The toughest problem in this set. Finer grids are usually needed for more complex $G(x,y)$ cases.

First tries at a new problem can fail due to exceeding the number of allowed secant "shoots" at a particular x or due to "real overflow". The latter occurs because of explosive (exponential?) growth of the spurious solution. The fix for either is simply to increase the number of y-steps or decrease the number of x-steps; either fix increases the aspect ratio AR and continues the dominance of the number of y-steps over the number of x-steps.

DFI code is interactive. The best "tracker" is to view interactively the approach of the shooter parameter to a limit; so long as it is approaching a limit, the code is succeeding. Here, four significant figures were used for this. In parallel with this technique, one needs to follow global PDE RMS errors for approach to a limit, say, to within $O(10^{-6})$ to $O(10^{-30})$, as has occurred in this work.

Sensible coding considerations, for any serious scientific worker, will include at least the items in the following list.

1. 64-bit minimum word size (DOUBLE PRECISION or REAL*8 in FORTRAN). Most 32-bit PCs have an 80-bit FPU (floating-point unit), which, in double precision, exceeds the accuracy of 64-bit main frames in single precision. Smaller word sizes are not adequate for serious scientific work.

2. Use binary grids; this can save orders of magnitude accuracy due to conversion errors of decimal to binary digits. $AR = 1\text{K}/128, \ldots,$ 8K/512 was the typical range of the number NY of y-intervals versus the aspect ratio $AR = NY/NX$, where NX is the number of x-intervals. Thus, grid sizes varied from 10K to 139K points. See Part IV [7] in this volume for grids up to 281 million grid points for 3D and 4D Dirichlet BCs.
3. Group terms to minimize machine operations and, hence, errors.
4. Avoid division if possible; this saves accuracy and CPU time.
5. In the code development stage, as here, copious outputs are useful, even essential.

For work reported in depth at the IMSE 2000/2002/2004 conferences (see [5]–[7]), RMS error measures included the following four items.

1. Validation of the PDE operator; one always has the PDE so one can always do this.
2. For exact solutions, as here, calculate the global RMS difference of exact values from actual computed values of $\Delta G - f$. For Neumann, both the derivative and the function (underlying the BC derivatives) fields were checked.
3. Maximum errors and their locations can be useful.
4. Copious comments, especially for large or complex codes, prevent confusion. The FORTRAN "!" in-line comment command can save many (about 100) code lines.

18.5 Closure

The theoretical bases for DFI are not complex. First, sophomore-level integration, partially along trajectories fixed in one (as here for 2D) or more (in dimensions $N \geq 3$) trajectories converts any DE to a Volterra IE or IDE. Such equations are usually implicit and require iteration. However, Lovitt application converts the IE/IDE to explicit form (under any 2-point quadrature such as trapezoid or Romberg) and removes the iteration requirement for all DEs of order $n \geq 2$ if they have no $(n-1)$-order derivative. The same massive numerical simplification occurs for any ODE.

Secondly, the questions of DFI existence and uniqueness are almost trivial for technological applications. For a linear Volterra IE/IDE, the only necessity is Tricomi's theorem [8], requiring Lebesque square-integrable functions. Nonlinear Volterra IE/IDE existence and uniqueness additionally require a pair of Lipschitz conditions [8], which are also easily satisfied for applications.

Another DFI facet, a serendipity, provides new physical and mathematical insights into any problem due to multiple but equivalent mathematical formulations. Consider Laplace and Poisson BVPs. From the three equivalent governing equations (18.1)–(18.3), the leading solution terms (the algebraic, nonintegral ones) are shown in Table 4.

Table 4. Leading terms of DFI solutions for the three BC classes.

1	$G_y(x,y) = G_y(x,0) - \cdots$	Robin, Neumann	(1R)
2	$G(x,y) = G(x,0) + yG_y(x,0) - \cdots$	Dirichlet	(2R)
3	$G_y(x,y) = G_y(x,0) - yG_{xx}(x,0) - \cdots$	Robin, Neumann	(3R)

This follows from formal y-integrations here; initial x-integrations over fixed y yield three similar forms with x interchanged with y in the G-derivatives. For any numericist this is ideal: eight (6 IDE, PDE, pure IE) mathematical descriptions of the same problem. Analysts also will benefit from having eight forms to analyze. Forms (18.2) and (18.3) appear best since the first two terms of the solution series are algebraic and are either known BC or "shooter" parameters. To change integration paths, switch two "DO-loop" indices by modifying 17 line pairs in a total of 290–370 code lines. Dirichlet [5] requires the fewest lines, Neumann the most lines of code.

DFI FORTRAN coding is rather easy; most of the work is in "DO-loops." The four code segments (main and three subroutines) are listed in Table 5. The main program also calculated errors beyond the three listed in Table 3, namely: 1) along an $x = y$ trajectory; 2) shooter "misses" of $\Delta G - f = 0$? at the top boundary $y = 1$; and 3) "misses" at the first and last x-values off the boundary. Code interaction is mostly governed by the main program; this includes tracking sweep number and an option to stop or continue the sweeps.

Table 5. F77 routines, purposes, and number of lines of 2D code.

Code segment	DO-loops and work load	$\sim$ lines of code
Main (DFI)	2 DFI (2D) and 5 error loops	200
Grid/BC	12 loops (set grid and BC)	90
PDE errors	2 loops (compute PDE operator)	40
Exact solution	3 loops (final errors; outputs)	40

The program totals to 24 "DO-loops" and $\sim$370 code lines. If the exact solution is unavailable (as in new applications), the last segment vanishes. The two DFI solver loops require 7 lines plus 33 lines of BC input. The secant shooter requires about 20 lines; the remaining 140 lines in the main program are error checking and I/O. Hence, the basic DFI solver is a tiny 40 lines. See Part IV for 3D/4D Dirichlet problems, which require only some 10–20 additional code lines.

This is the third in a series of four papers on the Laplace and Poisson equations treating three classes of boundary conditions: Part I, Dirichlet BCs, IMSE 2000 [5], Part II, Robin BCs, IMSE 2002 [6], Part III, Neumann BCs, IMSE 2004 (this paper), and Part IV, extensions to nonlinear Helmholtz equations and 3D/4D Laplace equation with Dirichlet BCs, IMSE 2004 [7].

The Numericists' Credo is

We compute numbers,
Not for their sake,
But to gain insights.

(John von Neumann)

Other DFI work can be found in [2] and [4]–[6]. The reader is invited to contact the author (frpdfi@airmail.net) for sample code (.FOR extension) and output (.TXT extension) files.

Added in proof: DFI onto Newtonian gravitation, that is,

$$\mathbf{F}^\alpha = m^\alpha \ddot{\mathbf{r}}^\alpha = -Gm^\alpha \sum_{k=1}^{n} \mathbf{S}_k^\alpha; \quad k \neq \alpha; \quad \alpha = 1, 2, \ldots, n \text{ (not summed)}$$

$$\mathbf{S}_k^\alpha = m_k \mathbf{e}_k^\alpha / \|\mathbf{R}_k^\alpha\|^2; \quad \mathbf{R}_k^\alpha = (\mathbf{r}^\alpha - \mathbf{r}_k); \quad \mathbf{e}_k^\alpha = \mathbf{R}_k^\alpha / \|\mathbf{R}_k^\alpha\|$$

yields $6n$ IEs; the $3n$ $\mathbf{v}^\alpha$ IEs are analogous to those for $\mathbf{r}^\alpha$, that is,

$$\mathbf{r}^\alpha(t) = \mathbf{r}^\alpha(0) + t\mathbf{v}^\alpha(0) - G\sum_{k=1}^{n} \int_0^t [t-s]\mathbf{S}_k^\alpha(s)ds; k \neq \alpha; \tag{18.4}$$

$\mathbf{v}$ equations are $\mathbf{r}$-coupled so $\mathbf{r}$ is solved first, or $\mathbf{r}, \mathbf{v}$ in an alternating sequence: $[\mathbf{r}(\Delta \mathrm{t}), \mathbf{v}(\Delta \mathrm{t})]$, repeat for $2\Delta t, \ldots,$ 4 DO-loops and 10 code lines solve $9n + 1$ equations; the number of passes is $n(9n - 8) \times$(number of t-steps).

1. Input $9n + 1$ ICs (initial positions, momenta, angular momenta, and total energy of the system). F77 "DATA" statement is useful.
2. Solve $\mathbf{r}^\alpha$ in (18.4) for $t = \Delta t, \ldots$; Lovitt decouples t from itself. Solve the similar velocity equations (or, much quicker, finite difference known $\mathbf{r}(\mathrm{t})$s). Get angular velocites, $\alpha = 1, \ldots, n$.
3. CHECKS for the entire system: (i) Do forces sum to 0? (ii), (iii), (iv) Conservation of linear and angular momenta and total energy? (v) Others, such as rotational energy for non-point masses?

DFI solves, in principle (numerically), the n-body problem limited only by computing machinery capacities. Note the universal DFI trademark of $(n-1)$ polynomial leading terms for all DE applications of order n.

References

1. F.R. Payne, Lect. Notes, UTA, 1980; AIAA Symposium, UTA, 1981 (unpublished).
2. F.R. Payne, A nonlinear system solver optimal for computer, in *Integral Methods in Science and Engineering,* C. Constanda (ed.), Longman, Harlow, 1994, 61–71.

3. W.V. Lovitt, *Linear Integral Equations,* Dover, New York, 1960.

4. F.R. Payne, Euler and inviscid Burger high-accuracy solutions, in *Nonlinear Problems in Aerospace and Aviation,* vol. 2, S. Sivasundaram (ed.), European Conf. Publications, Cambridge, 1999, 601–608.

5. F.R. Payne, Hybrid Laplace and Poisson solvers. I: Dirichlet boundary conditions, in *Integral Methods in Science and Engineering*, P. Schiavone, C. Constanda, and A. Mioduchowski (eds.), Birkhäuser, Boston, 2002, 203–208.

6. F.R. Payne, Hybrid Laplace and Poisson solvers. II: Robin BCs, in *Integral Methods in Science and Engineering*, C. Constanda, M. Ahues, and A. Largillier (eds.), Birkhäuser, Boston, 2004, 181–186.

7. F.R. Payne, Hybrid Laplace and Poisson solvers. Part IV: extensions, this volume, Chapter 19.

8. F.G. Tricomi, *Integral Equations,* Dover, New York, 1985, 10–15, 42–47.

19 Hybrid Laplace and Poisson Solvers. Part IV: Extensions

Fred R. Payne

19.1 Introduction

A new hybrid DE solver, "DFI" (direct, formal integration) [1], offers many analytic and numeric advantages; IMSE93 volume [2] lists 44. Improved CPU numerics and multiple analytic forms for analysis and coding are major ones inherent in this optimum DE solver. A series of IMSE papers between 1985–2004 (see [2]–[9]), suggest DFI applicability to any linear or nonlinear DE system. Prior work treated 2D Laplace and Poisson PDEs with Dirichlet BCs [7], Robin BCs [8], and Neumann BCs [9]; this extends the Dirichlet problem to 3D and 4D. Extension to n dimensions is almost trivial in regard to FORTRAN code modifications. This is yet another DFI advantage over conventional numeric solvers.

The author's 1980 discovery of DFI and its development was a major factor in his founding, in 1985, and chairing, in 1985 and 1990, of the first IMSE conferences.

DFI has two modes: "Simplex" integrates along one coordinate and "Multiplex" over two or more. In principle, one can formally integrate over all independent variables ("Full DFI") and eliminate all derivatives, but complexity may task a human. Simplex DFI, used here, has four stages.

1. Formally integrate DEs along a chosen trajectory (coordinate); Volterra IEs/IDEs result.
2. Study the new forms for insights; some will arise.
3. Analytically compute the solution near the initial point; one can usually do this, possibly approximately. This provides guidance for machine coding the full solution.
4. Compute error measures to validate results and for possible iteration or sweeping a global field. Such calculations are simple; e.g., difference current and prior run values and form global RMS values, etc. Convergence of errors denotes program success.

Results for several test analytic $G(x, y, z)$ for 3D and $G(x, y, z, t)$ for 4D Dirichlet BVPs serve as DFI accuracy checks. Other extensions herein are linear and nonlinear Helmholtz BVPs. DFI demonstrates ease of application to any system. 1980–2002 successes, with no failure, include (PDEs unless otherwise stated): Laplace and Poisson Dirichlet [7], Robin [8], and

Neumann BCs [9]; 2D Euler [10]; Bénard convection (sixth-order ODE) [11]; a Riccati equation (NLODE) and a turbulence NLODE model [11]; turbulent channel flow [12]; Lorenz "chaos" (three NLODEs) [13]; supersonic Prandtl boundary layer [14]; economics, and heat conduction (linear and nonlinear) [15]; stability of Burger's model [13]; flight mechanics, Maxwell and solid state physics [15]; Falkner–Skan NLODE [3]; Volterra predator-prey NLODE [16]; and others. DFI has succeeded, without failure, in many distinct DE systems.

19.2 Solution Methodologies

The 3D Poisson equation, where $f(x,y,z)=0$ yields Laplace's equation, is

$$G_{xx}+G_{yy}+G_{zz}=f(x,y,z).$$

Two successive integrations on $[0,y]$, $y\in[\Delta y,1]$, yield the pair of Volterra IDEs

$$G_y(x,y,z)=G_y(x,0,z)-\int_0^y [G_{xx}(x,s,z) \\ +G_{zz}(x,s,z)-f(x,s,z)]ds, \quad (19.1)$$

$$G(x,y,z)=G(x,0,z)+yG_y(x,0,z)-\int_0^y (y-s)[G_{xx}(x,s) \\ +G_{zz}(x,s,z)-f(x,s,z)]ds. \quad (19.2)$$

The "lag factor" $[y-s]$, unique to DFI, is due to Lovitt's "well-known" form for repeated integrals [17], easily proved through integration by parts and induction for $(k+1)$-repeated integrals with the same limits ($k=1$ above):

$$\int_0^y ds ... \int_0^s f(t)dt = \int_0^y [y-s]^k f(s)ds/k!$$

Either (19.1) or (19.2) can be used as the algorithm. (19.2) has the massive numerical advantage of bypassing the usual Volterra iterations for any 2-point quadrature; cost and time savings are huge. Elliptic DFI still requires sweeping but relaxation is automatic.

DFI easily extends to both higher dimensions and higher-order derivatives. A dimension increase adds another lagged integral and a new FORTRAN DO-loop in that variable. Unit increase in the DE order requires a new y-integration loop; a serendipity is increased dominance by the Lovitt "lag factor" effect, i.e., sequentially from $[y-s]$ for second-order, to $[y-s]^2/2, ...[y-s)^3/3!...[y-s]^n/n!$ for $(n+1)$-order DEs. This yields an ever-increasing dominance by the ICs/BCs as the DE order increases [13]. If the n-system contains a nonzero $(n-1)$-derivative, full Lovitt decoupling does not apply and some Volterra iteration is necessary for such implicit terms.

"Full" DFI adds two x- and two z-integrations and eliminates all derivatives as the $(n-1)$-derivative is missing in these systems. Full DFI applied to Euler (elliptic in subsonic flows) [10] obtained the solution in a single sweep. Both (19.1) and (19.2) are IDE and thus mixed quadratures and finite differences are required for numerics. Results for eight analytic trial $G(x, y, z)$ and $G(x, y, z, t)$ for 3D and 4D Laplace Dirchlet BCs serve as accuracy and CPU timing checks using form (19.2). The "shooter" parameters for 3D BVPs, $G(x, 1, z)$, and 4D BVPs, $G(x, 1, z, t)$, at the $y = 1$ upper boundary, are, respectively, $G_y(x, 0, z)$, which initiates dual "semi-sweeps" sequentially in x and z, whereas in 4D $G_y(x, 0, z, t)$ needs 3 "semi-sweeps" in x, z, t unless "full" DFI is employed.

The "lag factor" $[y - s]$ in (19.2) is Lovitt's form [10] for repeated integrals. Either (19.1) or (19.2) serves as the algorithm. (19.2) has the massive advantage of bypassing the usual Volterra iterations for any 2-point quadrature.

4D simply adds t as an argument in each term of (19.1), (19.2), and G_{tt}. DFI adds a lagged integral $G_{tt}(x, y, z, t)$ to (19.2) for 4D, namely,

$$G(x, y, z, t) = RHS(19.2) - \int_0^y [y - s] G_{tt}(x, s, z, t)\, ds. \qquad (19.3)$$

19.3 3D and 4D Laplace Dirichlet BVPs

3D test cases (2D BVPs are in [7]–[9]) including RMS errors in powers of 10 are listed in Table 1. PDE denotes RMS deviation of the Laplacian of the computed values from the exact solution. RMS denotes RMS deviation of the computed and exact solutions. MaxDG is the global maximum value of the difference of computed and exact values at a single point. NY/AR is the number NY of y-intervals; AR is the "aspect ratio" (the number of y-intervals divided by the number of x- or z-intervals). Sweeps denotes the number of sweeps over the entire computational field; CPU timing in seconds is calculated by a FORTRAN intrinsic. An Intel Pentium PRO, 200 MHz clock and 128MB RAM, was used for 3D problems but its RAM is too small for 4D problems. Results for 3D Laplace Dirchlet with zeroed initial interior fields (for $\Delta x = \Delta z$ and $= \Delta t$ in 4D) are given in Table 1.

Table 1. 3D Laplace errors (more sweeps needed for case 3).

Case	PDE	RMS	MaxDG	NY/AR	Sweeps	sec
1. xyz	-12	-11	-7	1K/128	98	60.7
2. $(x^2 + z^2)/2 - y^2$	-21	-17	-7	1K/128	96	58.3
3. $2x^2 - y^2 - z^2$	-7	-5	-2	1K/128	6	7.5

4D cases were tested and their errors, in powers of 10, are cited in Table 2. An Intel Pentium III, 733 MHz clock and 512MB RAM, was used.

Table 2. 4D Laplace results and errors.

Case	PDE	RMS	MaxDG	NY/AR	Sweeps	sec
41. $xyzt$	-9	-14	-11	1K/128	77	84.7
42. $x^2+z^2-y^2-t^2$	-5	-9	-10	1K/128	74	89.4
43. $x^2+y^2-z^2-t^2$	-15	-15	-11	1K/128	87	75.8
44. $y^2+z^2-x^2-t^2$	-5	-9	-10	1K/128	71	61.8
45. $3y^2-x^2-z^2-t^2$	-7	-14	-10	1K/128	94	100.2
46. Case 45 rerun	-9	-15	-11	1K/128	111	126.3

Every case set all initial interior G-field values to zero. There are two considerations: 1) accuracy as demonstrated by Table 2 results and 2) stability of the algorithm on that machine as discussed next.

In all cases (Parts I–IV in [7]–[9] and here), initial ("debugging") runs for a new case input the exact solution (and y-derivative for Neumann and Robin BCs) at all interior points; in nearly all cases, a single sweep yielded zero error (or a trivial one, such as 10^{-32}). Even so, usual practice was to perform more, usually 8, sweeps to validate stability; this procedure guarantees accuracy and stability. We note that an absurd number of sweeps $O(\geq 1000)$ may accumulate sufficient machine error to cause a drift in the results. If so, this is inconsequential. This is not a mathematical proof of algorithm stability but rather an engineering one.

In all cases, after the initial "debugging" run with exact field inputs, the initial fields were all zeroed; this worked well in all cases (an example of a best unbiased first estimator?). In a few cases, linear interpolation from the BCs was tried; those results were generally inferior, both in CPU timing and accuracy, to those with zeroed initial fields. We conjecture that Laplace operators "like to write on a blank page". That is, poor initial estimates force the operator to "work harder" for the solution.

Note the nonuniformity in errors from case to case. Generally, this increases as the functional complexity increases; compare cases 41, 42, and 45 with exact interior fields as inputs ("debugging" runs). This caution was universal when attacking any new problem or varying BCs; typical results are given in Table 3. Usually, such converged to minuscule or zero error in a single sweep (or trajectory for ODE), indicating that DFI faithfully reproduced the solution ($\Delta x = \Delta z = \Delta t$).

Table 3. Error variation for exact input G-fields (note the single sweeps).

Case	PDE	G	MaxDG	NY/AR	Sweeps	sec
41. $xyzt$	-22	-22	-15	1K/128	1	2.5
42. $x^2+z^2-y^2-z^2$	0	0	0	1K/128	1	0.9
45. $3y^2-x^2-z^2-t^2$	0	0	0	1K/128	1	1.1

Cases 42 and 45 are quadratic; case 41 is essentially quartic. All cases except 41 yield zero error (CPU likely set a tiny number to zero). In case 41, the maximum error of 10^{-15} is essentially zero on the computer; a computer zero output merely means that it cannot resolve the small number from a true zero value; this mostly depends upon machine "word size" and its builder. Other causes of nonuniform errors may include the rather primitive secant shooter. DFI expedites a Newton shooter since the derivative of the unknown function always appears in one DFI equation, namely, prior to the final IE in the DFI hierarchy which explicitly displays the solution.

For decades, the DFI procedure has been twofold: 1) insert (if known; otherwise bypass this) the exact solution into all interior domain grid points as code validation ("debugging"); 2) "record runs" estimate initial interior field values. The best estimate is a zeroed global field, less BCs. Linear extrapolation from the BCs was tried on occasion but was inferior to zeroed initial input fields in accuracy and execution times. Error checks included: 1) global RMS deviations of pertinent quantities; 2) absolute values of global maximum deviations at a field point.

19.4 Linear and Nonlinear Helmholtz Dirichlet BVPs

A generalized 2D Helmholtz equation, linear or nonlinear, is

$$\Delta H(x,y) = h(H(x,y),x,y).$$

Two successive y-integrations yield a form optimum for Dirichlet BVPs, namely,

$$H(x,y) = H(x,0) + yH_y(x,0) - \int_0^y [y-s][H_{xx}(x,s)ds - h(H(x,s))]\,ds, \qquad (19.4)$$

where $H_y(x,0)$ is the "shooting" parameter for the y-sweep at each fixed x. Note that (19.4) is the same form as (19.2) and (19.3) except 1) the dropping of z and 2) the forcing term is now dependent upon H. Results for both linear and nonlinear cases (denoted by L and N) are given in Table 4. NY, the number of y-intervals, and AR, the aspect ratio (the ratio of the number of y-steps to the number of x-steps), vary for the Helmholtz equation as shown in Table 4.

Table 4. Linear/nonlinear Helmholtz equation results and errors.

	$H(x,y)$	$h(H)$	DE	HRMS	MaxDH	NY/AR	Sweeps
L1	"cosh"	$25H$	-10	-6	-5	1K/128	66
L2	$e^x - e^{-y}$	H	-10	-7	-6	2K/512	6
N1	$1/(xy)$	$2(x^2+y^2)H^3$	-7	-6	-4	2K/128	28
N2	x/y	$2H^3/x^2$	-3	-11	-8	8K/128	14

Case L1 is from Cheney and Kincaid, *Numerical Mathematics and Computing*, 3rd ed., Booke/Cole, 1994, p. 481, who give the exact solution of

$\Delta H = h(H) = 25H$ as $[\cosh(5x) + \cosh(5y)]/(2\cosh(5))$. Note the few sweeps for L2; the y-grid is very fine.

The accuracy here is inferior to that for Laplace/Poisson BVPs (see [7]–[9]). Barring coding error (always possible), the reflexive nature of the Helmholtz nonlinear forcing function is likely a primary cause. Romberg iteration will greatly improve results. A 1985 unpublished result applied seven Romberg levels and resulted in $O(10^{-16})$ error with basic 0.01 steps. Modifying codes from trapezoid to Romberg is straightforward; simply add an "inner" loop at each y-value of the solver loop ($\sim$ 10 lines).

The search for the optimum grid pattern is an analog to the search for the optimum relaxation factor for elliptic problems. A "domain study" was run on the N2 case, $H = 1/xy$. Some results for varying domains (from $[1,2] \times [1,2]$ above to $[5,6] \times [5,6]$) with zero initial input H-fields are given in Table 5.

Table 5. Behavior of $H = 1/(xy)$ near its singularity at $(0,0)$.

Domain	PDE	HRMS	MaxDH	NY/AR
$[1,2]^2$	-2	-3	-4	1K/128
$[2,3]^2$	-3	-4	-4	1K/128
$[5,6]^2$	-7	-6	-4	2K/128

Note the improvement in PDE and HRMS errors as the computational domain moves farther from the singularity at $(0,0)$. These results indicate that beginning near $x = y = 0$ will likely fail to yield acceptable values.

Compare $\sim$ 5 sweeps for the $[5,6]$ domain here versus the large number (28) for the domain $[1,2]^2$ above. This is due to rapid functional growth near the origin for the Helmholtz case, $H = 1/(xy)$. All work used RMS error convergence of the DE to manually terminate the run.

On occasion, new optimizations were discovered. Time dictated that earlier codes not be retrofitted. A major improvement was the reduction of 24 accumulators (4D) to only one, with some reductions in run-times and RAM storage requirements but negligible improvement in accuracy.

19.5 Coding Considerations

All cases herein started at one wall BC and "shot" for the BC at the opposite wall. Since y-integration was used (x-integration may be preferable for some BVPs whose DE operator is not symmetric in x and y) and finite differences in the other directions, the y-loop must be the innermost one. Shooting for unknown BCs was quite successful, even though AR varied from 128 to 512 and NY from 1K to 8K for difficult problems.

In some earlier cases [8], experiment was required to continue the calculation; Neumann BCs were among such cases. An example: difference the BC on opposite walls and interpolate along adjacent walls. Subsequent passes correct these adjustments towards accurate values.

Virtually identical code structures were used for all problems since IMSE 1998. The four routines (with the number of code lines in parentheses) are:

Main: inputs, DFI solver, shooter, accuracy checks (field traverses compared to exact values), final outputs (140).

Grid: generated grids, BCs, a first estimate of the solution (60).

PDE: central differences validated the computed pointwise DE operator; RMS and maximum errors were also computed (35).

Exact: output final computed field, RMS, maximum deviations of the computed solution from the exact H (40).

The code totaled $\sim$ 275 lines; this varied about 10 percent for Dirichlet, Neumann, and Robin BCs, and 3D/4D elliptics. Five input FORTRAN parameters were used to begin all calculations:

NY: the number of y-intervals.

JSKIP: a print sampler of x and y outputs to reduce output (even so, some runs generated 200+ pages of output).

AR: (the number of y-intervals)/(the number of x-intervals).

EPS: shooter tolerance, usually $10^{-}8$; some increased to 10^{-14}; 64-bit machine noise level is $O(10^{-15})$.

EXACT= 1: exact input field (debugging and accuracy).

EXACT= 2: zeroed initial fields (all "record" runs).

EXACT= 3: linearily interpolated input fields (seldom used).

"Base" values were $NY = 1024$ (1K binary); JSKIP $= NY/4$ (yields 5 y-point output across the computational domain); $AR = 128$, though this varied from 16 to 512 for large grids; $NY/AR = 1024/128$ was optimum for many problems; this choice has 9225 field points.

19.6 Some Remarks on DFI Methodology

The Helmholtz algorithm is identical to other 2D cases; 3D and higher dimensions have identical structures excepting one more array and an extra sweep for each addition. FORTRAN77 DO-loops accommodate this easily.

1. Establish a [0, 1] grid for each variable and insert the BCs; this requires a 2D array for 2D problems; 3D problems require cubic arrays; 4D need hypercubic arrays; nD problems use n-dimensional arrays.
2. Sweep $x \in [\Delta x, 1-\Delta x]$ over the field, using a DFI y-trajectory at each fixed x. For nD problems, one can choose any for the DFI trajectory and cycle ("sweep" for elliptic) sequentially over the remaining variables, one at a time. Each such is termed a "semi-sweep". Thus, 4D requires a "nest" of 4 DO-loops for the three "semi-sweeps". Simple trapezoidal quadratures and 3-point central differences $O(\Delta x^2)$ were used; Romberg iteration (not needed here) can be incorporated by adding 10–20 lines of code.

3. Compute the global DE RMS error, which continually diminishes $\sim 1/e$ as the DE operator on the computed solution converges.
4. Repeat steps 2–3 until the desired accuracy is attained.

A curious result first noted for the Laplace equation in 1985 and recurring in IMSE work (see [7]–[9]) is the dependency of DFI on the grid "aspect ratio" (AR),

$$AR = NY/NX,$$

which is the ratio of the number of y-grid points to those of x. For a range of 1K to 4K (binary) y-points, a good AR value is 128 or nearby (in binary).

The "hammerhead" DFI stencil (see below) may exhibit Euler-like "column instability" for small values of AR. Why this is so is left to numerical analysts. ("A good piece of research leaves work for others." F.P., ca. 1970)

Consider the DFI "hammerhead" stencil. In Table 6, k is the known BCs or values already computed at previous y-steps starting at Δy off the wall. Let u be the unknown value to be computed at the current y-step. DFI/Lovitt decoupling to explicit dependency on already available values generates the "hammerhead" stencil, which for the first four y-steps is

Table 6. DFI "hammerhead" stencil for the first four steps off the boundary.

			u
		u	$k\ k\ k$
	u	$k\ k\ k$	$k\ k\ k$
u	$k\ k\ k$	$k\ k\ k$	$k\ k\ k$
$k\ k\ k$	$k\ k\ k$	$k\ k\ k$	$k\ k\ k$
$y = \Delta y$	$2\Delta y$	$3\Delta y$	$4\Delta y$

This pattern is repeated for each value of $x, z, \ldots$ "semi-sweeps" in those variables; nD requires $(n-1)$ semi-sweeps per global sweep.

As y traverses $[0,1]$, a narrow stencil develops. This may become unstable near the top of the trajectory for poor choices of grid sizes (AR and NY). Some experiment is required to find "good" AR, NY values ("relaxation"?).

Another DFI serendipity simulates the unsteady problem of zeroed initial internal field (at time $t = 0$) under given BCs and develops an "error wave" which is swept ever nearer a "steady" state of small or zero error. This facet will be valuable for unsteady problems. Each global sweep simulates a time step of some characteristic value.

19.7 Discussion

The usual grid of 1024 y-steps and $AR = 128$ implies that y-accuracy is about $O(10^{-6})$ and x-accuracy is $O(10^{-2})$. Accuracy of runs, generally less than, or equal to, $O(10^{-}6)$, is better than expected; this may be due to smoothing by the Laplace operator combined with similar quadrature effects. DFI has shown good success with a broad range of test functions. Global errors are dominated here by y step-size since 128–512 times as many y-steps as x-steps were taken for the ARs reported here.

First tries at a new problem can fail due to exceeding the number of allowed secant "shoots" at a particular x or "real overflow". The latter occurs because of explosive (exponential?) growth of spurious solutions. The fix for either is simply to increase the number of y-steps or decrease the number of x-steps; either fix increases AR, the aspect ratio, and continues dominance of the number of y-steps over those in x (and z, t for 3D/4D).

DFI code is interactive. The best "tracker" is to view interactively the approach of the shooter parameter to a limit; such behavior means the code is succeeding. Four significant figures were used for this. In parallel with this technique, one needs to follow global PDE RMS errors for approach to a limit, say $O(10^{-6})$–$O(10^{-30})$, as has occurred in this work.

Sensible coding considerations, for any serious scientific worker, will include the items in the following list.

1. 64-bit minimum word size. Smaller word sizes are not adequate for serious scientific work.
2. Use binary grids to preserve accuracy due to conversion errors of decimal to binary digits.
3. Group terms to minimize machine operations and, hence, errors.
4. Avoid division if possible to save much accuracy and CPU time.
5. Large outputs during code development are useful and essential.

For work reported in depth at the IMSE 2000/2002/2004 conferences (see [7]–[9]), RMS error measures included the following items.

1. Validation of the PDE operator; one always has the PDE.
2. For exact solutions, as here, calculate global RMS differences of exact from actual computed values of the DE.
3. Global maximum errors and their locations can be useful.

The theoretical bases for DFI are not complex. First, simply integrate by parts along trajectories fixed in one coordinate, or more in 3D or higher, and convert any DE to a Volterra IE or IDE. Such are usually implicit and require iteration. However, Lovitt application converts the IE/IDE to explicit form and removes all iteration requirements for DEs of order $n \geq 2$ with no $(n-1)$-order derivative. Similar simplification occurs for ODEs.

Second, the questions of DFI existence and uniqueness are almost trivial in technological application. For a linear Volterra IE/IDE, Tricomi's theorem [19] requires L_2-functions. Nonlinear Volterra IE/IDE existence and uniqueness additionally require a pair of Lipschitz conditions [19], which are also easily satisfied for applications.

Another DFI serendipity provides new insights into any problem due to multiple but equivalent mathematical formulations. Consider Laplace and Poisson BCs. From (19.1) and (19.2) above and (3) in [9], the leading solution terms (the nonintegral ones) are those listed in Table 7.

Table 7. Leading terms of DFI solutions for three BC classes.

1	$G_y(x,y) = G_y(x,0) - \cdots$	(Robin, Neumann)	(1R)
2	$G(x,y) = G(x,0) + yG_y(x,0) - \cdots$	(Dirichlet)	(2R)
3	$G_y(x,y) = G_y(x,0) - yG_{xx}(x,0) - \cdots$	(Robin, Neumann)	[9].

This follows from y-integrations; initial x-integrations over fixed y yield three similar forms with x interchanged with y in the G-derivatives. For numericists, this is ideal: eight (6 IDE, PDE, pure IE) mathematical descriptions of the problem. To change integration paths, switch two FORTRAN "DO-loop" indices by modifying a few of the 290–390 code lines.

DFI FORTRAN coding is rather easy; most of the work is in "DO-loops". The four code segments (main and three subroutines) are listed in Table 8. The main program also calculated errors beyond the three listed in Tables 1–5, namely 1) along an $x = y$ trajectory ($= z = t$ also in 3D/4D) and 2) shooter "misses" of $\Delta G - f = 0$? at $y = 1$, the top boundary. Code interaction is governed by the main program; this includes tracking sweep number and errors and an option to stop or continue the sweeps.

Table 8. F77 routines, purposes, and number of code lines.

Code segment	DO-loops and workload	~ lines of code
Main (DFI)	4 DFI (4D) and 9 error loops	210
Grid/BC	14 loops (set grid and BC)	100
PDE errors	4 loops (compute PDE operator)	40
Exact solution	4 loops (final errors; outputs)	40

If an exact solution is unknown, the last segment is null. The program totaled 35 "DO-loops" and ~390 code lines. The two DFI solver loops require 9 lines plus 37 lines for BCs. The secant shooter requires ~ 40 lines; 125 lines in the main program are error checkers and I/O. Hence, the basic DFI solver is a tiny 46 lines. This finalizes a series of four IMSE papers on elliptic BVPs covering three classes of boundary conditions. These are Part I, Dirichlet BCs, IMSE 2000 [7], Part II, Robin BCs, IMSE 2002 [8], Part III, Neumann BCs, IMSE 2004 [9], and Part IV, Extensions to Dirichlet 3D and 4D BVPs, this paper.

19.8 Some DFI Advantages

This section expands part of an IMSE93 paper [2], which was based on 12 years' work, four Ph.D. theses, and five MSAE theses.

19.8.1 DFI Conceptual Features

1. Nonlinear DEs are solved directly without linearization. All ICs and BCs are included explicitly in the Volterra formulation.
2. Imbeds arbitrary order predictor-correctors if $G \in C^{\infty}$.

3. New mathematical and physical insights from multiple formalisms.
4. Even nonlinear DE terms can integrate analytically to algebraic ones (e.g., udu/dx to $u^2/2$).
5. Any DE can usually be hand-integrated (approximately, perhaps) near the IP to provide insights for coding and analysis.
6. DFI embeds easily into other methods (FDM/FEM, spectral, etc.) or stands alone.
7. Symbolic manipulators can generate codes, useful for "full DFI", eliminating all derivatives. Such by hand can be a chore for high-order DEs. The author failed [12] in a system of 8 PDEs, 28 derivatives, and 21 formal integrations required by "full DFI", 1989.
8. PDEs allow multiple trajectories and sweeps via simple "DO-loop" changes.
9. All PDEs and any ODE of order ≥ 2 offer alternate algorithms.
10. Trapezoid quadratures are easily upgraded, by Romberg, to any desired accuracy, limited only by machine word size and arithmetic.
11. DFI solves IVPs as a sequence of smaller IVPs, and BVPs as the limit of a sequence of IVPs with "shooting". ABVPs are solved as the limit of a sequence of BVPs with ever-increasing upper limits until the solution changes by small, but acceptable, amounts.
12. Implicit, "organic shooter" for BVP/ABVP is available.
13. The infinite set of viscous fluid "wall compatibilities" ("no slip", etc.) is automatically satisfied; no other method does this.

19.8.2 Physical Features

1. DFI is a virtually error-free simulator. The *only* approximation is the quadrature rule (and difference formula if used). Computer error is not avoidable unless a more powerful machine is used.
2. A "cluster" property exhibits explicit causal relationships.
3. Modelling, especially of nonlinear processes, is simplified.
4. Three sets of physical mechanisms explicitly appear in all problems: the function and its slope values at the IP and diffusion normal to the trajectory (already explicit along the trajectory).
5. GLOBAL and LOCAL results are immediate, and "control volume" checks require only nominal effort.
6. The solution is displayed explicitly in the governing IEs.
7. Error calculation is virtually trivial; simply difference values at hand, square, and average for RMS errors.
8. Smoothes empiric and/or numeric data; these quadratures do best.

19.8.3 Mathematical Features

1. Only sophomore calculus and the concept of iteration are used.
2. Quadrature stability conditions are inverse to those of differencing methods. "With two arrows in the quiver, use the best one."
3. Volterra IEs/IDEs have unique solutions under rather weak physical conditions (L_2 and Lipschitz if nonlinear).
4. "Lovitt form" for repeated integrals decouples implicit Volterra forms for any 2-point quadrature and permits Romberg quadrature as needed. This produces massive CPU time savings.
5. Many little-used techniques fit naturally into DFI. "Lovitt" is the earliest and most important yet discovered.
6. "Micro-Picard" accelerator [16] is about nine times faster than classic Picard iteration (not needed here).
7. Does not, as FDM/FEM can, force the problem into linear molds.
8. "Full DFI" (natural anti-derivative (NAD)) eliminates all derivatives by sequential integration over all arguments. NAD implementation to incompressible Euler's flow equations (elliptic) required but a single global sweep [10].

19.8.4 Numerical Features

1. Uniquely compatible with computing machinery via minimal subtractions and divisions (none for ODEs). Digital computers "like" to integrate (sum, multiply) rather than differentiate, due to the "bad" operations of subtraction and division. The "bad" operations waste CPU time, 4 to 20+ times that of "good" operations.
2. Controls error propagation since global errors are the same order as pointwise errors.
3. Easy post-run checks confirm the DE solution.
4. DFI yields a simple rationale for "finding infinity" for ABVPs for that machine, grid, etc. Falkner–Skan, a nonlinear, third-order ODE, is the "similar" solution to incompressible Prandtl 2D boundary layer flows. For zero pressure gradient, "infinity" is order 15–20. One begins by integrating out to 1. Rerun to 2 or 3 and note results. Increase by stages to 20 and one nears the asymptotic BC of $u = 1$, e.g., $u = 0.9999$. Continue this empiric process until sufficiently close to "infinity". This is termed the "FAN" procedure [18].
5. The "FAN" procedure is quite powerful for ABVPs.
6. Any reasonable initial guess begins a convergent sequence of iterations.
7. "Lovitt decoupling" of Volterra IE/IDE for DEs of order two or greater is likely the largest CPU time and accuracy saver of all.

19.8.5 Coding and Execution Features

1. DFI is easy to code; solvers are often only 10–20 lines of FORTRAN.
2. Ideal for interactive computing upon microcomputers.
3. Minimal storage requirements allow update by overwriting data along the trajectory. To error check use a single "old" array.
4. "Bootstrap" recursion mimics the first step in all others.
5. Predictor-corrector (usually ODE): Taylor series predict values at the next grid point; iterate the Volterra system ("corrector") and repeat at subsequent grid points. An nth-order predictor (where n is the order of the DE) usually requires 2–3 iterations via trapezoid. Corrector iterates until $|f_n - f_o| \leq \varepsilon$, preset; thus the solution is uniformly convergent, so beloved by mathematicians.
6. Dirichlet, Neumann, and Robin BVPs yield identical IDEs and need only minor code changes to solve quite dissimilar problems.
7. Multiple integration trajectories are available for all PDEs and any ODE of order 2 or higher.
8. NO APPROXIMATION is made to this point. The only approximation is the quadrature formula (and difference formula if in "mixed mode", for IDEs).

The largest DFI numerical loads, prior to the work in [6]–[9], are listed below.

1. 1986, Blasius NLODE with 12,000,000 grid points (step-size of 10^{-6}), used 120 minutes on DEC20 (1 MFLOP, 72-bit word) [16].
2. 1989, turbulent channel flow large-scale structure predictions for 8 coupled NLPDEs with 28 distinct derivatives. DFI had a speed-up factor of $\sim$ 1000 over "psuedotime" CFD. It replaced 25,000 time iterations by $\sim$ 20 elliptic sweeps. This was "mixed mode" integration across the flow and FDM down- and cross-stream [12].
3. 1991, Lorenz's "chaos" (3 nonlinear and coupled ODEs) was solved with a range of time steps (0.1, 0.01, ..., 10^{-7}). Three integrations yielded three pure IEs; these were integrated to various times ranging from 100 to 10,000 (2.1 to 210 billion steps) [13]. Lorenz, 1963, used a 0.1 time step on a now obsolete computer.

19.9 Closure

DFI rests upon two of Tricomi's theorems [19]. Existence and uniqueness for linear Volterra IEs merely require functions to be L_2, that is, square integrable. Nonlinear IEs add two Lipschitz growth conditions; these are easily met for any technological application. DFI offers a sound alternative to classic DE solvers. For many problems, DFI is likely the optimum solver, especially if full Lovitt-decoupling prevails; then DFI's obviation of the usual Volterra iterations is massively advantageous.

The reader is invited to e-mail (frpdfi@airmail.net) for sample code and output files used herein.

DFI is the NATURAL THING TO DO and is EASY!

References

1. F.R. Payne, Lect. Notes, UTA, 1980; AIAA Symposium, UTA, 1981 (unpublished).
2. F.R. Payne, A nonlinear system solver optimal for computer, in *Integral Methods in Science and Engineering*, C. Constanda (ed.), Longman, Harlow, 1994, 61–71.
3. F.R. Payne, Direct, formal integration (DFI): an alternative to FDM/ FEM, in *Integral Methods in Science and Engineering*, F.R. Payne, C. Corduneanu, A. Haji-Sheikh, and T. Huang (eds.), Hemisphere, New York, 1986, 62–73; (with C.-S. Ahn) DFI solution of compressible laminar boundary layer flows, *ibid.*, 181–191. (with R. Mokkapati) Numeric implementation of matched asymptotic expansions for Prandtl flow, *ibid.*, 249–258.
4. F.R. Payne and M. Nair, A triad of solutions for 2D Navier-Stokes: global, semi-local, and local, in *Integral Methods in Science and Engineering*, A. Haji-Sheik, C. Corduneanu, J. Fry, T. Huang, and F.R. Payne, (eds.), Hemisphere, New York, 1991, 352–359.
5. F.R. Payne and K.R. Payne, New facets of DFI, a DE solver for all seasons, in *Integral Methods in Science and Engineering, vol. 2: Approximate Methods,* C. Constanda, J. Saranen, and S. Seikkala (eds.), Longman, Harlow, 1997, 176–180.
6. F.R. Payne and K.R. Payne, Linear and sublinear Tricomi via DFI, in *Integral Methods in Science and Engineering*, B. Bertram, C. Constanda, and A. Struthers (eds.), Chapman and Hall/CRC, Boca Raton, 2000, 268–273.
7. F.R. Payne, Hybrid Laplace and Poisson solvers. I: Dirichlet boundary conditions, in *Integral Methods in Science and Engineering,* P. Schiavone, C. Constanda, and A. Mioduchowski (eds.), Birkhäuser, Boston, 2002, 203–208.
8. F.R. Payne, Hybrid Laplace and Poisson solvers. II: Robin BCs, in *Integral Methods in Science and Engineering: Analytic and Numerical techniques,* C. Constanda, M. Ahues, and A. Largillier (eds.), Birkhäuser, Boston, 2004, 181–186.
9. F.R. Rayne, Hybrid Laplace and Poisson solvers. Part III: Neumann BCs, this volume, Chapter 18.
10. F.R. Payne, Euler and inviscid Burger high-accuracy solutions, in *Nonlinear Problems in Aerospace and Aviation,* vol. 2, S. Sivasundaram (ed.), European Conf. Publications, Cambridge, 1999, 601–608.

11. F.R. Payne, Global and local stability of Burger's analogy to Navier-Stokes, in *Proc. Internat. Conf. on Theory and Appl. of Diff. Equations,* Ohio Univ. Press, Athens, OH, 1988, 289–295.

12. F.R. Payne, NASA Ames Senior Fellowship Final Report, 1989 (unpublished).

13. F.R. Payne, Exact numeric solution of nonlinear DE systems, in *Dynamics of Continuous, Discrete, and Impulsive Systems,* vol. 5, Watam Press, Waterloo, ON, 1999, 39–51.

14. R. Mokkapati, Ph.D. Dissertation, UTA, 1989.

15. AE and Math graduate student term papers, UTA, 1982–2002.

16. F.R. Payne, Class notes, UTA, 1980–2002 (unpublished).

17. W.V. Lovitt, *Linear Integral Equations*, Dover, New York, 1960, 6–7.

18. F.-T. Ko and F.R. Payne, A simple conversion of two-point BVP to one-point BVP, in *Trends in the Theory and Practice of Nonlinear Differential Equations,* V. Lakshmikantham (ed.), Marcel Dekker, New York, 1985, 467–476.

19. F.G. Tricomi, *Integral Equations*, Dover, New York, 1985, 10–15, 42–47.

20 A Contact Problem for a Convection-diffusion Equation

Shirley Pomeranz, Gilbert Lewis, and Christian Constanda

20.1 Introduction

In this paper we consider a convection-diffusion equation with a piecewise constant coefficient, and indicate an iterative solution method for it. The usual numerical solution techniques are inadequate in this case because of the presence of a small coefficient and of a boundary layer. Here we propose a domain decomposition method that extends results published in [1] and [2]. After splitting the problem domain into two disjoint subdomains with an internal interface, we prove that the boundary value problem has a unique solution and then compute the solution numerically, using the finite element method in one subdomain and the method of matched asymptotic expansions in the other. These "local" solutions are related at the internal interface through continuity requirements.

20.2 The Boundary Value Problem

Consider the equilibrium distribution of temperature in heat conduction across the interface between two homogeneous plates with insulated faces, where convection is dominant in one and diffusion in the other. The model that best describes this problem mathematically is

$$\begin{aligned} -\nabla\cdot\big(a(x,y)(\nabla u)(x,y)\big) &+ b(\partial_y u)(x,y) \\ &+ cu(x,y) = f(x,y), \quad (x,y)\in\Omega, \qquad (20.1)\\ u(x,y) &= g(x,y), \quad (x,y)\in\partial\Omega, \qquad (20.2)\end{aligned}$$

where $\Omega = \big\{(x,y) : 0 < x < L,\ -H_1 < y < H_2\big\}$, L, H_1, H_2, b, and c are positive constants,

$$a(x,y) = \begin{cases} 1, & 0 < x < L,\ \ 0 \le y \le H_2,\\ \varepsilon, & 0 < x < L,\ \ -H_1 \le y < 0,\end{cases}$$

$\partial_y(\ldots) = \partial(\ldots)/\partial y$, and f and g are sufficiently smooth functions prescribed on Ω and $\partial\Omega$, respectively. The parameter ε, $0 < \varepsilon < 1$, is the relatively small diffusion coefficient.

We split Ω into two disjoint upper and lower subregions

$$\Omega_1 = \{(x,y) : 0 < x < L,\ 0 < y < H_2\},$$
$$\Omega_\varepsilon = \{(x,y) : 0 < x < L,\ -H_1 < y < 0\}$$

and make the notation

$$\begin{aligned}
\partial\Omega_0 &= \{(x,0) : 0 < x < L\},\\
\partial\Omega_1 &= \{(0,y) : 0 < y < H_2\} \cup \{(x,H_2) : 0 \le x \le L\}\\
&\qquad \cup \{(L,y) : 0 < y < H_2\},\\
\partial\Omega_\varepsilon &= \{(0,y) : -H_1 < y < 0\} \cup \{(x,-H_1) : 0 \le x \le L\}\\
&\qquad \cup \{(L,y) : -H_1 < y < 0\},\\
\Omega &= \Omega_1 \cup \Omega_\varepsilon \cup \partial\Omega_0, \quad \partial\Omega = \overline{\partial\Omega_1} \cup \overline{\partial\Omega_\varepsilon}.
\end{aligned}$$

If we also write

$$\begin{aligned}
u_1 &= u\big|_{\Omega_1}, & u_\varepsilon &= u\big|_{\Omega_\varepsilon}, & u &= \{u_1, u_\varepsilon\},\\
f_1 &= f\big|_{\Omega_1}, & f_\varepsilon &= f\big|_{\Omega_\varepsilon}, & f &= \{f_1, f_\varepsilon\},\\
g_1 &= g\big|_{\partial\Omega_1}, & g_\varepsilon &= g\big|_{\partial\Omega_\varepsilon}, & g &= \{g_1, g_\varepsilon\},
\end{aligned}$$

then problem (20.1), (20.2) splits into two separate boundary value problems, one in Ω_1 and one in Ω_ε; specifically,

$$\begin{aligned}
-\nabla \cdot (\nabla u_1) + b(\partial_y u_1) + cu_1 &= f_1 \quad \text{in } \Omega_1,\\
u_1 &= g_1 \quad \text{on } \partial\Omega_1,
\end{aligned} \tag{20.3}$$

and

$$\begin{aligned}
-\varepsilon\nabla \cdot (\nabla u_\varepsilon) + b(\partial_y u_\varepsilon) + cu_\varepsilon &= f_\varepsilon \quad \text{in } \Omega_\varepsilon,\\
u_\varepsilon &= g_\varepsilon \quad \text{on } \partial\Omega_\varepsilon.
\end{aligned} \tag{20.4}$$

The boundary conditions in (20.3) and (20.4) are augmented with the natural conditions of continuity of the solution and of the normal component of its flux across the internal interface $\partial\Omega_0$:

$$u_1 = u_\varepsilon, \quad \partial_y u_1 = \varepsilon(\partial_y u_\varepsilon) \quad \text{on } \partial\Omega_0. \tag{20.5}$$

For an arbitrary function $v \in C_0^\infty(\Omega)$, we write $v = \{v_1, v_\varepsilon\}$. Multiplying the equations in (20.3) and (20.4) by v_1 and v_ε, integrating over Ω_1 and Ω_ε, respectively, adding the results, and using Gauss's formula and the transmission conditions (20.5), we find that

$$A(u,v) = (f,v),$$

where

$$A(u,v) = \int_{\Omega_1} \big[(\nabla u_1)\cdot(\nabla v_1) - bu_1(\partial_y v_1) + cu_1 v_1\big]\,d\sigma + \int_{\Omega_\varepsilon} \big[\varepsilon(\nabla u_\varepsilon)\cdot(\nabla v_\varepsilon) - bu_\varepsilon(\partial_y v_\varepsilon) + cu_\varepsilon v_\varepsilon\big]\,d\sigma$$

is a bilinear form associated with the internal energy of the system and

$$(f,v) = \int_{\Omega_1} f_1 v_1\,d\sigma + \int_{\Omega_\varepsilon} f_\varepsilon v_\varepsilon\,d\sigma$$

is the $L^2(\Omega)$-inner product of f and v. Hence, in a Sobolev space (distributional) setting [3], the variational problem corresponding to (20.3)–(20.5) consists in finding $u \in H_1(\Omega)$, $u = \{u_1, u_\varepsilon\}$, satisfying

$$\begin{aligned} &A(u,v) = (f,v) \quad \forall v \in \mathring{H}_1(\Omega), \\ &\gamma u = g, \quad \gamma_{10}u_1 = \gamma_{\varepsilon 0}u_\varepsilon, \end{aligned} \tag{20.6}$$

where $f \in H_{-1}(\Omega)$ and $g \in H_{1/2}(\partial\Omega)$ are prescribed, $\gamma u = \{\gamma_1 u_1, \gamma_\varepsilon u_\varepsilon\}$, γ_1 and γ_ε are the continuous trace operators from $H_1(\Omega_1)$ and $H_1(\Omega_\varepsilon)$ to $H_{1/2}(\partial\Omega_1)$ and $H_{1/2}(\partial\Omega_\varepsilon)$, respectively, and γ_{10} and $\gamma_{\varepsilon 0}$ are the continuous trace operators from $H_1(\Omega_1)$ and $H_1(\Omega_\varepsilon)$ to $H_{1/2}(\partial\Omega_0)$. The continuity of the normal component of the flux across the internal interface is accounted for by the fact that the variational equation in (20.6) has no term defined on $\partial\Omega_0$.

Theorem 1. *The variational problem* (20.6) *has a unique solution, which satisfies the estimate*

$$\|u\|_{H_1(\Omega)} \le c(\|f\|_{H_{-1}(\Omega)} + \|g\|_{H_{1/2}(\partial\Omega)}), \quad c = \text{const} > 0.$$

20.3 Numerical Method

We denote by $u_1^{(k)}$ the kth iterate approximating u in Ω_1, and by $u_\varepsilon^{(k)}$ the kth iterate approximating u in Ω_ε, $k = 1, 2, \ldots$, and write

$$u^{(k)}(x,y) = \begin{cases} u_1^{(k)}(x,y), & (x,y) \in \Omega_1, \\ u_\varepsilon^{(k)}(x,y), & (x,y) \in \Omega_\varepsilon. \end{cases}$$

We are using a version of the Dirichlet/Neumann (DN) and Adaptive Dirichlet/Neumann (ADN) methods (see [1] and [2]). The difference is that here, the interface condition of continuity of the normal derivative [1] is replaced by the continuity of the normal component of the flux. In addition, our method is applied to a domain where the decomposition into subregions arises from the discontinuity of the diffusion coefficient. Our technique is described below to $O(\varepsilon)$.

Let

$$u_1^{(1)}(x,0+) = \lambda^{(0)}(x)$$

be an initial guess for $u(x,0+)$, $0 \le x \le L$. For $k = 1,2,\dots$, we use the Dirichlet boundary condition on $\partial\Omega_0$ $(y = 0+)$ to solve

$$-\nabla^2 u_1^{(k)} + b(\partial_y u_1^{(k)}) + c u_1^{(k)} = f(x,y), \quad (x,y) \in \Omega_1, \tag{20.7}$$

$$\begin{aligned} u_1^{(k)}(0,y) &= g(0,y), \quad u_1^{(k)}(L,y) = g(L,y), \quad 0 \le y \le H_2, \\ u_1^{(k)}(x,H_2) &= g(x,H_2), \quad u_1^{(k)}(x,0+) = \lambda^{(k-1)}(x), \quad 0 \le x \le L, \end{aligned} \tag{20.8}$$

then the Neumann boundary condition on $\partial\Omega_0$ $(y = 0-)$ to solve

$$-\varepsilon\, \nabla^2 u_\varepsilon^{(k)} + b(\partial_y u_\varepsilon^{(k)}) + c u_\varepsilon^{(k)} = f(x,y), \quad (x,y) \in \Omega_\varepsilon, \tag{20.9}$$

$$\begin{aligned} u_\varepsilon^{(k)}(0,y) &= g(0,y), \quad u_\varepsilon^{(k)}(L,y) = g(L,y), \quad -H_1 \le y \le 0, \\ u_\varepsilon^{(k)}(x,-H_1) &= g(x,-H_1), \\ &\varepsilon(\partial_y u_\varepsilon^{(k)})(x,0-) = (\partial_y u_1^{(k)})(x,0+), \quad 0 \le x \le L. \end{aligned} \tag{20.10}$$

In Ω_1 we now update the Dirichlet boundary condition at the interface $(y = 0+)$ by setting, for $k = 1,2,\dots$,

$$\lambda^{(k)}(x) = (1-\theta)\lambda^{(k-1)}(x) + \theta u_\varepsilon^{(k)}(x,0-), \quad 0 \le x \le L. \tag{20.11}$$

The continuity across the interface of the normal flux component of each iterate is imposed by the Neumann boundary condition at $y = 0-$. But the iterates $u^{(k)}(x,y)$, $k = 1,2,\dots$, are generally not continuous across $\partial\Omega_0$. However, if the method converges, then the limit as $k \to \infty$ of these iterates is continuous across the interface.

The problem in Ω_1 is solved numerically, for example, by means of finite elements. Our method has the advantage that, at each iteration, the problem in Ω_ε does not need to be solved explicitly for $u_\varepsilon^{(k)}(x,y)$, $k = 1,2,\dots$. This needs to be done only at the final iteration.

The problem on Ω_ε is a singular perturbation problem whose solution can be approximated by means of the boundary layer method with matched asymptotic expansions [4]. The main boundary layer is at $y = 0-$. For the kth iteration, the input Neumann boundary data on Ω_ε at $y = 0-$ are obtained by approximating $\partial_y u_1^{(k)}$ at $y = 0+$ using the output from the problem on Ω_1 and invoking the continuity of the normal flux iterates at the interface; that is,

$$\varepsilon(\partial_y u_\varepsilon^{(k)})(x,0-) = (\partial_y u_1^{(k)})(x,0+), \quad 0 \le x \le L, \quad k = 1,2,\dots.$$

Matched asymptotic expansions to $O(\varepsilon)$ then lead to

$$
\begin{aligned}
u_\varepsilon^{(k)}(x,y) &= b^{-1}(\partial_y u_1^{(k)})(x,0+)e^{by/\varepsilon} \\
&\quad + b^{-1}\int_{-H_1}^{y} e^{(c/b)(s-y)} f(x,s)\,ds + e^{-(c/b)(y+H_1)} g(x,-H_1) \\
&\quad + O(\varepsilon), \quad 0 \le x \le L,\ -H_1 \le y \le 0, \quad k = 1,2,\ldots,
\end{aligned}
$$

so

$$
\begin{aligned}
u_\varepsilon^{(k)}(x,0-) &= b^{-1}(\partial_y u_1^{(k)})(x,0+) \\
&\quad + b^{-1}\int_{-H_1}^{0} e^{(c/b)(s)} f(x,s)\,ds + e^{-(c/b)(H_1)} g(x,-H_1) \\
&\quad + O(\varepsilon), \quad 0 \le x \le L, \quad k = 1,2,\ldots.
\end{aligned}
\tag{20.12}
$$

The boundary layer solutions at $x = 0+$ and/or $x = L-$ can be obtained by means of the Laplace transformation and can be added at the end of the computations in order to obtain the complete solution.

The first term on the right-hand side in (20.12) is known from the previous step in Ω_1. The second and third terms are independent of the iteration process and depend only on the data, so they need to be computed only once. Hence, when $(\partial_y u_1^{(k)})(x,0+)$ has been computed in Ω_1, we substitute it into (20.12) to obtain $u_\varepsilon^{(k)}(x,0-)$, use this in (20.11) to obtain $\lambda^{(k)}(x)$, and so on.

20.4 Convergence

The error iterates in Ω_1 and Ω_ε, defined by

$$
\left.
\begin{aligned}
e_1^{(k)}(x,y) &= u(x,y) - u_1^{(k)}(x,y), \\
e_\varepsilon^{(k)}(x,y) &= u(x,y) - u_\varepsilon^{(k)}(x,y),
\end{aligned}
\right. \quad k = 1,2,\ldots,
$$

are found by separation of variables in the associated homogeneous problems. We write

$$
\beta^{(k-1)}(x) = e_1^{(k)}(x,0+) = u(x,0+) - \lambda^{(k-1)}(x), \quad 0 \le x \le L, \quad k = 1,\ldots.
$$

For $k = 1,2,\ldots$, in Ω_1 we have

$$
\begin{aligned}
-\nabla^2 e_1^{(k)} + b(\partial_y e_1^{(k)}) + c e_1^{(k)} &= 0, \quad (x,y) \in \Omega_1, \\
e_1^{(k)}(0,y) = 0, \quad e_1^{(k)}(L,y) &= 0, \quad 0 \le y \le H_2, \\
e_1^{(k)}(x,H_2) = 0, \quad e_1^{(k)}(x,0+) &= \beta^{(k-1)}(x), \quad 0 \le x \le L;
\end{aligned}
$$

the error is

$$e_1^{(k)}(x,y) = \sum_{n=1}^{\infty} C_{1,n}^{(k)} e^{b(y-H_2)/2} \sinh\left\{\left[\left(\frac{b}{2}\right)^2 + c + \left(\frac{n\pi}{L}\right)^2\right]^{1/2}(-H_2+y)\right\} \times \sin\left(\frac{n\pi x}{L}\right), \quad k=1,2,\dots, \tag{20.13}$$

where

$$C_{1,n}^{(k)} = \frac{e^{bH_2/2}}{\sinh\left\{\left[\left(\frac{b}{2}\right)^2 + c + \left(\frac{n\pi}{L}\right)^2\right]^{1/2}(-H_2)\right\}} \times \frac{2}{L}\int_0^L \beta^{(k-1)}(x)\sin\left(\frac{n\pi x}{L}\right)dx, \quad n=1,2,\dots,\ k=1,2,\dots.$$

For $k=1,2,\dots,$ in Ω_ε we have

$$\begin{gathered}
-\varepsilon\,\nabla^2 e_\varepsilon^{(k)} + b(\partial_y e_\varepsilon^{(k)}) + c e_\varepsilon^{(k)} = 0, \quad (x,y)\in\Omega_\varepsilon, \\
e_\varepsilon^{(k)}(0,y)=0, \quad e_\varepsilon^{(k)}(L,y)=0, \quad -H_1\le y\le 0, \\
e_\varepsilon^{(k)}(x,-H_1)=0, \quad \varepsilon(\partial_y e_\varepsilon^{(k)})(x,0-) = (\partial_y e_1^{(k)})(x,0+), \quad 0\le x\le L,
\end{gathered}$$

and the error is

$$e_\varepsilon^{(k)}(x,y) = \sum_{n=1}^{\infty} C_{\varepsilon,n}^{(k)} e^{b(y+H_1)/(2\varepsilon)} \sinh\left\{\left[\left(\frac{b}{2\varepsilon}\right)^2 + \frac{c}{\varepsilon} + \left(\frac{n\pi}{L}\right)^2\right]^{1/2}(H_1+y)\right\} \times \sin\left(\frac{n\pi x}{L}\right), \quad k=1,2,\dots, \tag{20.14}$$

where

$$C_{\varepsilon,n}^{(k)} = C_{1,n}^{(k)}\, e^{-b(H_1+\varepsilon H_2)/(2\varepsilon)}\, \frac{2r_n\cosh(r_nH_2) - b\sinh(r_nH_2)}{2\varepsilon s_n\cosh(s_nH_1) + b\sinh(s_nH_1)}, \quad n=1,2,\dots,\ k=1,2,\dots. \tag{20.15}$$

Here

$$\left.\begin{aligned}
r_n &= \left[\left(\frac{b}{2}\right)^2 + c + \left(\frac{n\pi}{L}\right)^2\right]^{1/2}, \\
s_n &= \left[\left(\frac{b}{2\varepsilon}\right)^2 + \frac{c}{\varepsilon} + \left(\frac{n\pi}{L}\right)^2\right]^{1/2},
\end{aligned}\right\} \quad n=1,2,\dots.$$

Applying the update formula (20.11), we find that

$$\beta^{(k)}(x) = (1-\theta)\beta^{(k-1)}(x) + \theta e_\varepsilon^{(k)}(x,0-), \quad 0 \le x \le L, \quad k = 1,2,\ldots.$$

Using (20.13) and (20.14) and replacing $\beta^{(k-1)}(x) = e_1^{(k)}(x,0+)$ and $e_\varepsilon^{(k)}(x,0-)$, $k = 1,2,\ldots$, by their Fourier sine series representations on $[0,L]$, we see that the kth error iterate $\beta^{(k)}(x)$ at $y = 0+$ is given by

$$\begin{aligned}\beta^{(k)}(x) &= (1-\theta)\beta^{(k-1)}(x) + \theta e_\varepsilon^{(k)}(x,0-)\\ &= (1-\theta)\sum_{n=1}^{\infty} C_{1,n}^{(k)} e^{-bH_2/2}\sinh(-r_nH_2)\sin\left(\frac{n\pi x}{L}\right)\\ &\quad + \theta\sum_{n=1}^{\infty} C_{\varepsilon,n}^{(k)} e^{bH_1/(2\varepsilon)}\sinh(s_nH_1)\sin\left(\frac{n\pi x}{L}\right),\\ &\qquad 0 \le x \le L, \quad k = 1,2,\ldots. \end{aligned} \tag{20.16}$$

Expressing $\beta^{(k)}$ on the left-hand side in (20.16) as a Fourier sine series, we have

$$\begin{aligned}&\sum_{n=1}^{\infty} C_{1,n}^{(k+1)} e^{-bH_2/2}\sinh(-r_nH_2)\sin\left(\frac{n\pi x}{L}\right)\\ &\quad = (1-\theta)\sum_{n=1}^{\infty} C_{1,n}^{(k)} e^{-bH_2/2}\sinh(-r_nH_2)\sin\left(\frac{n\pi x}{L}\right),\\ &\qquad + \theta\sum_{n=1}^{\infty} C_{\varepsilon,n}^{(k)} e^{bH_1/(2\varepsilon)}\sinh(s_nH_1)\sin\left(\frac{n\pi x}{L}\right),\\ &\qquad\qquad 0 \le x \le L, \quad k = 1,2,\ldots. \end{aligned} \tag{20.17}$$

We now make use of the orthogonality of the functions $\sin(n\pi x/L)$, $n = 1,2,\ldots$, on $[0,L]$ in (20.17) to determine a connection between the nth Fourier sine series coefficients of $\beta^{(k)}$ and those of $\beta^{(k-1)}$ of the form

$$\begin{aligned}&C_{1,n}^{(k+1)} e^{-bH_2/2}\sinh(-r_nH_2)\\ &\quad = (1-\theta)C_{1,n}^{(k)} e^{-bH_2/2}\sinh(-r_nH_2) + \theta\, C_{\varepsilon,n}^{(k)} e^{bH_1/(2\varepsilon)}\sinh(s_nH_1)\\ &\qquad\qquad n = 1,2,\ldots,\ k = 1,2,\ldots. \end{aligned} \tag{20.18}$$

From (20.15) and (20.18) it follows that

$$\begin{aligned}C_{1,n}^{(k+1)} &= (1-\theta)C_{1,n}^{(k)}\\ &\quad + \theta\, C_{1,n}^{(k)} \frac{2r_n\cosh(r_nH_2) - b\sinh(r_nH_2)}{2\varepsilon s_n\cosh(s_nH_1) + b\sinh(s_nH_1)}\,\frac{\sinh(s_nH_1)}{\sinh(-r_nH_2)}\\ &= \big[1-\theta\big(1-\gamma(\varepsilon,b,c,L,H_1,H_2,n)\big)\big]C_{1,n}^{(k)},\\ &\qquad\qquad n = 1,2,\ldots,\ k = 1,2,\ldots, \end{aligned} \tag{20.19}$$

where

$$\gamma(\varepsilon,b,c,L,H_1,H_2,n)=\frac{1}{\varepsilon}\left(\frac{(b/2)-r_n\coth(r_nH_2)}{(b/(2\varepsilon))+s_n\coth(s_nH_1)}\right),\qquad n=1,2,\dots. \tag{20.20}$$

From (20.19) and (20.20) we obtain

$$C_{1,n}^{(k+1)}=\alpha(\theta,\varepsilon,b,c,L,H_1,H_2,n)C_{1,n}^{(k)},\quad n=1,2,\dots,\ k=1,2,\dots,$$

where the reduction factors α are given by

$$\alpha(\theta,\varepsilon,b,c,L,H_1,H_2,n)=1-\theta\big(1-\gamma(\varepsilon,b,c,L,H_1,H_2,n)\big),\quad n=1,2,\dots.$$

Theorem 2. (i) *The iterative method* (20.7)–(20.11) *converges in* $L^2(0,L)$ *to the unique continuous solution of* (20.1), (20.2) *if and only if*

$$|\alpha(\theta,\varepsilon,b,c,L,H_1,H_2,n)|<1,\quad n=1,2,\dots.$$

(ii) *If convergence occurs for a specific value of* θ*, then it also does for all smaller positive values of* θ*.*

(iii) *If there are positive proper fractions* $\bar\theta$ *and* $\bar\alpha$ *such that*

$$|\alpha(\bar\theta,\varepsilon,b,c,L,H_1,H_2,n)|\le\bar\alpha,\quad n=1,2,\dots,$$

then the method converges pointwise on $[0,L]$ *for* $\bar\theta$*, with convergence rate at least* $\bar\alpha$*.*

For example, in the test problem (20.1), (20.2) with $b=c=L=H_1=H_2=1$ and $\varepsilon=0.1$, if we choose $\theta=0.15$, then we can take $\bar\alpha=0.65$; in this case,

$$\|\beta^{(k)}\|_{L^2(0,L)}=(0.65)^k\|\beta^{(0)}\|_{L^2(0,L)}\to 0\quad\text{as }k\to\infty.$$

20.5 Computational Results

The numerical solution in Ω_1 was computed by means of the finite element method with piecewise linear triangular elements. The uniform finite element grid had 11 nodes in both the x and y directions.

The method of matched asymptotic expansions to $O(\varepsilon^2)$ was used in Ω_ε. Numerically, this modified the expression of the Dirichlet boundary condition for the next iteration in Ω_1 (at $y=0+$) from that given by (20.11) to

$$\lambda^{(k)}(x)=(1-\theta)\lambda^{(k-1)}(x)+\theta(1-\varepsilon)u_\varepsilon^{(k)}(x,0-),\ 0\le x\le L,\ k=1,2,\dots.$$

The numerical results were obtained with Mathematica Version 4 for the test problem (20.1), (20.2) with $H_1 = H_2 = L = b = c = 1$, $\theta = 0.25$, $\varepsilon = 0.01$, and data functions

$$f_1(x,y) = 2(1+y) + x(1-x)(2+y),$$

$$f_\varepsilon = \frac{1}{\varepsilon^2}\, e^{y/\varepsilon}\big[2\varepsilon^2 + (\varepsilon^2 + \varepsilon - 1)x(1-x)\big],$$

$$g_1(x,y) = \begin{cases} 2x(1-x), & 0 \le x \le 1,\ y = 1, \\ 0, & x = 0,\ 0 < y < 1, \\ 0, & x = 1,\ 0 < y < 1, \end{cases}$$

$$g_\varepsilon(x,y) = \begin{cases} x(1-x)e^{-1/\varepsilon}, & 0 \le x \le 1,\ y = -1, \\ 0, & x = 0,\ -1 < y < 0, \\ 0, & x = 1,\ -1 < y < 0, \end{cases}$$

The exact solution is

$$u_1(x,y) = x(1-x)(1+y),$$
$$u_\varepsilon(x,y) = x(1-x)e^{y/\varepsilon}.$$

Both the exact solution and its normal flux are continuous across $y = 0$, $0 \le x \le 1$. The iterations were started using the linear interpolant of the given boundary values at $(x,y) = (0,0)$ and $(x,y) = (1,0)$ as an initial guess for $\lambda^{(0)}$; in other words, $\lambda^{(0)}(x) = 0$, $0 \le x \le 1$.

The values of the interface error

$$e^{(10)}(x,0+) = x(1-x) - u^{(10)}(x,0+)$$

after 10 iterations can be found in Table 1.

Table 1. The error at $y = 0+$ after 10 iterations.

x	$e^{(10)}(x,0+)$
0.0	0.0
0.1	0.00768254
0.2	−0.00533406
0.3	0.000561597
0.4	0.00116317
0.5	0.00075847
0.6	0.00106993
0.7	0.000327891
0.8	−0.00486947
0.9	0.00744013
1.0	0.0

20.6 Conclusions

We have proposed an efficient iterative domain decomposition method that solves general convection-diffusion singular perturbation problems. Our specific application involves a piecewise constant diffusion coefficient. We have established sufficient conditions for the convergence of the method, identified suitable values of the relaxation parameter θ, and started investigations into the rate of convergence. The method was implemented to $O(\varepsilon^2)$ to solve test problems.

Full details of the proofs of these assertions will appear in a future publication.

References

1. C. Carlenzoli and A. Quarteroni, Adaptive domain decomposition methods for advection-diffusion problems, in *Modeling, Mesh Generation, and Adaptive Numerical Methods for Partial Differential Equations,* IMA Vol. Math. Appl. **75**, Springer-Verlag, New York, 1995, 165–186.
2. A. Quarteroni and A. Valli, *Domain Decomposition Methods for Partial Differential Equations,* Clarendon Press, Oxford, 1999.
3. I. Chudinovich and C. Constanda, *Variational and Potential Methods in the Theory of Bending of Plates with Transverse Shear Deformation,* Chapman & Hall/CRC, Boca Raton-London-New York-Washington, DC, 2000.
4. C. Constanda, *Solution Techniques for Elementary Partial Differential Equations,* Chapman & Hall/CRC, Boca Raton-London-New York-Washington, DC, 2002.

21 Integral Representation of the Solution of Torsion of an Elliptic Beam with Microstructure

Stanislav Potapenko

21.1 Introduction

The theory of micropolar elasticity [1] was developed to account for discrepancies between the classical theory and experiments when the effects of material microstructure were known to significantly affect a body's overall deformation. The problem of torsion of micropolar elastic beams has been considered in [2] and [3]. However, the results in [2] are confined to the simple case of a beam with circular cross section while the analysis in [3] overlooks certain differentiability requirements that are essential to establish the rigorous solution of the problem (see, for example, [4]). In neither case is there any attempt to quantify the influence of material microstructure on the beam's deformation.

The treatment of the torsion problem in micropolar elasticity requires the rigorous analysis of a Neumann-type boundary value problem in which the governing equations are a set of three second-order coupled partial differential equations for three unknown antiplane displacement and microrotation fields. This is in contrast to the relatively simple torsion problem arising in classical linear elasticity, in which a single antiplane displacement is found from the solution of a Neumann problem for Laplace's equation [5]. This means that in the case of a micropolar beam with noncircular cross section it is extremely difficult (if not impossible) to find a closed-form analytic solution to the torsion problem.

In this paper, we use a simple, yet effective, numerical scheme based on an extension of Kupradze's method of generalized Fourier series [6] to approximate the solution of the problem of torsion of an elliptic micropolar beam. Our numerical results demonstrate that the material microstructure does indeed have a significant effect on the torsional function and the subsequent warping of a typical cross section.

21.2 Torsion of Micropolar Beams

Let V be a domain in the real three-dimensional space $\mathbb{R}^3$, occupied by a homogeneous and isotropic linearly elastic micropolar material with elastic constants λ, μ, α, β, γ, and κ, whose boundary is denoted by ∂V. The deformation of a micropolar elastic solid can be characterized by a displacement field of the form

$$U(x) = (u_1(x), u_2(x), u_3(x))^T$$

and a microrotation field of the form

$$\Phi(x) = (\varphi_1(x), \varphi_2(x), \varphi_3(x))^T,$$

where $x = (x_1, x_2, x_3)$ is a generic point in $\mathbb{R}^3$ and a superscript T indicates matrix transposition. We consider an isotropic, homogeneous, prismatic micropolar beam bounded by plane ends perpendicular to the generators. A typical cross section S is assumed to be a simply connected region bounded by a closed C^2-curve ∂S with outward unit normal $n = (n_1, n_2)^T$. Taking into account the basic relations describing the deformations of a homogeneous and isotropic, linearly elastic micropolar solid [7], we can formulate the problem of torsion of a cylindrical micropolar beam (see, for example, [2] and [3]) as an interior Neumann problem of antiplane micropolar elasticity [8]:

Find $u \in C^2(S) \cap C^1(S \cup \partial S)$ satisfying

$$L(\partial x)u(x) = 0, \quad x \in S, \tag{21.1}$$

such that

$$T(\partial x)u(x) = f(x), \quad x \in \partial S. \tag{21.2}$$

Here, $L(\partial x)$ is the (3×3)-matrix partial differential operator corresponding to the governing equations of torsion of a micropolar beam [3], $u(x_1, x_2) = (\varphi_1(x_1, x_2), \varphi_2(x_1, x_2), u_3(x_1, x_2))^T$, $T(\partial x)$ is the boundary stress operator [3], and $f = (\gamma n_1, \gamma n_2, \mu(x_2 n_1 - x_1 n_2))^T$.

In [8], the boundary integral equation method is used to prove existence and uniqueness results in the appropriate function spaces for the boundary value problem (21.1), (21.2). As part of this analysis, it is shown that the solution of (21.1), (21.2) can be expressed in the form of an integral potential.

21.3 Generalized Fourier Series

Let ∂S_* be a simple closed Liapunov curve such that ∂S lies strictly inside the domain S_* enclosed by ∂S_*, and let $\{x^{(k)} \in \partial S_*,\ k = 1, 2, \dots\}$ be a countable set of points densely distributed on ∂S_*. We set $S_*^- = \mathbb{R}^2 \backslash \bar{S}_*$, denote by $D^{(i)}$ the columns of the fundamental matrix D [8] and by $F^{(i)}$ the columns (of matrix F) that form the basis of the set of rigid displacement and microrotations associated with (21.1) and (21.2); that is,

$$F = \begin{bmatrix} 0 & 0 & 0 \\ 0 & 0 & 0 \\ 0 & 0 & 1 \end{bmatrix}. \tag{21.3}$$

The following result is fundamental to the numerical scheme used to approximate the solution of the micropolar torsion problem. Its proof proceeds as in [6].

Theorem 1. *The set*

$$\{F^{(3)}, \theta^{(jk)},\ j=1,2,3,\ k=1,2,\ldots\}, \tag{21.4}$$

where $F^{(3)}$ is the third column of matrix (21.3) and

$$\theta^{(jk)}(x) = T(\partial x) D^{(j)}(x, x^{(k)}),$$

is linearly independent on ∂S and fundamental in $L^2(\partial S)$.

If we now introduce the new sequence $\{\eta^{(n)}\}_{n=1}^{\infty}$ obtained from (21.4) by means of a Gram–Schmidt orthonormalization process, and use the integral representation formula for the solution of a boundary value problem (Somigliana formula) [8], then, as in [9], we can derive the approximate solution for the torsion problem in the form of a generalized Fourier series, that is,

$$u^{(n)}(x) = \tilde{q}_3 \tilde{F}^{(3)} - \sum_{r=1}^{n} q_r \int_{\partial S} P(x,y)\eta^{(r)}(y)ds(y) + G(x), \quad x \in S. \tag{21.5}$$

Here, the first term on the right-hand side is a rigid displacement independent of n, the Fourier coefficients q_r are computed by means of the procedure discussed in [6] and [9], $P(x,y)$ is a matrix of singular solutions [8], and $G(x)$ is given by

$$G(x) = \int_{\partial S} D(x,y) f(y) ds(y), \quad x \in \mathbb{R}^2 \backslash \partial S.$$

Since $\tilde{q}_3$ cannot be determined in terms of the boundary data of the problem, we can conclude that the solution is unique up to an arbitrary rigid displacement/microrotation, which is consistent with the results obtained in [8].

This numerical method is extremely attractive in that it inherits all the advantages of the boundary integral equation method and, as in the following example, can be shown to produce accurate, fast-converging, and effective results.

21.4 Example: Torsion of an Elliptic Beam

To verify the numerical method, it is a relatively simple matter to show that for the problem of a *circular* micropolar beam, the numerical scheme produces results that converge rapidly to the exact solution established in [2] (that the cross section does not warp, i.e., that the material microstructure is insignificant in the torsion of a *circular* micropolar bar). Of more interest, however, is the case of an elliptic micropolar bar [5], which, to the author's knowledge, remains absent from the literature.

As an example, consider the torsion of a micropolar beam of elliptic cross section in which the elastic constants take the values $\alpha = 3$, $\beta = 6$, $\gamma = 2$, $\kappa = 1$, and $\mu = 1$. The domain S is bounded by the ellipse

$$x_1 = \cos t, \quad x_2 = 1.5 \sin t.$$

As the auxiliary contour ∂S_* we take the confocal ellipse

$$x_1 = 1.1 \cos t, \quad x_2 = 1.6 \sin t.$$

Using the Gauss quadrature formula with 16 ordinates to evaluate the integrals over ∂S and following the computational procedure discussed in [6] and [9], the approximate solution (21.5) is found to converge to eight decimal places for $n = 62$ terms of the series. Numerical values are presented in Table 1 for representative points $(0,0)$, $(0.25, 0.25)$, $(0.5, 0.5)$, and $(0.5, 0.75)$ inside the elliptic cross section.

Table 1. Approximate solution of micropolar beam with elliptic cross section for $n = 62$ in (21.5).

Point in cross section	$(0,0)$	$(0.25, 0.25)$	$(0.5, 0.5)$	$(0.5, 0.75)$
φ_1	0.74431942	1.17355112	1.24343810	1.82784247
φ_2	0.48152259	0.97222035	1.11246544	1.36181203
u_3	0.00006160	0.02139392	0.08461420	0.12380739

Here φ_1 is the microrotation about the x_1-axis, φ_2 the microrotation about the x_2-axis, and u_3 the antiplane displacement.

Note that if we compare the values of the out-of-plane displacement or torsional function u_3 with those obtained in the case of a classical elastic elliptic beam (which are 0, 0.02403812, 0.09615251, and 0.14422874 at the same points, based on the exact solution for the warping function [5]), we conclude that there is a difference of up to 15% between them at certain points. (In addition to the results in Table 1, we considered several other interior points of the ellipse and arrived at a similar conclusion).

In contrast to the case of a circular micropolar beam, for which the cross section remains flat [2] (as in the classical case [5]), there is a significant difference in the torsional function for an elliptic beam made of micropolar material when compared to the same beam in which the microstructure is ignored (i.e., the classical case [5]).

The method used here is easily extended, with only minor changes of detail, to the analysis of the torsion of micropolar beams of any (smooth) cross section, where we again expect a significant contribution from the material microstructure.

References

1. A.C. Eringen, Linear theory of micropolar elasticity, *J. Math. Mech.* **15** (1966), 909–923.
2. A.C. Smith, Torsion and vibrations of cylinders of a micropolar elastic solid, in *Recent Adv. Engng. Sci.* **5**, A.C. Eringen (ed.), Gordon and Breach, New York, 1970, 129–137.
3. D. Iesan, Torsion of micropolar elastic beams, *Int. J. Engng. Sci.* **9** (1971), 1047–1060.
4. P. Schiavone, On existence theorems in the theory of extensional motions of thin micropolar plates, *Int. J. Engng. Sci.* **27** (1989), 1129–1133.
5. S. Timoshenko and J. Goodier, *Theory of Elasticity*, McGraw-Hill, New York, 1970.
6. V.D. Kupradze, T.G. Gegelia, M.O. Basheleishvili, and T.V. Burchuladze, *Three-Dimensional Problems of the Mathematical Theory of Elasticity and Thermoelasticity,* North-Holland, Amsterdam, 1979.
7. W. Nowacki, *Theory of Asymmetric Elasticity,* Polish Scientific Publ., Warsaw, 1986.
8. S. Potapenko, P. Schiavone, and A. Mioduchowski, Antiplane shear deformations in a linear theory of elasticity with microstructure, *Z. Angew. Math. Phys.* **56** (2005), 516–528.
9. C. Constanda, *A Mathematical Analysis of Bending of Plates with Transverse Shear Deformation*, Longman-Wiley, Harlow-New York, 1990.

22 A Coupled Second-order Boundary Value Problem at Resonance

Seppo Seikkala and Markku Hihnala

22.1 Introduction

We continue the study started in [1] of the boundary value problem (BVP)

$$\begin{pmatrix} x_1'' \\ x_2'' \end{pmatrix} + A \begin{pmatrix} x_1 \\ x_2 \end{pmatrix} = \begin{pmatrix} f_1(ax_1 + bx_2) \\ f_2(cx_1 + dx_2) \end{pmatrix} + \begin{pmatrix} b_1(t) \\ b_2(t) \end{pmatrix}, \tag{22.1}$$
$$x_1(0) = x_2(0) = x_1(\pi) = x_2(\pi) = 0,$$

now considering the case where the null space of the differential operator on the left-hand side in (22.1), with the given Dirichlet boundary conditions, is two dimensional.

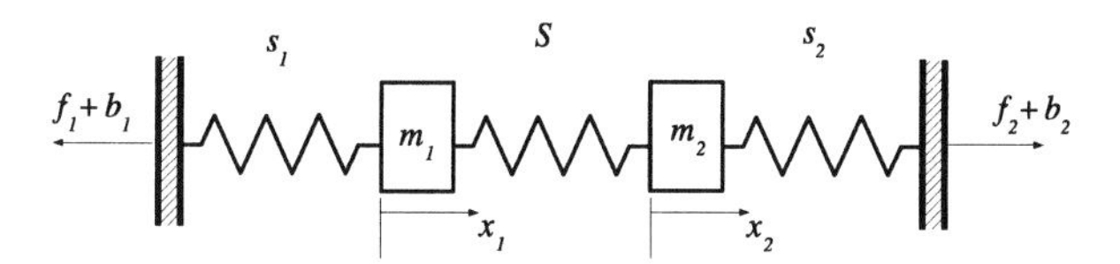

Fig. 1. A coupled spring-mass system.

Problems of this type arise, for example, in mechanics (coupled oscillators) or in coupled circuits theory [2]. Thus, for the spring-mass system in Fig. 1 (m_1 and m_2 are the masses and s_1, s_2, and S the stiffnesses) we have

$$A = \begin{pmatrix} a_1 & -\alpha \\ -\beta & a_2 \end{pmatrix}, \tag{22.2}$$

where $a_1 = s_1/m_1 + S/m_1$, $a_2 = s_2/m_2 + S/m_2$, $\alpha = S/m_1$, and $\beta = S/m_2$. The special case where $s_1/m_1 = s_2/m_2 = 1$ and, hence, where

$$A = \begin{pmatrix} \alpha + 1 & -\alpha \\ -\beta & \beta + 1 \end{pmatrix},$$

was studied in [3]. The case $s_1 = s_2 = m_1 = m_2$,

$$A = \begin{pmatrix} \alpha+1 & -\alpha \\ -\alpha & \alpha+1 \end{pmatrix},$$

was studied in [4]. In [1], we considered more general problems (22.1) but still having a one-dimensional null space for the differential operator with the given Dirichlet boundary conditions. If, for example,

$$m_1 = \frac{64}{81}, \quad m_2 = 1, \quad s_1 = \frac{32}{81}, \quad s_2 = \frac{13}{9}, \quad S = \frac{32}{9},$$

that is,

$$A = \begin{pmatrix} 5 & -\frac{9}{2} \\ -\frac{32}{9} & 5 \end{pmatrix},$$

then the null space is two dimensional.

The change of variables $z = T^{-1}x$, where T is a matrix that diagonalizes A and has the eigenvectors of A as columns, transforms problem (22.1) into a system

$$\begin{pmatrix} z_1'' + d_1 z_1 \\ z_2'' + d_2 z_2 \end{pmatrix} = T^{-1}\begin{pmatrix} f_1(ax_1+bx_2) \\ f_2(cx_1+dx_2) \end{pmatrix} + T^{-1}\begin{pmatrix} b_1(t) \\ b_2(t) \end{pmatrix}, \tag{22.3}$$
$$z_1(0) = z_2(0) = z_1(\pi) = z_2(\pi) = 0,$$

where $x = \begin{pmatrix} x_1 \\ x_2 \end{pmatrix} = Tz$ and d_1 and d_2 are the eigenvalues of A. For the matrix A in (22.2) we may choose

$$T = \begin{pmatrix} \alpha & \gamma \\ \gamma & -\beta \end{pmatrix}, \quad T^{-1} = \frac{1}{\alpha\beta+\gamma^2}\begin{pmatrix} \beta & \gamma \\ \gamma & -\alpha \end{pmatrix},$$

where either $\gamma = \frac{1}{2}(a_1 - a_2) + \frac{1}{2}\left[(a_1-a_2)^2 + 4\alpha\beta\right]^{1/2}$ or $\gamma = \frac{1}{2}(a_1 - a_2) - \frac{1}{2}\left[(a_1-a_2)^2 + 4\alpha\beta\right]^{1/2}$.

If $d_1 = {l_1}^2$ and $d_2 = {l_2}^2$, where l_1 and l_2 are positive integers, then we have a resonance case, and the null space of the differential operator E in the coupled system (22.1), defined by

$$Ex = \begin{pmatrix} x_1'' \\ x_2'' \end{pmatrix} + A\begin{pmatrix} x_1 \\ x_2 \end{pmatrix},$$

with the given Dirichlet boundary conditions, is spanned by the functions

$$\Phi(t) = T\begin{pmatrix} 1 \\ 0 \end{pmatrix}\sin(l_1 t) \quad \text{and} \quad \eta(t) = T\begin{pmatrix} 0 \\ 1 \end{pmatrix}\sin(l_2 t).$$

We study the existence of the solutions of (22.1) in terms of the parameters $\bar{b}_1$ and $\bar{b}_2$ in the decomposition

$$b(t) = \begin{pmatrix} b_1(t) \\ b_2(t) \end{pmatrix} = \bar{b}_1\Phi(t) + \bar{b}_2\eta(t) + T\begin{pmatrix} \tilde{b}_1(t) \\ \tilde{b}_2(t) \end{pmatrix}, \tag{22.4}$$

where $\tilde{b}_1$ is orthogonal to $\psi_1(t) = (2/\pi)^{1/2}\sin(l_1 t)$ and $\tilde{b}_2$ is orthogonal to $\psi_2(t) = (2/\pi)1/2\sin(l_2 t)$. The motivation for decomposition (22.4) is that the BVP

$$Ex = b, \quad x(0) = x(\pi) = 0$$

has a solution if and only if $\bar{b}_1 = \bar{b}_2 = 0$.

22.2 Results

Let X be the set of all continuous functions $x : [0, T] \to \mathbb{R}^2$, and suppose that f_1 and f_2 are continuous and bounded. For a fixed $\lambda = (\lambda_1, \lambda_2) \in \mathbb{R}^2$, consider the integral equation system

$$z = \lambda\psi + KNz, \tag{22.5}$$

where

$$Nz = N\begin{pmatrix} z_1 \\ z_2 \end{pmatrix} = T^{-1}\begin{pmatrix} f_1(ax_1 + bx_2) + b_1 \\ f_2(cx_1 + dx_2) + b_2 \end{pmatrix},$$

$x = Tz$, $\psi = (\psi_1, \psi_2)^T$, $\lambda\psi = (\lambda_1\psi_1, \lambda_2\psi_2)^T$, and the linear operator $K : X \to X$ is defined by

$$Kz = \begin{pmatrix} \int_0^\pi k_1(t,s)z_1(s)\,ds \\ \int_0^\pi k_2(t,s)z_2(s)\,ds \end{pmatrix};$$

here the k_i are the solutions of the problems

$$L_i k_i(t,s) = \delta(t-s) - \psi_i(t)\psi_i(s),$$
$$k_i(0,s) = k_i(\pi,s) = 0,$$
$$\int_0^\pi k_i(t,s)\psi_i(t)\,dt = 0, \quad i = 1, 2,$$

$L_1 u = u'' + {l_1}^2 u$ and $L_2 u = u'' + {l_2}^2 u$. Since f_1 and f_2 are continuous and bounded, it follows that for a fixed $\lambda \in \mathbb{R}^2$, system (22.5) has at least one solution. For any such solution z^λ, we write $x^\lambda = Tz^\lambda$, $\delta(\lambda) = (\delta_1(\lambda), \delta_2(\lambda))$,

$$\delta_i(\lambda) = \int_0^\pi (Nz)_i(t)\psi_i(t)\,dt,$$

and

$$\tilde{\delta}_i(\lambda) = \delta_i(\lambda) - \bar{b}_i, \quad i = 1, 2.$$

We now easily deduce that the BVP (22.3) is equivalent to the pair of equations

$$\begin{aligned} z &= \lambda\psi + KNz, \\ \delta(\lambda) &= 0, \end{aligned}$$

from which it follows that the BVP (22.1) is equivalent to the pair

$$\begin{aligned} x &= \lambda\Psi + Fx, \\ \delta(\lambda) &= 0, \end{aligned}$$

where $\Psi = T\psi$ and $F = TKNT^{-1}$.

For simplicity, assume that the limits $f_i(\infty)$ and $f_i(-\infty)$ exist and that $f_i(-\infty) < 0 < f_i(\infty)$, $i = 1, 2$. We make the notation $D_i{}^+ = \{t \in [0, \pi] : \psi_i(t) > 0\}$ and $D_i{}^- = \{t \in [0, \pi] : \psi_i(t) < 0\}$, $i = 1, 2$.

Theorem. *Suppose that $ad - bc \neq 0$ and that for $i = 1, 2$,*

$$\begin{aligned} f_i(-\infty) \int\limits_{D_i{}^+} \psi_i(t)\,dt + f_i(\infty) \int\limits_{D_i{}^-} \psi_i(t)\,dt & \\ < \int\limits_0^\pi b_i(t)\psi_i(t)\,dt & \\ < f_i(\infty) \int\limits_{D_i{}^+} \psi_i(t)\,dt + f_i(-\infty) \int\limits_{D_i{}^-} \psi_i(t)\,dt. & \end{aligned} \tag{22.6}$$

Then the BVP (22.1) has at least one solution.

Proof. The change of variables

$$\begin{pmatrix} z_1 \\ z_2 \end{pmatrix} = BT^{-1} \begin{pmatrix} x_1 \\ x_2 \end{pmatrix},$$

where B is the matrix

$$B = \begin{pmatrix} a & b \\ c & d \end{pmatrix},$$

transforms problem (22.1) into the equivalent problem

$$\begin{gathered} \begin{pmatrix} z_1'' + d_1 z_1 \\ z_2'' + d_2 z_2 \end{pmatrix} = BT^{-1} \begin{pmatrix} f_1(z_1) \\ f_2(z_2) \end{pmatrix} + BT^{-1} \begin{pmatrix} b_1(t) \\ b_2(t) \end{pmatrix}, \\ z_1(0) = z_2(0) = z_1(\pi) = z_2(\pi) = 0. \end{gathered}$$

This, in turn, is equivalent to the system of equations

$$\begin{aligned} z &= Pz + KNz, \\ PNz &= 0, \end{aligned} \tag{22.7}$$

where the projection operator $P : C^2[0,\pi] \times C^2[0,\pi] \to C[0,\pi] \times C[0,\pi]$ is defined by

$$Pz = \begin{pmatrix} \lambda_1 \psi_1 \\ \lambda_2 \psi_2 \end{pmatrix}, \quad \lambda_i = \int_0^\pi z_i(t)\psi_i(t)\,dt, \quad i = 1, 2,$$

and N is defined by

$$Nz = BT^{-1} \begin{pmatrix} f_1(z_1) + b_1(t) \\ f_2(z_2) + b_2(t) \end{pmatrix}.$$

Let $Q : C^2[0,\pi] \times C^2[0,\pi] \to C[0,\pi] \times C[0,\pi]$ be defined by

$$Qz = Pz + KNz - B^{-1}PNz.$$

If Q has a fixed point, then system (22.7) has a solution since $PNz = 0$ if and only if $PB^{-1}PNz = A^{-1}PNz = 0$. On the other hand, if z is a solution of (22.7), then z is a fixed point of Q. By the Leray–Schauder theory, to show that Q has a fixed point, it suffices to show that $(I - \tau Q)z \neq 0$ for all $z \in \partial\Omega$ and $\tau \in (0, 1)$, where

$$\Omega = \{z \mid z = z^\star + \tilde{z},\ ||z^\star|| < R,\ ||\tilde{z}|| < M\}, \quad z^\star = Pz,$$

for some chosen $R > 0$ and $M > 0$. Now, $(I - \tau Q)z \neq 0$ if $P(I - \tau Q)z \neq 0$, that is, if

$$(1-\tau)\begin{pmatrix} \lambda_1 \psi_1 \\ \lambda_2 \psi_2 \end{pmatrix} + \tau B^{-1}BT^{-1}\begin{pmatrix} \delta_1 \psi_1 \\ \delta_2 \psi_2 \end{pmatrix} \neq \begin{pmatrix} 0 \\ 0 \end{pmatrix}, \tag{22.8}$$

where

$$\delta_i = \int_0^\pi f_i[\lambda_i\psi_i(t) + \tilde{z}_i(t)]\psi_i(t)\,dt + \int_0^\pi b_i(t)\psi_i(t)\,dt, \quad i = 1, 2.$$

Let $z \in \partial\Omega$. Then either 1) $||z^\star|| = R$ and $||\tilde{z}|| \leq M$, or 2) $||z^\star|| \leq R$ and $||\tilde{z}|| = M$. In the case 1), we have $\max\{|\lambda_1|, |\lambda_2|\} = R$. By condition (22.6), we can choose R so that $\delta_1 > 0$ if $\lambda_1 = R$, $\delta_1 < 0$ if $\lambda_1 = -R$, $\delta_2 > 0$ if $\lambda_2 = R$, and $\delta_2 < 0$ if $\lambda_2 = R$; hence, (22.8) holds, so $(I - \tau Q)z \neq 0$.

In the case 2), we choose $M > ||K||\ ||Nz||$. Then $(I - \tau Q)z = 0$ is equivalent to $z = \tau(Pz - B^{-1}PNz) + \tau KNz$, which implies that $\tilde{z} = \tau KNz$ and, hence, that $||\tilde{z}|| < M$: a contradiction. Thus, $(I - \tau Q)z \neq 0$, and the proof is complete.

References

1. S. Seikkala and M. Hihnala, A resonance problem for a second-order vector differential equation, in *Integral Methods in Science and Engineering,* C. Constanda, M. Ahues, and A. Largillier (eds.), Birkhäuser, Boston, 2004, 233–238.
2. I.G. Main, *Vibrations and Waves in Physics*, Cambridge Univ. Press, Cambridge, 1993.
3. S. Seikkala and D. Vorobiev, A resonance problem for a system of second-order differential equations, *Mathematiscs*, Oulu, preprint series, 2003.
4. A. Canada, Nonlinear ordinary boundary value problems under a combined effect of periodic and attractive nonlinearities, *J. Math. Anal. Appl.* **243** (2001), 174–189.

23 Multiple Impact Dynamics of a Falling Rod and Its Numerical Solution

Hua Shan, Jianzhong Su, Florin Badiu, Jiansen Zhu, and Leon Xu

23.1 Introduction

When an electronic device drops to the floor, it usually comes down at an inclination angle. After an initial impact at an uneven level, a clattering sequence occurs in rapid succession. There has been growing recognition that the entire impact sequence, rather than just the initial impact, is important for the shock response of circuits, displays, and disk drives. In a pioneering study by Goyal et al. (see [1] and [2]), it was found that when a two-dimensional rod was dropped at a small angle to the ground, the second impact might be as large as twice the initial impact. This raised the issue of how adequate the current testing procedure and simulation analysis are. Standard fragility tests, which typically involve a single impact with no rotation, are not adequate for that type of drop. Angled dropping tests have been seen in the literature, but they are apparently more complicated in nature. Similarly, including subsequent impacts in numerical simulations increases computational costs. So there is a need for further study of multiple impacts from both analytic and numerical analysis and experimental points of view.

Problems of a single impact or first impact are discussed in many articles in mechanics and mathematical literature, (see, for example, [3]–[5] for rigid-body collisions). Even in single-impact cases, the topic remains the focus of much discussion (see [6] and [7]) as many theoretical issues of contact dynamics with friction have started to be resolved. Recent attention has been paid to detecting and calculating the micro-collisions that occur in a short time interval, when the bodies are allowed to be flexible (see [8] and [9]). A whole range of commercial software (ANSYS, etc.) is available to study the single-impact problem, with a rather detailed approach, such as finite element analysis, implemented (see [10] and [11]).

The study of multiple impacts, however, is an emerging area. Goyal et al. used in [1] and [2] a transition matrix method to calculate the clattering sequence and its impacts. There, the contact times of impacts are assumed to be instantaneous and time intervals between the impacts are also brief, as

The authors are indebted to Nokia for the support of this work.

a two-dimensional rod is dropped to the ground at a very small angle. We note here that these impacts are still relatively far apart in times of collision, by comparison to the micro-collision situation [8]. These collisions occur in different parts of the body according to their movement after the collisions, and are not a consequence of elastic oscillations. Rather, the dynamics between the impacts play a very significant role in determining the velocity and location of the next impact, particularly when the inclination angle of the initial dropping is moderate.

In this study, we provide a comprehensive analysis of the clattering dynamics. The work is therefore a continuation of the study by Goyal et al. in [1] and [2]. Our results in the multiple impact of a uniform rod are consistent with that of Goyal's. We further explore the scenario where a larger initial dropping angle leads to a different impact sequence. Our methodology and numerical tools allow us to consider the problem of a three-dimensional body with nonuniform density. Substantial new knowledge of the multiple impacts is gained with numerical experiments. Such research is the first step towards a simulation tool for the design and optimization of electronic components.

We outline our article as follows. In Section 23.2, we state the basic rigid-body dynamics equation. Section 23.3 introduces the continuous contact model that will be used in numerical simulation. Section 23.4 contains theoretical results for a two-dimensional uniform rod based on the discrete contact model. We present the numerical simulation of the multiple impacts of a three-dimensional rod in Section 23.5. The discussion and conclusion are presented in Section 23.6.

23.2 Rigid-Body Dynamics Model

Two sets of coordinate systems are used to describe the displacement and rotation of a rigid body. The global coordinate system (x, y, z) is fixed to the ground, as shown in Fig. 1. The local coordinate system (x', y', z'), or rigid-body coordinate system, is a body-fixed frame with its origin located at the mass center of the rigid body. In the present work, only the gravitational force $\mathbf{F_g}$ and the impact contact force $\mathbf{F_c}$ are considered.

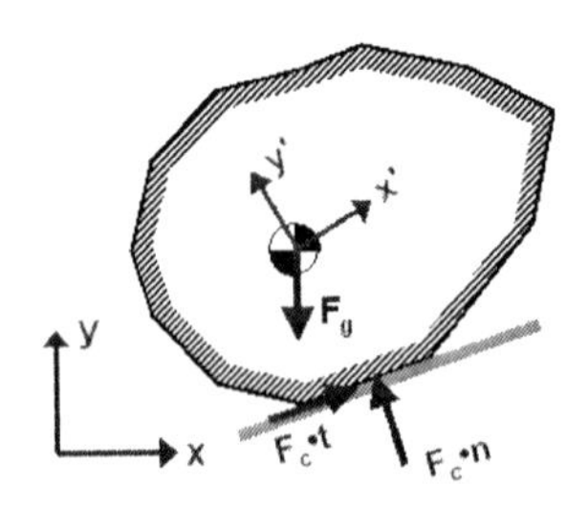

Fig. 1. Rigid body and coordinate systems.

The equation of unconstrained motion for the rigid body can be written as a set of ordinary differential equations in the matrix form [12]

$$\mathbf{M}\,\ddot{\mathbf{q}} = \mathbf{Q_v} + \mathbf{Q_e}\,, \tag{23.1}$$

where $\mathbf{q} = \{\mathbf{R^T}, \beta^\mathbf{T}\}$ denotes the vector of generalized coordinates, $\mathbf{R} = \{x_c, y_c, z_c\}^\mathbf{T}$ is the coordinate of the mass center, and $\beta = \{\beta_0, \beta_1, \beta_2, \beta_3\}^\mathbf{T}$ represents the vector of Euler parameters. The inertia matrix $\mathbf{M}$ is given by

$$\mathbf{M} = \int_\Omega \rho \begin{bmatrix} \mathbf{I} \\ \mathbf{G'^T}\,\tilde{\mathbf{u}}'^\mathbf{T}\,\mathbf{A^T} \end{bmatrix} [\mathbf{I} - \mathbf{A}\,\tilde{\mathbf{u}}'\,\mathbf{G}']\,d\Omega\,,$$

where the integration is over the entire rigid body, ρ is the density, $\mathbf{I}$ is the identity (3×3)-matrix, $\mathbf{A}$ is the transformation (4×4)-matrix, and $\tilde{\mathbf{u}}'$ is the skew-symmetric matrix obtained from the local coordinates vector $\mathbf{u}'$. The matrix $\mathbf{G}'$ is a (3×4)-matrix expressed in terms of the Euler parameters.

$\mathbf{Q_v}$, the first term on the right-hand side of (23.1), represents the vector that absorbs the quadratic velocity terms, that is,

$$\mathbf{Q_v} = -\int_\Omega \rho \begin{bmatrix} \mathbf{I} \\ \mathbf{G'^T}\,\tilde{\mathbf{u}}'^\mathbf{T}\,\mathbf{A^T} \end{bmatrix} \alpha_\mathbf{v}\,d\Omega,$$

where $\alpha_\mathbf{v} = \mathbf{A}\,\tilde{\omega}'\,\tilde{\omega}'\,\mathbf{u}' - \mathbf{A}\,\tilde{\mathbf{u}}'\,\dot{\mathbf{G}}'\,\dot{\beta}$, $\tilde{\omega}'$ is the skew-symmetric matrix corresponding to the angular velocity vector ω' of the rigid body in the local coordinate system. The superposed dot denotes the derivative with respect to time. The second term on the right-hand side of (23.1) is the vector of generalized forces

$$\mathbf{Q_e} = \begin{bmatrix} (\mathbf{Q_e})_\mathbf{R} \\ (\mathbf{Q_e})_\beta \end{bmatrix} = \begin{bmatrix} \mathbf{F_g} + \mathbf{F_c} \\ \mathbf{G^T}\,\mathbf{M_c} \end{bmatrix},$$

where $\mathbf{G} = \mathbf{A}\,\mathbf{G}'$. The general force includes the gravitational force $\mathbf{F_g}$ and the impact contact force $\mathbf{F_c}$, and $\mathbf{M_c}$ is the vector of the moment of the contact force with respect to the mass center. The modeling of the impact contact force will be given in Section 23.3.

The matrix $\mathbf{M}$ is assumed to be positive definite, and (23.1) can be written as

$$\ddot{\mathbf{q}} = \mathbf{M^{-1}}\,(\mathbf{Q_v} + \mathbf{Q_e}). \tag{23.2}$$

We introduce the state vector

$$\mathbf{U} = \begin{bmatrix} \dot{\mathbf{q}} \\ \mathbf{q} \end{bmatrix}$$

and the load vector

$$\mathbf{R} = \begin{bmatrix} \mathbf{M^{-1}}\,(\mathbf{Q_v} + \mathbf{Q_e}) \\ \dot{\mathbf{q}} \end{bmatrix}.$$

Then (23.2) can be written as

$$\dot{\mathbf{U}} = \mathbf{R},$$

that is, a set of ordinary differential equations, which can be solved for given initial conditions. The time integration is accomplished by means of the third-order total-variation-diminishing (TVD) Runge-Kutta method (see [13]).

23.3 Continuous Contact Model

As mentioned earlier, the purpose of the present work is to study the collision of a rigid body with a horizontal floor. The continuous contact model, also known as the compliant contact model, is used to model the impact contact force. This model is well suited for the problems in discussion. First, the model allows us to record specific impact and forces at any particular moment; second, the viscoelastic parameters in the model can be used to describe the energy dissipation and elastic reconstitution of the floor. As we are more concerned with the trajectory of the impacts rather than micro-collisions, we consider this continuous contact model adequate. Below we give a brief description of the numerical procedure; full details of the proofs of these assertions will appear in a future publication.

The horizontal ground is modeled as a distributed viscoelastic foundation that consists of a layer of continuously distributed parallel springs and dampers, as shown in Fig. 2. The surface stiffness is represented by the spring coefficient k_G, and c_G is the ground damping coefficient.

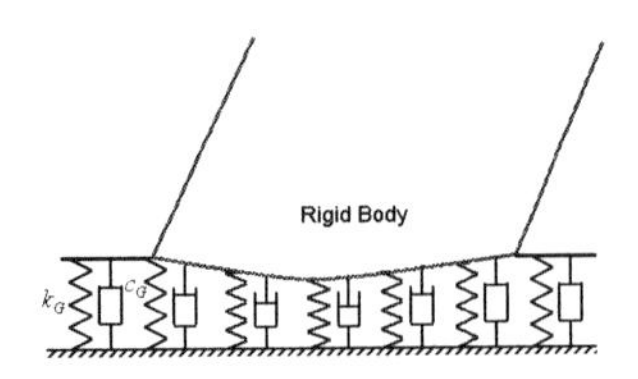

Fig. 2. Distributed viscoelastic foundation.

The impact contact force is calculated as the integral of a distributed load over the contact area S, in the form

$$\mathbf{F_c} = \int_S \mathbf{f_c}\, dS,$$

where $\mathbf{f_c} = f_n\, \mathbf{n} + f_t\, \mathbf{t}$ is the vector of the distributed load and $\mathbf{n}$ and $\mathbf{t}$ represent the unit vectors in the normal and tangential directions, respectively. The normal distributed contact load f_n is determined explicitly by the local indentation δ and its rate $\dot{\delta}$: $f_n = (k_G + c_G\dot{\delta})\,\delta$. The local tangential contact load f_t is determined from Coulomb's law, that is, $f_t \le \mu_s f_n$ when sticking occurs, and $f_t = \mu_k f_n$ when sliding occurs. Here μ_s and μ_k are the coefficients of static friction and sliding friction, respectively. The moment of the impact contact force with respect to the mass center is computed as

$$\mathbf{M_c} = \int_S \mathbf{m_c}\, dS,$$

where $\mathbf{m_c} = \tilde{\mathbf{u}}\, \mathbf{f_c}$ and $\tilde{\mathbf{u}}$ is the skew-symmetric matrix corresponding to the local coordinate vector expressed in the global coordinate system.

The continuous contact model will be used to evaluate the impact contact force in the rigid-body dynamics model of Section 23.2. The numerical simulation results of the falling rod problem will be given in Section 23.5.

23.4 Discrete Contact Model for a Falling Rod

This section presents a discrete contact dynamics model for the falling rod problem. This model is based on the linear impulse-momentum principle, the angular impulse-momentum principle for a rigid body, and some impact parameters that relate the pre- and post-impact variables, such as the coefficient of restitution, which is defined as the ratio of the post-impact relative normal velocity to the pre-impact relative normal velocity at the impact location. The discrete model assumes that the impact occurs instantaneously and the interaction forces are high, and, thus, that the change in position and orientation during the contact duration are negligible and the effects of other forces (for example, the gravitational force) are disregarded [4]. This model is able to predict post-impact status for given pre-impact information and predefined impact parameters. The theoretical solutions for the discrete model in this section will be compared with the numerical results for the continuous impact model in Section 23.5.

The falling rod problem is depicted in Fig. 3, which shows a schematic sketch of the collision of a falling rod with the horizontal ground. The length of the rod is L. The rod forms an angle θ with the floor when one of its ends hits the ground. Unit vectors $\mathbf{n}$ and $\mathbf{t}$ define the normal and tangential directions. The vector pointing from the mass center of the rod to the impact location is defined as $\mathbf{d} = d_n\, \mathbf{n} + d_t\, \mathbf{t}$, and therefore $d_n = -1/2\, L \sin\theta$, and $d_t = -1/2\, L \cos\theta$.

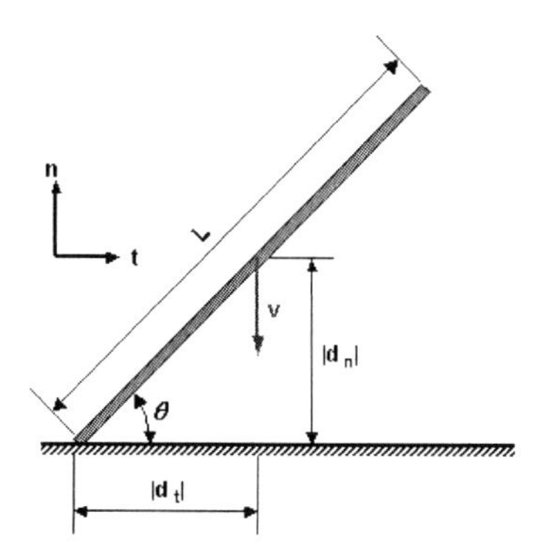

Fig. 3. Falling rod colliding with a massive horizontal surface.

The impact dynamics equations are

$$\begin{aligned} m(V_n - v_n) &= P_n, \\ m(V_t - v_t) &= \mu_i P_n, \\ \tfrac{1}{12} mL^2(\Omega - \omega) &= \mu_i P_n d_n - P_n d_t, \\ V_n - \Omega d_t &= -e(v_n - \omega d_t). \end{aligned} \tag{23.3}$$

The first two equations (23.3) represent the linear impulse-momentum law in the normal and tangential directions. The third equation follows from the angular impulse-momentum law. The last equation gives the relationship between the relative normal velocities at the impact location before and after impact. The mass of the rod is given by m. The pre- and post-impact vectors of velocity at the center of mass of the rod are defined as $\mathbf{v} = v_n\mathbf{n} + v_t\mathbf{t}$ and $\mathbf{V} = V_n\mathbf{n} + V_t\mathbf{t}$, respectively. The variables ω and Ω are the pre- and post-impact angular velocities of the rod. The impact impulse vector is given by $\mathbf{P} = P_n\mathbf{n} + P_t\mathbf{t}$, with normal component P_n and tangential component P_t. The relationship between these two components is given by $P_t = \mu_i P_n$, where μ_i is the impulse ratio. The restitution coefficient is denoted by e. When the pre-impact variables and the impact parameters (μ_i and e) are known, (23.3) gives a closed system of algebraic equations for the four unknowns V_n, V_t, Ω, and P_n. Setting $m = 1$, $L = 1$, $\omega = 0$, and $e = 1$, we find that the solutions of (23.3) are

$$\begin{aligned} V_n &= \frac{1 - 3\cos^2\theta + 3\mu_i \sin\theta\cos\theta}{1 + 3\cos^2\theta - 3\mu_i \sin\theta\cos\theta} v_n, \\ V_t &= v_t - \frac{2\mu_i}{1 + 3\cos^2\theta - 3\mu_i \sin\theta\cos\theta} v_n, \\ \Omega &= \frac{12(\cos\theta - \mu_i\sin\theta)}{1 + 3\cos^2\theta - 3\mu_i \sin\theta\cos\theta} v_n, \\ P_n &= -\frac{2v_n}{1 + 3\cos^2\theta - 3\mu_i \sin\theta\cos\theta} v_n. \end{aligned}$$

Note that the third equation (23.3) is based on the assumption of point contact, and becomes invalid when $\theta = 0$. In this case, (23.3) is reduced to the set of equations

$$\begin{aligned} m(V_n - v_n) &= P_n, \\ m(V_t - v_t) &= \mu_i P_n, \\ V_n &= -ev_n. \end{aligned} \tag{23.4}$$

When $e = 1$, the solutions of (23.4) are

$$\begin{aligned} V_n &= -v_n, \\ V_t &= -2\mu_i v_n + v_t, \\ P_n &= -2v_n. \end{aligned}$$

23.5 Numerical Simulation of a Falling Rigid Rod

The discrete contact dynamics model presented in the previous section is illustrative of the qualitative features in a clattering impact, but it certainly has several shortcomings. First of all, the discrete model is based on the assumption that the impact occurs instantaneously and the contact duration is negligible. Second, it also assumes that interaction forces are high, and thus the effects of other forces (for example, the gravitational force) are disregarded. Third, it assumes that the impact is only with point-contact, and therefore the contact point must be known in advance. The discrete contact model needs to be modified to handle the case with multiple contact points or with surface-contact impact. The continuous contact model presented in Section 23.3 is able to overcome these problems. In this section, the rigid-body dynamics equations given in Section 23.2 and the continuous contact model are applied to the numerical simulations of a falling rigid rod colliding with a horizontal ground. The numerical results on the first impact are compared to the theoretical solutions of the discrete model for the purpose of code validation. Then the numerical results involving a sequence of multiple impacts of the falling rigid rod with the ground are displayed.

Figure 4 shows the geometry and surface meshes of the rigid rod used in all test cases in this section. The parameters of the rod are given as mass $m = 1$, length $L = 1$, and radius $r_d = 0.01$.

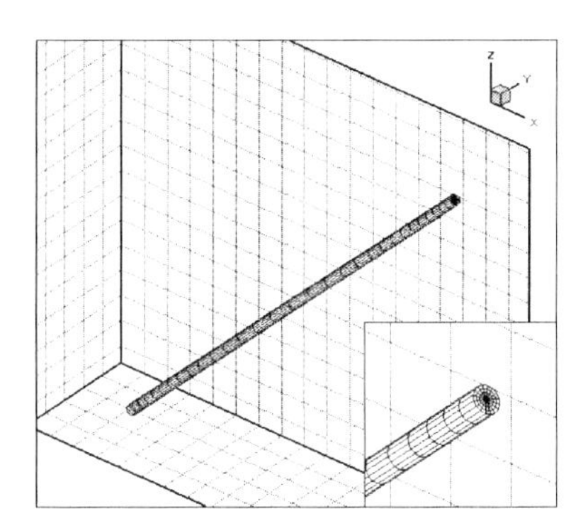

Fig. 4. Meshes at the surface of the falling rod.

In the first test case, the falling rigid rod collides with the horizontal ground at different contact angles θ ranging from $0°$ to $90°$. The initial angular velocity is $\omega = 0$, the pre-impact normal velocity is $v_n = -1$, the pre-impact tangential velocity is $v_t = 0$, the ground stiffness coefficient is $k_G = 10^{11}$, and the ground damping coefficient is $c_G = 0$. The reason that k_G takes such a large value is to make the contact surface sufficiently small, so that the numerical results will be comparable to the theoretical solution of the discrete contact model where a point-contact is assumed. The friction is not considered in this case, so $\mu_s = 0$ and $\mu_k = 0$. The discrete contact model in Section 23.4 is also applied to this test case for comparison. In the discrete contact model, for the case of $\theta \neq 0°$, the post-impact values are calculated by setting $v_n = -1$, $v_t = 0$, and $\mu_i = 0$ in (23.3). When $\theta = 0°$, (23.4) is used to calculate the solutions. These solutions from the discrete contact model are referred to as the theoretical solutions in this section.

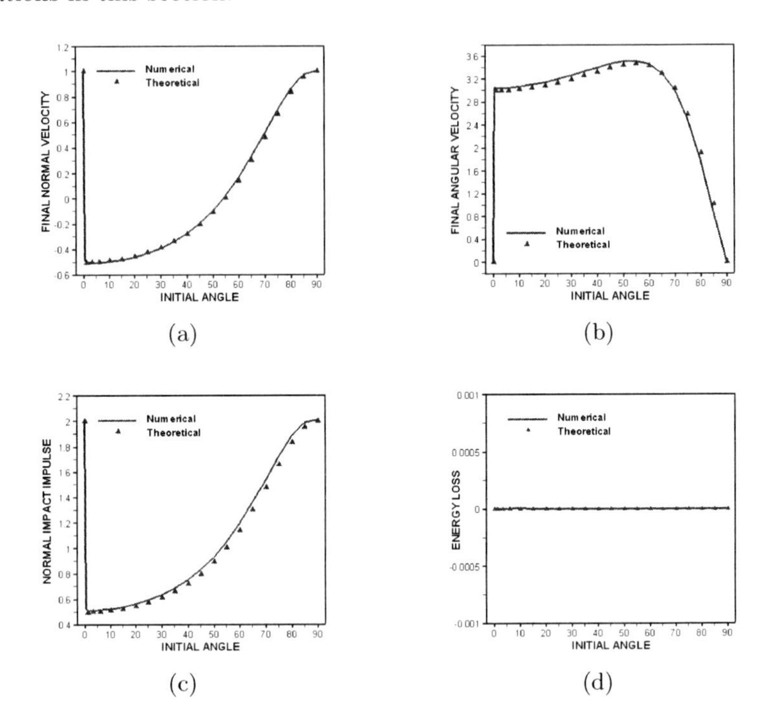

Fig. 5. Comparison between numerical and theoretical results for the first collision of the rigid rod in the first test case, with initial angular velocity $\omega = 0$, pre-impact normal velocity $v_n = -1$, ground stiffness $k_G = 10^{11}$, ground damping coefficient $c_G = 0$, and no friction. (a) The post-impact normal velocity of the center of mass as a function of the impact contact angle. (b) The post-impact angular velocity as a function of the impact contact angle. (c) The normal impact impulse as a function of the impact contact angle. (d) The energy loss as a function of the impact contact angle.

The post-impact status of the rigid rod for the first collision given by numerical simulation is compared to the theoretical solutions, as shown in Fig. 5. The post-impact normal (z-direction) velocity at the mass center, the angular velocity, the normal impact impulse, and the energy loss are shown as functions of the initial contact angles in Fig. 5(a), (b), (c), and (d). Both the numerical simulation results and the theoretical solutions indicate a sudden change of velocity, angular velocity, and normal impact impulse when θ changes from zero to nonzero values. The energy loss is the difference in the total energy before and after the collision. In this test case, there is no friction and floor damping: Fig. 5(d) shows clearly the conservation of the total energy. Overall, Fig. 5 indicates that the numerical results of the rigid-body dynamics simulation with the continuous contact impact model agree very well with the theoretical solutions.

The second test case demonstrates the collision between a falling rigid rod and the horizontal "rough" floor with a sliding friction coefficient μ_k ranging from 0 to 0.8. The static friction coefficient μ_s takes the same value as μ_k. The initial impact contact angle is $\theta = 45°$. The other parameters are the same as in the first test case.

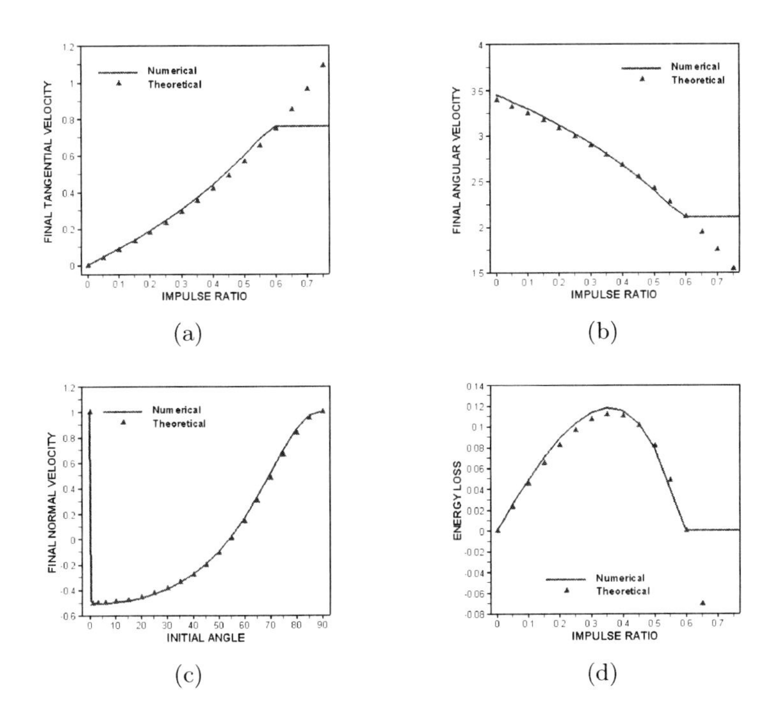

Fig. 6. Comparison between numerical and theoretical results for the first collision of the rigid rod in the second test case, with initial impact contact angle $\theta = 45°$, initial angular velocity $\omega = 0$, pre-impact normal velocity $v_n = -1$, ground stiffness $k_G = 10^{11}$, ground damping coefficient $c_G = 0$, sliding friction coefficient $\mu_k = 0$–0.8, and static friction coefficient $\mu_s = \mu_k$. (a) The post-impact normal velocity of the center of mass as a function of the impact contact angle. (b) The post-impact angular velocity as a function of the impact contact angle. (c) The normal impact impulse as a function of the impact contact angle. (d) The energy loss as a function of the impact contact angle.

Setting $\theta = 45°$, we see that the theoretical solutions in (23.4) give the post-impact value of different variables for the first collision. The comparisons between the theoretical solutions and the numerical results are given in Fig. 6, where the impulse ratio μ_i is calculated in the same way in both numerical results and theoretical solutions. The post-impact values of the normal velocity and tangential velocity at the mass center are shown in Fig. 6(a) and (b) as a function of the impulse ratio. Fig. 6(c) shows the normal component of the impact impulse for the first impact. The total energy loss in the first collision is shown in Fig. 6(d). When $\mu_i < 0.6$, the numerical results agree very well with the theoretical solutions, but large deviations are observed for $\mu_i > 0.6$. We note that the theoretical solution based on the discrete contact model gives a negative energy loss when $\mu_i > 0.6$, as shown in Fig. 6(d). Indeed, as pointed by Brach [4], the impulse ratio μ_i in the discrete contact dynamics model must satisfy $|\mu_i| \leq \min(|\mu_O|, |\mu_T|)$ to ensure the conservation of energy. In our case, $\mu_O = \mu_T = 0.6$, which gives the upper limit of μ_i where the theoretical model is applicable. In the numerical results, all the curves turn into horizontal lines when $\mu_i > 0.6$, as shown in Fig. 6, indicating that the post-impact motion of the rod is independent of μ_i. The explanation will be given later in this section. Unlike the discrete contact model, the numerical results of the rigid-body dynamics simulation with the continuous contact model are able to give the correct answer.

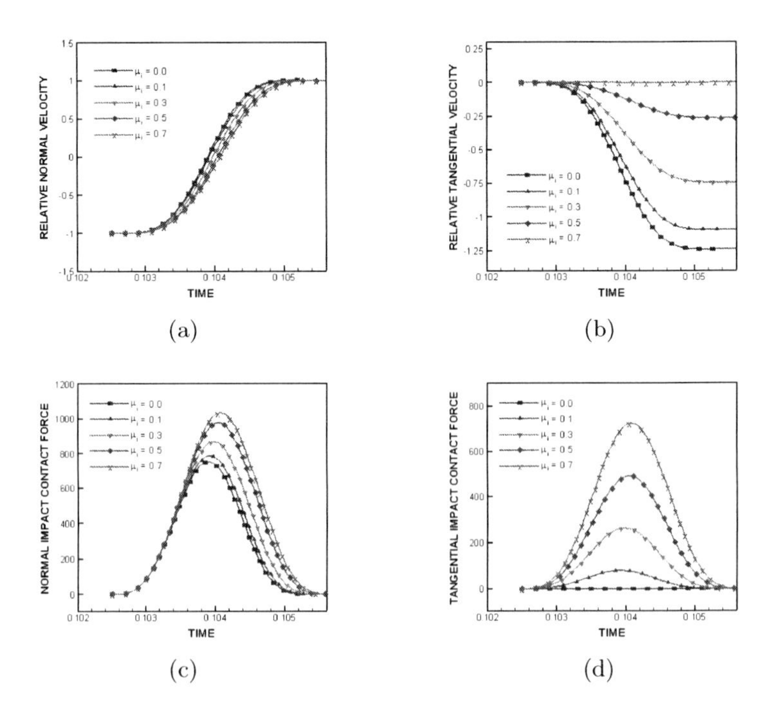

Fig. 7. Numerical results for the first collision of the rigid rod with the ground. All the parameters are the same as for Fig. 6. (a) The relative normal velocity at the impact location as a function of time. (b) The relative tangential velocity at the impact location as a function of time. (c) The normal impact force as a function of time. (d) The tangential impact force as a function of time.

One of the advantages of the continuous contact dynamics model over a discrete model is that it reveals the impact contact process explicitly and therefore is able to give details regarding the time-varying variables during the impact process. From the numerical simulation results in this test case, the variations of different quantities as functions of time during the first collision are presented in Fig. 7. Thus, Fig. 7(a) shows the normal component of the rod velocity at the contact point, each curve corresponding to a different impulse ratio. The normal velocity is -1 as the impact starts. During the impact process, the normal velocity at the contact point gradually increases and reaches 1 at the end of the collision. When the ground damping coefficient is $c_G = 0$, the numerical results are in agreement with the theoretical solutions with restitution coefficient $e = 1$. Fig. 7(b) shows the variation of the tangential velocity at the contact point during the first collision at different impulse ratio μ_i representing different friction effects. Initially the tangential velocity is zero, and it becomes nonzero for all impulse ratios except $\mu_i = 0.7$, indicating that the contact point is sliding along the ground surface during the collision if the impulse ratio is small. As μ_i increases, the magnitude of the tangential velocity becomes smaller and smaller. For the case $\mu_i = 0.7$, the friction force grows large enough to prevent the rod from sliding along the ground surface; therefore, the tangential velocity at the contact point becomes zero, indicating that the end of the rod is sticking to the ground during the impact process. Once sticking occurs, the post-impact motion of the rod is independent of the impulse ratio, which provides an explanation why all the curves turn into horizontal lines when $\mu_i > 0.6$ in Fig. 6. Fig. 6(d) shows that there is no energy loss when $\mu_i > 0.6$, because the damping effect of the ground is not included in this test case and no work is done by the friction force when sticking occurs. Fig. 7(c) and (d) display the variation of the normal and tangential impact contact force, indicating that larger friction induces larger impact contact force in both the normal and tangential directions.

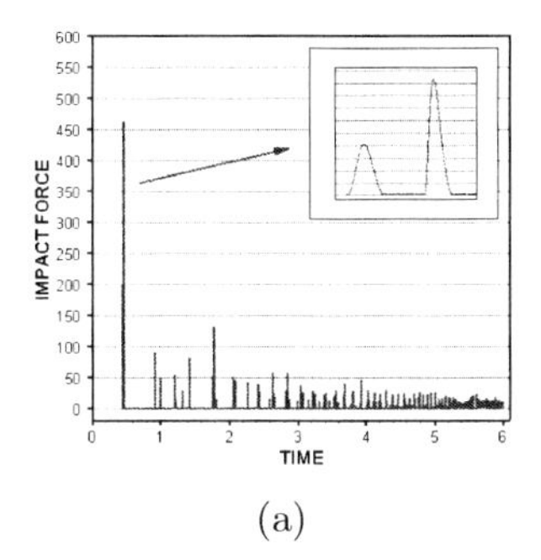

(a)

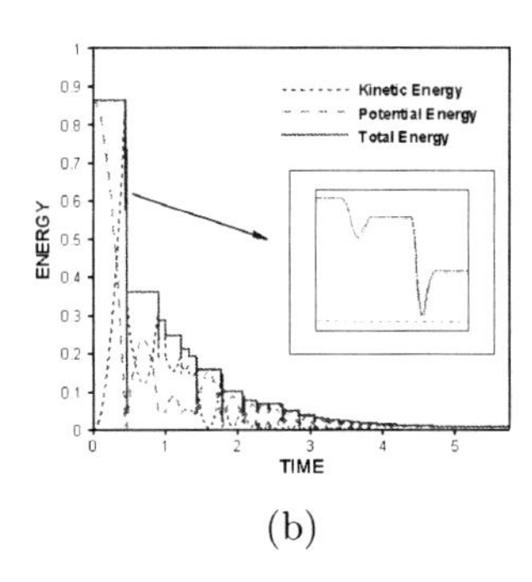

(b)

Fig. 8. Numerical results for multiple collisions of the rigid rod with the ground, with mass $m = 1$, length $L = 1$, radius $r_d = 0.01$, initial impact contact angle $\theta = 45°$, initial angular velocity $\omega = 0$, initial elevation of the center of mass $z_c|_{t=0} = 2$, ground stiffness $k_G = 10^9$, ground damping coefficient $c_G = 10^8$, sliding friction coefficient $\mu_k = 0.3$, and static friction coefficient $\mu_s = 0.35$. (a) The normal impact contact force as a function of time. (b) The total energy, kinetic energy, and potential energy as a function of time.

The numerical simulation code developed in this paper can be used to study multiple impacts of the rigid body. The third test case is designed to demonstrate the numerical simulation of a sequence of collisions of the falling rod with a horizontal ground, with initial contact angle $\theta = 45°$, initial angular velocity $\omega = 0$, pre-impact normal velocity $v_n = -1$, pre-impact tangential velocity $v_t = 0$, ground stiffness $k_G = 10^9$, damping coefficient of the ground $C_G = 10^8$, static friction coefficient $\mu_s = 0.35$, and sliding friction coefficient $\mu_k = 0.3$. The numerical simulation results are shown in Fig. 8. Fig. 8(a) shows the normal impact force as a function of time, each spike corresponding to one impact (some of the spikes are too close to each other and cannot be distinguished from the figure). The second impact is larger than first one. The third impact, apparently separated from the first two impacts by a larger time interval, does not belong to the same clattering series. The change of energy in this series of impacts is given in Fig. 8(b). The horizontal line segments on the total energy curve indicate the conservation of the total energy during the airborne time between the impacts, as the energy loss is attributed to ground damping and friction effects.

23.6 Discussion and Conclusion

The overall aim of this article is to analyze both analytically and through numerical simulation the issues surrounding clattering. Our discussions are limited to a rod and a rectangular model cell phone with uniformly distributed mass. Our three-dimensional computational dynamics model, however, allows us to study a rigid body of any shape and arbitrary distribution of mass for its multiple-impact sequence. For the falling rod case, the numerical simulation based on the continuous contact model predicts the sliding and sticking that occur at the contact point correctly, whereas the theoretical solution based on the discrete contact model fails when sticking occurs.

Our study confirms the results of Goyal et al. (see [1] and [2]) that if a rod falls to the ground at a small angle, then its clattering impact series has a much larger second impact than the initial one. Furthermore, our analytic study finds that this same phenomenon is happening at angles as large as 54°. In realistic situations, the range might be small when the energy dissipation and the friction effect of the ground are taken into account, as our three-dimensional numerical model study shows. Clattering is also common in the falling of three-dimensional objects such as cell phones, notebook computers, and other mobile devices. The dynamics can be more complicated because the second impact can be more than twice the initial one, as our numerical study shows.

The true implication of clattering remains an interesting topic of study. On the one hand, it means that second impacts cannot be ignored in analysis. On the other hand, our calculation shows that when the dropping inclination angle is small, the summation of the first three impacts equals the first impact of zero angle drop. This seems to suggest that clattering spreads the impact. Of course, in application to an electronic device, one needs to determine the effect of the impact on the internal contents; the amount of impact is only one of the contributing quantities. A flexible model incorporating the detailed structure of devices is helpful to test the effects of micro-collisions, vibrations, and shock wave propagation in these devices.

In our numerical study we find that the kinetic energy transfers between its rotational part and translational part are important to clattering. Larger impacts typically occur at times from fast rotation to fast translation. Our numerical model has been proven to be a realistic tool for a multiple-impact study.

References

1. S. Goyal, J.M. Papadopoulos, and P.A. Sullivan, The dynamics of clattering I: equation of motion and examples, *ASME J. Dynamic Systems, Measurement and Control* **120** (1998), 83–93.
2. S. Goyal, J.M. Papadopoulos, and P.A. Sullivan, The dynamics of clattering II: global results and shock protection, *ASME J. Dynamic Systems, Measurement and Control* **120** (1998), 94–101.
3. J.B. Keller, Impact with friction, *ASME J. Appl. Mech.* **53** (1986), 1–4.
4. R.M. Brach, Rigid-body collision, *ASME J. Appl. Mech.* **56** (1989), 133–139.
5. W.J. Stronge, Rigid-body collision with friction, *Proc. Roy. Soc. London A* **431** (1990), 169–181.
6. D.E. Stewart, Rigid-body dynamics with friction and impact, *SIAM Review* **42** (2000), 3–39.
7. G. Gilardi and I. Sharf, Literature survey of contact dynamics modeling, *Mechanism and Machine Theory* **37** (2002), 11213–11239.
8. D. Stoianovici and Y. Hurmuzlu, A critical study of the applicability of rigid-body collision theory, *ASME J. Appl. Mech.* **63** (1996), 307–316.
9. A. Yigit, A. Ulsoy, and R. Scott, Dynamics of radially rotating beam with impact, part 1: theoretical and computational model, *J. Vibration and Acoustics* **12** (1990), 65–70.
10. J. Wu, G. Song, C. Yeh, and K. Waytt, Drop/impact simulation and test validation of telecommunication products, in *Intersociety Conference on Thermal Phenomena,* IEEE, 1998, 330–336.
11. T. Tee, H. Ng, C. Lim, E. Pek, and Z. Zhong, Application of drop test simulation in electronic packaging, in *ANSYS Conference,* 2002.
12. A. Shabana, *Computational Dynamics,* 2nd ed., Wiley, New York, 2000.
13. C.W. Shu and S. Osher, Efficient implementation of essentially non-oscillatory shock-capturing schemes, *J. Comput. Phys.* **75** (1988), 439–471.

24 On the Monotone Solutions of Some ODEs. I: Structure of the Solutions

Tadie

24.1 Introduction

The aim of this work is to investigate some properties of the solutions of the ODE $u'' = g(t)f(u(t))$ in certain domains in $\mathbb{R}$. The case when the coefficient g is regular and bounded in those domains need not be considered. The initial value $u'(R)$, $R > 0$, is prescribed. That value and the energy function $E(u) := u'(r)^2 - u'(R)^2 - 2\{F(u(r)) - F(u(R))\}$, where $F(t) := \int_0^t f(s)ds$, play a crucial role in the study.

For any $R > 0$, define $E(R) = [R, \infty)$ and $I(R) = (R, \infty)$. For some $\delta, m > 0$, consider the problem

$$\begin{gathered} u'' = f(u) \quad \text{in } I(R), \quad u(R) = m, \\ f \in C^1(I(R)) \cap C(E(R)), \quad f(0) = f(\delta) = 0, \\ f(t) > 0 \quad \text{in } (0,\delta), \quad f(t) < 0 \quad \text{in } (\delta, \infty). \end{gathered} \tag{E}$$

When f is replaced by f_1 in (E), that is,

$$f_1 \in C(E(R)) \cap C^1(I(R)), \quad f_1(t) > 0 \quad \forall t > 0,$$

the problem is denoted by (E1). It is obvious that because $f(t) < 0$ for $t > \delta$, any solution u of (E) with finite $u'(R)$ is bounded above. Our main focus is on the existence and structure of the solutions $u \in C(E(R)) \cap C^2(I(R))$ of (E) *whose positive parts are strictly decreasing.* We start by studying the associated problems

$$v'' = f(v) \quad \text{in } E(R), \quad v(R) = m, \quad v'(R) = \nu, \tag{Eν}$$

where $\nu \in \mathbb{R}$. By the classical theory of ordinary differential equations, (Eν) has a unique solution $v \in C([R, R+\tau)) \cap C^2((R, R+\tau))$ for some small $\tau > 0$. In fact, considering the operator T defined for $r > R$ by

$$Tu(r) := m + \nu(r - R) + \int_R^r (r-s)g(s)f(u(s))ds, \tag{T}$$

The author is indebted to the Department of Mathematics of the Danish Technical University, Lyngby, for the support of this work.

where g is a regular and bounded function, we know that various existence theories for ODEs show that T has a fixed point in some space $C([R, R+\tau))$, $\tau > 0$. That fixed point turns out to be in $C^2([R, R+\tau))$ and solves

$$\begin{aligned} u'' = g(r)f(u), \quad & r \in [R, R+\tau), \\ u(R) = m, \quad & u'(R) = \nu. \end{aligned} \tag{T1}$$

Any such local solution has an extension in a maximal interval of existence $(R, \rho) \subseteq (R, +\infty)$, say, in which it remains nonnegative. We are interested in those solutions that are decreasing wherever they are positive. *Our interest in this part will be on problem* (E) *and mainly on autonomous problems.*

Such an extension of u could be of several types, as listed below.

(SP): strictly positive, if there is $a > 0$ such that $u(r) > a$ in $r > R$;

(SC): strictly crossing, if there is $T > R$ such that $u > 0$ in (R, T), $u(T) = 0$, and $u'(T) < 0$;

(C): crossing, if it is (SC) but not necessarily with $u'(T) < 0$;

(D): decaying, if $u \geq 0$, is nonincreasing in (R, ∞), and $\lim_\infty u(r) = 0$;

(OS): oscillatory, if for all $T > 0$ there is $T < r < t$ such that $(r, u(r))$ and $(t, u(t))$ are, respectively, a local minimum and a local maximum and $u(r) < u(t)$.

We define $F(t) := \int_0^t f(s)ds$ and denote by θ its positive zero.

Lemma 1. *Any solution $v := v_\nu$ of* (Eν) *satisfies, for $r > t \geq R$,*

$$v'(r)^2 = v'(t)^2 + 2\{F(v(r)) - F(v(t))\}, \tag{24.1}$$

$$v'(r)^2 = \nu^2 + 2\{F(v(r)) - F(m)\}, \tag{24.2a}$$

$$v'(t)^2 = v'(r)^2 \quad \Leftrightarrow \quad F(v(t)) = F(v(r)). \tag{24.2b}$$

Consequently, for $r > t \geq R$,

$$\begin{aligned} &\text{(i)} \quad v'(r) = 0 \quad \Rightarrow \quad \nu^2 = 2[F(m) - F(v(r))], \\ &\text{(ii)} \quad v'(r) = v'(t) = 0 \quad \Rightarrow \quad F(v(r)) = F(v(t)), \\ &\text{(iii)} \quad v'(r) = 0,\ v(r) \geq \delta,\ v''(r) < 0 \quad \Rightarrow \quad v(t) \leq v(r) \quad for R \leq t < r, \\ &\text{(iv)} \quad v'(r) = 0,\ v(r) < \delta,\ v''(r) > 0 \quad \Rightarrow \quad v(t) \geq v(r) \quad for R \leq t < r. \end{aligned} \tag{24.3a}$$

Also, by (24.2a), *any solution of* (Eν) *is bounded above for any given finite ν such that $\nu^2 \geq 2[F(m) - F(u(r))]$.*

If v is a ground-state solution (i.e., $v \geq 0$ and $\lim_\infty v = 0$) or if v has an extremum $(r_1, v(r_1))$ with $F(v(r_1)) = 0$, then $v'(R)^2 = 2F(m)$, so $m \in (0, \theta)$.

Furthermore, if $(r, v(r))$ and $(s, v(s))$ with $v'(r) = v'(s) = 0$ are local maxima, then $v(r) = v(s) > \delta$. Similarly, if they are local minima, then $v(r) = v(s) < \delta$.

From (24.3a)(i), *it is obvious that if a solution v of* (E) *satisfies*

$$v(\rho) = v'(\rho) = 0 \quad \text{for some } \rho > R, \tag{24.3b}$$

then $v'(R)^2 = 2F(m)$*, so* $m \in (0, \theta]$*. If a solution* u *has* $(r_1, u(r_1))$ *and* $(r_2, u(r_2))$ *as its first local minimum and first local maximum with* $r_1 < r_2$*, then* $u(r_1) < \delta < u(r_2) \leq \theta$ *and* $u(r) \in [u(r_1), u(r_2)]$ *for all* $r > r_2$*.*

Proof. Multiplying the equation by v' and integrating over (t, r) yields (24.1) and (24.2a).

Statements (24.3a)(i),(ii) follow from (24.1) and (24.2a).

For (24.3a)(iii), if there is $R < t < r$ such that $v(t) > v(r)$, then $v'(t)^2 + 2[F(v(r)) - F(v(t))] = 0$, so $F(v(r)) \leq F(v(t))$, which, according to the properties of F, can happen only if $v(r) \geq v(t) \geq \delta$.

Statement (24.3a)(iv) follows from the same type of argument. The proof is completed by means of (24.2b).

The last statement follows from (24.3a)(ii) and the properties of F.

Lemma 2. *Let* $\nu \in R$ *and* $m > 0$ *be given (and finite), and let* v *be the corresponding solution of* (Eν)*. If there is a finite* $\rho > R$ *such that* $v(\rho) = k > 0$ *and* $v'(\rho) \neq 0$*, then* $\lim_{r \nearrow \infty} v(r) = k$ *cannot hold. In other words, if* $\lim_\infty v(r) = k > 0$*, then* $v(r) = k$ *with* $v'(r) = 0$ *has no solution in* $[R, \infty)$*.*

Proof. Suppose that such a limit $k > 0$ exists. Then for r sufficiently large we can have $v'(\rho)^2 - v'(r)^2 > v'(\rho)^2/4$, while $F(v(\rho)) - F(v(r)) = 0$, which violates (24.1).

Theorem 1. *Let* $m > \delta$ *and* $\nu \in R$*, and suppose that the corresponding solution* v*, say, of* (Eν) *has no strictly positive limit at infinity. Then one of the following situations holds:*

(i) *v is strictly decreasing;*

(ii) *v has exactly one local extremum;*

(iii) *v is oscillatory.*

If $0 < m < \delta$*, then* v *is decreasing if it has no local maximum at* $r > R$*, and oscillatory if it has any local positive minimum there.*

Generally, if v *has a local minimum* $(t, v(t))$ *with* $t > R$ *and* $0 < v(t) < \delta$*, then* v *is oscillatory.*

Proof. It is enough to notice from the above results that if v has one local maximum and one local minimum $(r, v(r))$ and $(s, v(s))$, respectively, with $r < s$, then (1) $v(t)$ remains between $v(r)$ and $v(s)$ for any $t \geq s$, and (2) any other local maximum $(x, v(x))$ and local minimum $(y, v(y))$ with $x, y > s$ would satisfy $v(x) = v(r)$ and $v(y) = v(s)$. Furthermore, because $\lim_{r \nearrow \infty} v(r)$ does not exist, it follows that for any $T > R$ there are $t, y > T$ such that $(t, v(t))$ is a local maximum and $(y, v(y))$ is a local minimum.

24.2 Some Comparison Results

If $u, v \in C^2(I(R))$ are, respectively, solutions of

$$\begin{gathered} \phi'' = f(\phi) \quad \text{in } I(R), \\ u(R) = m_u, \;\; v(R) = m_v, \quad u'(R) = \alpha_u, \;\; v'(R) = \alpha_v, \end{gathered} \tag{24.4}$$

then

$$[uv' - vu']' = uv\left\{\frac{f(v)}{v} - \frac{f(u)}{u}\right\} \quad \text{in } I(R), \tag{24.5}$$
$$(uv' - vu')(R) = m_u\alpha_v - m_v\alpha_u.$$

Theorem 2. *Suppose that $f(t)/t$ is increasing in $K := [0, m]$, and let u and v be the solutions of (24.4) with range in K. Then $v > u$ for $r > R$ if $u > 0$ and*
(i) *$m_u = m_v = m > 0$ and $\alpha_v > \alpha_u$, or*
(ii) *$0 < m_u < m_v = m$ and $0 \geq \alpha_v \geq \alpha_u$, or*
(iii) *$\alpha_u = \alpha_v < 0$ and $m_u < m_v$.*

Proof. (i) If $v = u$ at some $T > R$, then in $(R, T) := J_T$, (24.5) reads

$$[uv' - vu']' = uv\{f(v)/v - f(u)/u\} > 0 \quad \text{in } J_T,$$
$$(uv' - vu')(R) = m(\alpha_v - \alpha_u) > 0;$$

hence, v/u is increasing with value 1 at R, which conflicts with its value at T. The other cases are handled similarly.

Corollary. *If $m > 0$ and $\alpha = 0$, then the corresponding solution v*
(i) *is oscillatory if $m < \delta$;*
ii) for $m > \delta$, it is crossing if it has no local minimum, and is oscillatory otherwise.

Lemma 3. *For any $m > 0$, $m \neq \delta$, the α_v can be chosen so that the corresponding solution v of (24.4) is strictly crossing.*

Proof. Let $f_* := \max_{[0,\delta]} f(t)$. From the equation, $v'(r) \leq \alpha_v + f_*(r - R)$ for all $r > R$, so $v(4R) \leq m + 3\alpha_v R + (9/2)R^2 f_*$. This is negative for any negative α_v sufficiently large in absolute value.

Theorem 3. *Let f be as in problem* (E), *let $\rho > R$, and define the function $\Psi(y, z)(r) := (y'z - yz')(r)$.*
(a) *Suppose that $f(t)/t$ is increasing in some interval $(0, M]$. Then*
(i) *if two decreasing and distinct functions u, v with range in $[0, M]$ satisfy $\phi'' = f(\phi)$ in $J := (R, \rho)$ and $\phi(\rho) = 0$, then $\Psi(v, u) \neq 0$ in J;*
(ii) *if the problem*

$$u'' = f(u), \quad in\ (R, \rho), \quad u(R) = m \leq M, \quad u(\rho) = 0 \tag{ρ/m}$$

has a decreasing solution, then this solution is unique.
(b) *Suppose that $f(t)/t$ is decreasing in some interval $(0, M]$. If the problem*

$$u'' = f(u) \quad in\ (R, \rho), \quad u(\rho) = 0, \quad u(R) \leq M \tag{ρ_*}$$

has a decreasing solution, then this solution is unique.

The results in (a)(ii) *and* (b) *establish that if $f(t)/t$ is monotone in some $(0, M]$ then the decreasing solution of $u'' = f(u)$ in (R, ρ); $u(\rho) = 0$ has at most one solution whose maximum is less than or equal to M.*

Proof. Let u and v be two such solutions.

(a)(i) If there is $R \le T < \rho$ such that $\Psi(v,u)(T) = 0$, then in some (T,τ) where $v < u$, say, v/u is decreasing (by (24.5)) from a value less than 1; thus, we would not have $u(\rho) = v(\rho)$. Hence, $\Psi(v,u)(T) \neq 0$ in (R,ρ).

(ii) Suppose that $\Psi(v,u) < 0$ in $(R,\rho) := J$. Then $v'(R) \le u'(R)$ and $v < u$ in some $(R,T) := J_T$. Using (24.5), we find that v/u is strictly decreasing in J starting with the value 1 at R; this violates the fact that $u = v$ at ρ. The case $uv' - vu' > 0$ in J yields the same conclusion. Thus, the two solutions must coincide.

(b) Let u, v be two such solutions. If there is $T \in [R,\rho)$ such that $\Psi(v,u)(T) = 0$, $v > u$ in $(T,\rho) := J(T)$, then $\{\Psi(v,u)\}'(r) = uv\{f(v)/v - f(u)/u\} < 0$ in $J(T)$, so Ψ is strictly negative and decreasing in $J(T)$. This conflicts with its value 0 at ρ. Thus, $uv' - u'v \neq 0$ in $J = [R,\rho)$ unless $u \equiv v$ there.

If $\Psi(v,u)(r) < 0$ in J, it would be negative and strictly decreasing, hence conflicting with its value at ρ.

Theorem 4. *For f as in* (E), *consider the problem*

$$u'' = f(u) \quad in \quad J := (R,\rho); \quad u(\rho) = 0. \tag{ρ}$$

(A) *If $\chi(t) := f(t)/t$ is decreasing in $(0,\lambda)$ and increasing for $t > \lambda$ with $\chi(\lambda) = 0$ for some $\lambda > 0$, then if there are two distinct solutions u and v of (ρ) with $v > u$ in J, $uv' - vu' < 0$ cannot hold in the whole interval J. Furthermore, if two distinct solutions u and v satisfy $v > u$ in (r_1,ρ) for some $r_1 \in J$, then $(uv' - vu')(r_1) > 0$.*

(B) *If $\chi(t) := f(t)/t$ is increasing in $(0,\lambda)$ and decreasing for $t > \lambda$ for some $\lambda > 0$, then if there are two distinct solutions u and v of (ρ) with $v > u$ in J, $uv' - vu' > 0$ cannot hold in the whole interval J. Furthermore, if two distinct solutions u and v satisfy $v > u$ in some (r_1,ρ), then $(uv' - vu')(r_1) < 0$.*

Proof. Let u and v be two such solutions.

(A) Suppose that $v > u$ in J, say. If $uv' - vu' < 0$ in J, then there is $t \in J$ such that $0 < v(t) < \lambda$. So, for $r > t$, $uv' - vu' < 0$ is negative and strictly decreasing, contradicting its value 0 at ρ.

(B) Here, if $v > u$ in J and $0 < u(T) < v(R) \le \lambda$ with $(uv' - vu')(T) > 0$, then $uv' - vu'$ is positive and strictly increasing in (T,ρ); hence, it could not be 0 at ρ.

Further details can be found in [1] and [2].

24.3 Problem (E1). Blow-up Solutions

Consider positive and increasing solutions of

$$\begin{gathered} u'' = h(u), \quad r > R, \\ u(R) = m \ge 0, \quad u'(R) = \nu \neq 0, \\ h \in C([0,\infty)) \cap C^1((0,\infty)), \quad h'(t), h(t) > 0 \quad \text{for } t > 0. \end{gathered} \tag{E1}$$

From the property of h, there is $r_1 > R$ such that $u'(r) > 0$ for all $r > r_1$ for any solution of (E1). So, in what follows we assume that $u' \geq 0$. For $H(t) := \int_0^t h(s)ds$, we see from the equation that for a solution of (E1), $u'(r)^2 = \nu^2 + 2\{H(u(r)) - H(m)\}$ and, as $u' \geq 0$, for all $r > R$

$$\int_m^{u(r)} \frac{du}{\sqrt{\nu^2 + 2\{H(u) - H(m)\}}} = r - R. \tag{24.6}$$

Theorem 5. *If for any $m \geq 0$ and $\nu \neq 0$*

$$R_\infty - R := \int_m^\infty \frac{dy}{\sqrt{\nu^2 + 2\{H(y) - H(m)\}}} < \infty, \tag{24.7}$$

then any such solution of (E1) *blows up at the finite value $R_\infty := R_\infty(m, \nu)$ i.e., $\lim_{r \nearrow R_\infty} u(r) = +\infty$.*

Proof. Any such solution is strictly increasing and, by (T), satisfies $u(r) \geq h(m)(r-R)^2/2$ for $r > R$. Hence, u cannot be bounded. Equalities (24.6) and (24.7) now complete the proof.

Either of the formulas (24.6) and (24.7) indicates that we would have problems if we set $\nu = 0$. In fact, in that case, under the given assumptions, we might have

$$\int_m^{u(r)} \frac{dy}{\sqrt{2\{H(y) - H(m)\}}} = +\infty,$$

which would create some difficulties.

The way around such a difficulty is shown in the second part of this work (the next chapter in this volume). When it comes to evaluating integral (24.7) or, more generally, (24.6) with $\nu = 0$, the expression under the square root can be written in the form

$$A(m) + G(u) := (u - m)^\theta G_1(u), \quad \theta \geq 0, \quad G_1(t) \neq 0 \quad \text{in } t \geq m. \tag{G}$$

The cases to be considered are

(c0) $\theta = 0$;
(c1) $\theta \in (0, 2)$;
(c2) $\theta \geq 2$.

For $Y := u - m$, (G) becomes $A(m) + G(Y) := Y^\theta G_1(Y + m) := \Gamma(Y)$. Suppose that there exist constants g_1, g_2, $g_3 > 0$ and a polynomial of the form $p(Y) := a + bY^\gamma$ with strictly positive coefficients such that

$$\begin{aligned} g_1 Y^\theta < \Gamma(Y) < g_2 Y^\theta \quad & \text{for Y in } (0, 1], \\ g_3 Y^\theta < \Gamma(Y) < Y^\theta p(Y) \quad & \text{for } Y > 1. \end{aligned} \tag{G1}$$

Then, as $\tau \searrow 0$, $\int_\tau^u dY/\sqrt{\Gamma(Y)}$ converges for (c0) and (c1) and diverges for (c2). Also, as $u \nearrow +\infty$, $\int_1^u dY/\sqrt{\Gamma(Y)}$ tends to $+\infty$ for (c0) and (c1) and converges for (c2). This leads to the following assertion.

Theorem 6. *Consider* (E1) *with* $m = \nu = 0$, *i.e.,*

$$u'' = h(u), \quad r > R, \quad u(R) = u'(R) = 0. \tag{E'1}$$

(i) *If* $h(t) = O(t^\gamma)$ *near* $t = 0^+$, *then if* $\gamma \in (0, 1)$, *problem* (E′1) *has a positive increasing solution in* $C^2([R, \infty))$, *which is large at infinity.*

(ii) *If* $h(t) = O(t^{1+\gamma})$, $\gamma \geq 0$, *near* 0^+, *then no solution* $u \in C([R, R + \eta))$, $\eta > 0$, *of* $u'' = h(u)$, $r > R$, *can have the value zero at* R.

Proof. (i) Multiplying both sides of the equation by u' and integrating over (R, r) yields

$$\int_0^{u(r)} \frac{dt}{\sqrt{H(t)}} = \sqrt{2}\{r - R\}, \tag{G2}$$

which leads to the desired conclusion.

(ii) If we assume that $u(\tau) > 0$, $\tau \in (R, R + 1)$, then, as in (G2),

$$\int_{u(\tau)}^{u(R+1)} \frac{dt}{\sqrt{H(t)}} = \sqrt{2}\{R + 1 - \tau\},$$

and this time the integral on the left-hand side diverges if $u(\tau) = 0^+$. Hence, no solution $u \in C([R, R + \tau))$ of the equation can satisfy $u(R) = 0$.

References

1. Tadie, On uniqueness conditions for decreasing solutions of semilinear elliptic equations, *Z. Anal. Anwendungen* **18** (1999), 517–523.

2. Tadie, Monotonicity and boundedness of the atomic radius in the Thomas-Fermi theory: mathematical proof, *Canadian Appl. Math. Quart.* **7** (1999), 301–311.

25 On the Monotone Solutions of Some ODEs. II: Dead-core, Compact-support, and Blow-up Solutions

Tadie

25.1 Introduction

In this chapter we focus on some problems motivated by physical situations. The first case is about some compact-support ground-state solutions, i.e., solutions that are strictly positive in a bounded domain and identically zero together with their derivatives outside that domain (see [1] and [2]). Dead-core solutions are solutions that are identically zero together with their derivatives inside a bounded, nonempty subdomain and positive outside it. The compact case applies, for example, to the existence of "magnetic islands" [3] and the dead-core one to some chemical reactions in which the reactant reacts only at the lateral region of the basin. Here we point out how some equations can lead to such types of situations. We also bring in some cases of explosive or large solutions.

For any $R > 0$, define $E(R) = [R, \infty)$ and $I(R) = (R, \infty)$. For some $\delta, m > 0$ we consider the problem

$$\begin{gathered} u'' = f(u) \quad \text{in } I(R), \qquad u(R) = m, \\ f \in C^1(I(R)) \cap C(E(R)), \quad f(0) = f(\delta) = 0, \\ f(t) > 0 \quad \text{in } (0, \delta), \quad f(t) < 0 \quad \text{in } (\delta, \infty). \end{gathered} \tag{E}$$

We have seen in Part I of this work that, using the associated problem

$$v'' = f(v) \quad \text{in } E(R), \quad v(R) = m, \quad v'(R) = \nu, \tag{Eν}$$

we can easily prove the existence of decreasing solutions of (E) by choosing (if necessary) suitable values for $\nu \in \mathbb{R}$. In this part of the work, we will focus on compact-support decreasing solutions of (E). Later, the dead-core and blow-up solutions of the problems, this time in $(0, R)$, with $f(u)$ replaced by $g(r)f(u)$ in certain cases, will be considered. We recall that a solution u of (E) is a *compact-support* solution if there is $\rho > R$ such that

$$u'' = f(u), \quad u > 0 \quad \text{in } (R, \rho), \quad u(r) = u'(r) = 0 \quad \forall r \geq \rho. \tag{CS}$$

Note that if a solution of (E) satisfies $u(\rho) = u'(\rho) = 0$, $\rho > R$, then, since for any $r > R$ we must have $u'(r)^2 = 2F(u(r))$, it follows that $u(r) \in [0, \theta]$ (see (24.2a) in Part I).

A solution is said to be a *blow-up* (or *large*) solution if it tends to infinity at some point or at infinity.

In this context, w is a *dead-core* solution of the equation

$$w'' = g(r)f(w) \text{ in } (0, R) := I_R, \quad w(R) = m$$

if there is an open interval $K \subset I_R$ such that

$$w'' = g(r)f(w) \text{ in } I_R, \quad w \equiv w' \equiv 0 \text{ in } K, \quad w > 0 \text{ in } I_R \setminus \bar{K}. \qquad \text{(DCS)}$$

25.2 Compact-support Solutions

Lemma 1. (Comparison results) (a) *Suppose that a C^1-function K satisfies $K > f$ with $K(t)/t > f(t)/t$ and either f or K is increasing for $t > 0$. Let u and v be, respectively, decreasing and crossing C^2-solutions of the initial value problems*

$$\begin{gathered} u'' \leq f(u), \quad v'' \geq K(v) \quad \text{in } r > R, \\ u(R) = v(R) = m > 0, \quad u'(R) \leq v'(R). \end{gathered}$$

Then $u < v$ as long as $u > 0$ in (R, ∞).

(b) *Suppose that f or K is increasing and that $f < K$. For the boundary value problems*

$$u'' \leq f(u), \quad v'' \geq K(v) \quad \text{in } r > R, \quad u(\rho) = v(\rho) = 0$$

we have $u > v$ in (R, ρ).

Proof. (a) It is easy to see that

$$\begin{gathered} (vu' - uv')(R) \leq 0, \\ vu'' - uv'' \leq (vu' - uv')' = v^2\{u/v\}' = uv\{f(u)/u - K(v)/v\}. \end{gathered}$$

At R, we have

$$f(u)/u - K(v)/v = f(u)/u - f(v)/v + f(v)/v - K(v)/v < 0;$$

hence, $vu'' - uv''$ is negative in some $(R, R + \tau)$. Thus, u/v is strictly decreasing there with the value 1 at R, and the assertion follows.

(b) Suppose that $u < v$ in some nonempty $A \subset (R, \rho)$; then $u'' - v'' \leq f(u) - K(v) = f(u) - f(v) + f(v) - K(v) = f(u) - K(u) + K(u) - K(v) < 0$ in A, conflicting with its value at the local minimum of $(u - v)$. Thus, $u > v$ in (R, ρ).

To simplify the notation, we write $(A, B) \leq (C, D)$ for $A \leq C$ and $B \leq D$. The reverse inequality is defined similarly.

Corollary 1. (c) *Under the assumptions in Lemma* 1 *but with the inequalities in* (a) *and* (b) *reversed, that is,*

(a′) $u > v$ *in the conclusion of* (a),

(b′) $u < v$ *in the conclusion of* (b),

and

(d) *if there are piecewise C^2-functions ϕ_1, ϕ_2, w_1, and w_2 such that in* (R,∞)

$$\begin{aligned} \phi_2'' - f(\phi_2) \le 0 \le \phi_1'' - f(\phi_1), \quad \phi_2(\rho_2) = \phi_1(\rho_1) = 0, \quad \rho_2 \ge \rho_1, \\ w_1'' - f(w_1) \le 0 \le w_2'' - f(w_2), \quad (w_1, w_1') \le (w_2, w_2') \quad \text{at } R, \end{aligned} \tag{25.1}$$

then, under the above assumptions, the solutions of

$$\begin{aligned} \phi'' = f(\phi) \;\text{ in } (R,\rho), \quad \phi(\rho) = 0, \quad \rho \in [\rho_1, \rho_2], \\ w'' = f(w) \;\text{ in } r > R, \quad (w_1, w_1') \le (w, w') \le (w_2, w_2') \;\text{ at } R, \end{aligned} \tag{25.2}$$

satisfy $\phi_1 \le \phi \le \phi_2$ and $w_1 \le w \le w_2$ as long as they are positive.

Remarks. (1) In Lemma 1, it is the comparison between f and K that matters as long as they are locally C^1. In fact, if $M > 0$ is such that each of $f_1(t) := f(t) + Mt$ and $K_1(t) := K(t) + Mt$ is increasing, we can consider instead the equations $u'' + Mu = f_1(u)$ and $v'' + Mv = K_1(v)$ and get the same conclusions.

(2) If the functions ϕ_i and w_i in (25.1) are found, then, according to the classical theory of supersolution-subsolution methods, solutions similar to those mentioned in (25.2) exist.

More generally, for initial value problems, if $h \in C^1$ is an increasing function and there are $\phi, \psi \in C^1$ such that

$$\begin{aligned} &\text{(i)} \quad (\psi - \phi)(R) \ge 0, \quad (\psi - \phi)'(R) \ge 0, \\ &\text{(ii)} \quad \psi'' - h(\psi) \ge 0 \ge \phi'' - h(\phi) \quad \text{for } r > R, \end{aligned} \tag{R1}$$

then $u'' = h(u) r > R$ has a solution $u \in C^2([R,\infty))$ such that

$$(\phi, \phi') \le (u, u') \le (\psi, \psi') \quad \text{in } [R, \infty).$$

Note that for final value problems we arrive at the same conclusion, with the following modifications.

a) In (i), R is replaced by the endpoint ρ, say;

b) in (ii), the inequalities are reversed and the domain is a finite interval $(0,\rho)$.

We say that

(i) $u \in U_m(\nu)$ if u is decreasing and solves (Eν) with compact support;

(ii) $\nu \in U_m$ if (Eν) has a decreasing solution with compact support.

Theorem 1. *Let $F(t) := \int_0^t f(s)\, ds$, and let $m > 0$ be given. If*

$$\rho_* := \int_0^m \frac{dt}{\sqrt{F(t) + \nu^2/2 - F(m)}} < \infty, \tag{C}$$

then for any monotone decreasing and crossing solution u of (Eν), u^+ has its support in $[R, \rho_* + R]$.

Furthermore, if (C) *holds for any ν such that* (Eν) *has a decreasing solution with $u(R) = m$, then there is $\nu_* > 0$ such that* (Eν_*) *has a compact-support solution. This solution is unique if the map $t \mapsto f(t)/t$ is monotone in $(0, m)$.*

Proof. If v is such a decreasing solution of (Eν), then $v' < 0$ as long as $v > 0$ and, for $r > R$, we get $-v'(r) = [2\{F(v(r)) + A_{m\nu}\}]^{1/2}$, where $A_{m\nu} := \nu^2/2 - F(m)$. Thus,

$$-\int_m^{v(r)} \frac{du}{[2\{F(u) + A_{m\nu}\}]^{1/2}} + R = r, \tag{25.3}$$

and (C) implies that the left-hand side of (25.3) is uniformly bounded above for $v(r) \in [0, m]$. It is obvious from (24.2a) in Part I that for such a solution, $\{F(u) + A_{m\nu}\} \geq 0$; more precisely, for those solutions that have a positive zero a, say, $v'(a)^2 = \nu^2 - 2F(m)$; hence we arrive at the following necessary condition.

$$\text{If the solution } v \text{ has a zero in } r > R, \text{ then } \nu^2 \geq 2F(m). \tag{C1}$$

The left-hand side of (25.3) is uniformly bounded if we have, for example, $f(t) = O(t^\alpha)$ near $t = 0$ with $\alpha \in (0, 1)$. In fact, for $\beta = \alpha + 1$,

(i) if $A := A_{m\nu} > 0$, then $\int_0^m \{t^\alpha + A\}^{-1/2}\, dt \leq \int_0^m t^{-\beta/2}\, dt < \infty$, and

(ii) if $A = -\tau^\beta$, $\tau > 0$, then $t^\beta + A = t^\beta - \tau^\beta$ in $(0, \tau)$ and, by the classical inequality $t^\beta - \tau^\beta > \beta\tau^\alpha(t - \tau)$, if $t > \tau > 0$ and $\beta > 1$, then $\int_0^\tau [t^\beta - \tau^\beta]^{-1/2}\, dt < c \int_0^\tau [t - \tau]^{-1/2}\, dt$, which is finite.

If (C) holds for any $\nu \in U_m$, we define

$$R_* := \max\{\rho > R;\ \operatorname{supp}(u^+) = [R, \rho],\ \ u \in U_m(\nu)\},$$

where ν_m and u_* are the corresponding ν and solution of (Eν_m), that is,

$$u'' = f(u) \quad \text{in } E(R_*), \quad u(R) = m, \quad u'(R) = \nu_m, \quad u(R_*) = 0.$$

If we assume that $u'(R_*) \neq 0$, then, by the implicit function theorem [4], there should be $I := (\nu_* - \tau, \nu_* + \tau)$, $\tau > 0$, such that for all $\nu \in I$, (Eν) has such a solution. Some solutions would then have the corresponding ρ greater than R_*, violating the definition of R_*; hence, $u(R_*) = u'(R_*) = 0$.

Theorem 3 of Part I now completes the proof.

Theorem 2. *Let $g \in C([R,\infty); R_+)$ be strictly positive and decreasing or bounded in $r \geq R$, let $m > 0$, and consider decreasing positive solutions of*

$$u'' = g(r)f(u), \quad r > R, \quad u(R) = m > 0, \tag{g}$$

where f and m are as in Theorem 1, with (C) holding for all $\nu \in U_m$. Then all those solutions have their supports in a fixed and finite interval $[R, \rho_*]$.

Proof. Let $g_0 := \max_{r>R} g(r)$. It suffices to compare the above solutions with that of $v'' = g_0 f(v)$, $r > R$, $v(R) = m > 0$.

When v is a decreasing solution corresponding to the data $v(R) = m > 0$ and $v'(R) = \nu$, the identity (24.2a) in Part I implies that

$$F(v(r)) + \tfrac{1}{2}\nu^2 - F(m) \geq 0 \quad \text{whenever } v(r) > 0 \text{ in } r > R. \tag{25.4}$$

Theorem 3. *Consider*

$$u'' = f(u) \quad \text{in } J := (R, \rho), \quad u(\rho) = 0, \tag{25.5}$$

and suppose that

(i) *$\chi(t) := f(t)/t$ is decreasing in $(0,\lambda)$ and increasing in $t > \lambda$ with $\chi(\lambda) = 0$ for some $\lambda > 0$;*

(ii) *f is increasing in some $(0, \lambda_1)$ and either $|\chi'(t)/f'(t)|$ is locally bounded, and bounded away from 0 in $t > 0$, or $[\chi'(t)/f'(t) + 1/t^2] > 0$ for small $t > 0$.*

Then (25.5) has at most one decreasing solution.

Proof. Let u and v be two such solutions, and let $U(r) := u(r) + c$ and $V(r) := v(r) + c$, where $c > 0$ is small. For $X(r) := U$ or V,

$$\begin{aligned}
&\text{(i)} \quad X'' = f(X-c) \quad \text{in } (R,\rho), \quad X(\rho) = c;\\
&\text{(ii)} \quad [UV' - VU']' = UV\{\chi_c(V) - \chi_c(U)\}\\
&\qquad = uv\left\{\frac{f(v)}{v} - \frac{f(u)}{u}\right\} + c(f(v) - f(u));\\
&\text{(iii)} \quad \chi_c'(t) = \frac{1}{t^2}\{(t-c)^2\chi'(t-c) + cf'(t-c)\}\\
&\qquad = \left[\frac{(t-c)}{t}\right]^2 f'(t-c)\left\{\frac{\chi'(t-c)}{f'(t-c)} + \frac{c}{(t-c)^2}\right\},
\end{aligned} \tag{χ}$$

where $\chi(t) := f(t)/t$ and $\chi_c(t) := f(t-c)/t$. So, for $r > T > R$,

$$\begin{aligned}
(UV' - VU')(r) &:= (UV' - VU')(T) + \int_T^r UV\{\chi_c(V) - \chi_c(U)\}\,ds\\
&= (uv' - vu')(r) + c\left\{(v' - u')(T) + \int_T^r (f(v) - f(u))\,ds\right\}\\
&= (uv' - vu')(r) + c(v' - u')(r).
\end{aligned} \tag{$\chi 1$}$$

For some $c > 0$ and U and V defined above, if $|\chi'/f'|$ is bounded away from 0 with $f' > 0$, then, by (χ)(iii), $\chi_c'(t) > 0$ when t approaches c or

$$[UV' - VU']' = UV\{\chi_c(V) - \chi_c(U)\} > 0$$

when U and V approach c.

Let $\Lambda^+ := \{r \in [R, |uv' - vu' > 0\}$ and $U^+ := u(\Lambda^+)$. Then there is $c \in U^+$ such that for some $r \in \Lambda^+$ close to ρ we have

$$(UV' - VU')(r) = (uv' - vu')(r) + c(v' - u')(r) \geq 0.$$

In fact, as $u' \neq 0$ in J and $u, v \in C^1(\overline{J})$, since $uv' - vu' > 0$ in some (r_1, ρ), we only need to take a sufficiently small $c > 0$. Taking $T \in \Lambda^+$ close to ρ and $c > 0$ close to $u(T)$ so that $(UV' - VU')(T) \geq 0$, we see that $UV' - VU'$ is positive and increasing in (T, ρ), which implies that V/U is strictly increasing there with $(V/U)(T) > 1$. This conflicts with $V = U = c > 0$ at ρ (see also [5] and [6]).

25.3 Dead-core and Blow-up Solutions

We now consider possibilities of dead-core solutions for problems of the type

$$v'' = g(r)f(v), \quad r \in (0, R), \quad v(R) = m > 0, \quad v(0) = 0. \tag{DC}$$

Theorem 4. *Let $g \in C^1((0, R]; R_+)$ be strictly decreasing, and let $m > 0$ be such that $F(t) := \int_0^t f(s)ds > 0$ in $(0, m)$ and $f > 0$ in $(0, \delta)$, $\delta < m$. Consider the overdetermined problem*

$$v'' = g(r)f(v) \quad in\ (0, R), \quad v(R) = m, \quad v(0) = v'(0) = 0, \tag{V}$$

and suppose that

$$\lim_{r \searrow 0} \int_r^R \sqrt{g(s)}\, ds = +\infty, \quad M(m) := \int_0^m \frac{du}{\sqrt{F(u)}} < +\infty. \tag{25.6}$$

Then for any increasing C^2-solution u of (V), there is $\rho := \rho_m \in (0, R)$ such that

$$\begin{gathered} u \equiv u' \equiv 0 \quad in\ [0, \rho], \quad u > 0 \quad for\ r > \rho, \\ \int_\rho^R \sqrt{2g(s)}\, ds < M(m). \end{gathered} \tag{25.7}$$

If $g(r) = O(r^{-k})$, $k > 2$, at 0, then for $\mu := (k - 2)/2$,

$$\rho_m \geq R\{1 + \sqrt{2}\mu R^\mu M(m)\}^{-1/\mu}. \tag{25.8}$$

Proof. If u is such a solution, then

$$((u')^2)' = 2g(r)u'f(u) = 2g(r)F(u)'.$$

So, from the properties of g, for all $r_1 \in (0, \delta)$,

$$u'(r)^2 - u'(r_1)^2 = 2\int_{r_1}^{r} g(s)F(u(s))'\,ds \geq 2g(r)[F(u(r)) - F(u(r_1))],$$

from which

$$u'(r)^2 \geq u'(r_1)^2 + 2\sqrt{g(r)\{F(u(r)) - F(u(r_1))\}}.$$

As u and F are increasing, we get

$$u'(r) \geq u'(r_1) + \sqrt{2g(r)\{F(u(r)) - F(u(r_1))\}},$$

so

$$\int_{r_1}^{r} \sqrt{2g(s)}\,ds < \int_{u(r_1)}^{u(r)} \frac{dt}{\sqrt{\{F(t) - F(u(r_1))\}}}, \tag{25.9}$$

and the conclusion follows. In fact, (25.7) implies that r_1 on the left-hand side of (25.9) cannot be allowed to be arbitrarily small.

For some sufficiently large $\mu > 0$, the problem

$$U'' = g(r)f(U), \quad r \in (0, R), \quad U(R) = m, \quad U'(R) = \mu \tag{25.9a}$$

has a unique decreasing solution $u := U_\mu$ that has a positive zero, $a := a_\mu \in (0, R)$, say. To any such *admissible* μ there corresponds a unique a_μ. Thus, (25.6) implies that $a_0 := \min\{a_\mu | \mu \text{ is admissible}\} > 0$.

If we denote the solution of (25.9a) by $U(a, r)$ when its positive zero is $a := a_\mu$, then the implicit function theorem yields $\{d/dr\}U|_{(a_0,a_0)} = 0$. Otherwise, there would be $\eta > 0$ such that to any element $\alpha \in (a_0 - \eta, a_0 + \eta)$ there corresponds an admissible μ, and a solution whose positive zero is α.

If $g(r) = r^{-k}$, then, by (25.9), for $R > r_1 > 0$

$$\int_{r_1}^{R} g(r)\,dr = \frac{2\{r_1^{(2-k)/2} - R^{(2-k)/2}\}}{(k-2)} < \int_{u(r_1)}^{m} \frac{dt}{\sqrt{F(t) - F(u(r_1))}}. \tag{25.10}$$

When $r_1 \searrow 0$, we have $F(u(r_1)) \searrow 0$; hence, by (25.6), r_1 has a minimum positive value, T, say. From (25.10) we then obtain

$$T^{(2-k)/2} < R^{(2-k)/2} + \tfrac{1}{2}(k-2)\int_0^m \frac{dt}{\sqrt{F(t)}},$$

and the estimate for $\rho_m := T$ follows.

For $\tau > 0$, consider a solution u_τ of the problem

$$\begin{gathered} u'' = g(r)h(u), \quad u' > 0 \quad \text{for } r > \tau, \quad u(\tau) = u'(\tau) = 0, \\ h \in C^1 \text{ and is increasing}, \\ g \in C(0,\infty) \text{ is strictly positive for } r > 0. \end{gathered} \tag{T}$$

Definition. A function u is said
(i) to *blow up at* $T > 0$ if $\lim_{r \nearrow T} u(r) = \infty$;
(ii) to be *above* another function v if $u(r) > v(r)$ for all $r \in \operatorname{supp}(u) \cap \operatorname{supp}(v)$, or if their supports are disjoint and u blows up at some $T < \inf\{\operatorname{supp}(v)\}$.

Lemma 2. (1) *If $\tau_1 < \tau_2$ and $u_i \equiv u_{\tau_i}$, then u_1 is above u_2.*
(2) *Suppose that $t \mapsto h(t)/t$ is increasing in $t > 0$. If $u, v \in C^2$ satisfy*

$$\begin{gathered} u'' - g(r)h(u) \le 0 \le v'' - g(r)h(v), \quad r > \tau, \\ (vu' - uv')(\tau) \le 0, \quad u < v \text{ in some interval } (\tau, \tau_1), \end{gathered}$$

then v is above u for $r > \tau$.

Proof. (1) We have $(u_1 - u_2)'' = h(u_1) - h(u_2) > 0$ in some (τ_2, r), $(u_1 - u_1)'(\tau_2) > 0$, and $(u_1 - u_2)(\tau_2) > 0$. This implies that $u_1 > u_2$ at (and beyond) any such t.
(2) In some (τ, t), we have $(vu'' - v''u)(r) \le uvg(r)\{h(u)/u - h(v)/v\} < 0$; hence, $(vu' - uv') = v^2(u/v)' < 0$ in (τ, t), which means that u/v is strictly decreasing with a value less than 1 at τ.

This easily leads to the next assertion.

Theorem 5. *Let $h \in C([0,\infty); \mathbb{R}_+) \cap C^1((0,\infty))$ be increasing, let $g \in C([0,\infty); \mathbb{R}_+)$ be decreasing or bounded below for $r > R > 0$, and suppose that $H(t) := \int_0^t h(s)ds$ satisfies*

$$\int_0^\infty \frac{dt}{\sqrt{H(T) - H(m)}} := H_0 < +\infty, \quad \lim_{r \nearrow \infty} \int_R^r \sqrt{2g(s)}\, ds = \infty. \tag{25.11}$$

Then for any $\nu > 0$, the problem

$$u'' = h(u), \quad r > R, \quad u(R) = m \ge 0, \quad u'(R) = \nu$$

has a solution that blows up at a finite $T_\nu > R$.

Proof. Let $\nu > 0$, and let u be its corresponding solution in a maximal interval of existence $[R, T)$, say. From the equation, u, u', and u'' are strictly increasing and positive, so u cannot be bounded. Furthermore, $[(u')^2]' = 2g(r)H(u)'$, therefore,

$$u'(r)^2 = \nu^2 + 2\int_R^r g(s)H(u(s))'\, ds > 2g(r)\{H(u(r)) - H(m)\}$$

and $u'(r) > \sqrt{2g(r)\{H(u) - H(m)\}}$, or

$$\int_m^{u(r)} \frac{dt}{\sqrt{H(T) - H(m)}} > \int_R^r \sqrt{2g(s)}\, ds.$$

Since, by (25.11), the left-hand side is bounded whereas the right-hand side is not, T_ν has to be finite.

Theorem 6. *Let $h \in C([0,\infty);\mathbb{R}_+) \cap C^1((0,\infty))$ be increasing, let $g \in C([0,\infty);\mathbb{R}_+)$ be decreasing in some interval $(0,R)$ and decreasing or bounded below for $r > R$, let $H(t) := \int_0^t h(s)ds$, and suppose that*

$$\int_0^\infty \frac{dt}{\sqrt{H(T)}} := H_0 < +\infty,$$
$$\lim_{\rho \nearrow \infty} \int_R^\rho \sqrt{2g(s)}\, ds = \infty, \quad \lim_{r \searrow 0} \int_r^R \sqrt{2g(s)}\, ds = \infty. \tag{25.12}$$

In this case,

(a) *for any $\tau > 0$, any increasing solution u_τ of the problem*

$$u'' = h(u), \quad r > 0, \quad u(\tau) = 0, \quad u'(\tau) = 0 \tag{25.13}$$

blows up at some finite $R_\tau > 0$ such that

$$\text{(i)} \quad R_\tau \geq \frac{H_0}{\sqrt{2g(\tau)}} + \tau,$$
$$\text{(ii)} \quad R_\tau \leq \tau\left\{\frac{2\sqrt{2}}{2\sqrt{2} - (k-2)H_0\tau^{(k-2)/2}}\right\}; \tag{25.14}$$

(b) *if $\tau > \sigma > 0$, then u_σ is above u_τ;*

(c) *$R_\tau \searrow 0$ as $\tau \searrow 0$, so there is no such solution for $\tau = 0$.*

(An example of this type of function h is $h(t) = t^\alpha + t^\gamma$, $\alpha \in (0,2)$, $\gamma > 2$.)

Proof. (a) First, we show that u_τ exists.

Let $R_1 > \tau$, and let w and v be large solutions of

$$w'' = h(w), \quad r > 0, \quad w(0) = \mu > 0, \quad w'(0) = \nu,$$
$$v'' = h(v), \quad r > R_1, \quad v(R_1) = 0, \quad v'(R) = 0.$$

By taking $\nu > v'(R_1)$ and extending v and v' by 0 to $[0, R_1)]$, we ensure that $(v, v') \leq (w, w')$ for $r > \tau$. By the comparison method (see the above Remarks), (25.13) has a solution lying between w and v.

From the equation, in $(0,R)$ we have

$$\{u'(r)^2\}' = 2g(r)H(u)', \quad u'(r)^2 = 2\int_\tau^r g(s)H(u(s))'ds \leq 2g(\tau)H(u(r)).$$

Consequently,

$$\frac{u'(r)}{\sqrt{H(u)}} \leq \sqrt{2g(\tau)}, \quad \sqrt{2g(\tau)}(r-\tau) \geq \int_0^{u_\tau(r)} \frac{dt}{\sqrt{H(T)}} = H_0$$

for any such increasing solution u_τ. This yields estimate (i).

If $g(r) = O(r^{-k})$ for small $r > 0$, then $u'(r)^2 = 2\int_\tau^r g(s)H(u(s))'ds \geq 2g(r)H(u(r))$, from which $u'(r) \geq \sqrt{2g(r)H(u(r))}$ and

$$H_0 = \int_0^\infty \frac{dt}{\sqrt{H(u)}} \geq \sqrt{2}\int_\tau^R r^{-k/2}\,dr = \frac{2\sqrt{2}}{k-2}\{\tau^{(2-k)/2} - R^{(2-k)/2}\};$$

therefore, (ii) follows.

(b) At $r = \tau > \sigma$, the function $u_\sigma - u_\tau$ and its derivative are strictly positive; so, by the comparison method, $u_\sigma > u_\tau$ for $r > \tau$.

References

1. J.I. Diaz and M.A. Herrero, Estimates on the support of the solutions of some nonlinear elliptic and parabolic problems, *Proc. Royal Soc. Edinburgh* **89A** (1981), 249–258.
2. P. Pucci, J. Serrin, and H. Zou, A strong maximum principle and a compact support principle for singular elliptic inequalities, *J. Math. Pures Appl.* **78** (1999), 769–789.
3. H.G. Kaper and M.K. Kwong, Free boundary problems for Emden-Fowler equations, *Diff. Integral Equations* **3** (1990), 353–362.
4. Tadie, An ODE approach for $\triangle U + \lambda U^p - U^{-\gamma} = 0$ in $\mathbb{R}^n$, $\gamma \in (0,1)$, $p > 0$, *Canadian Appl. Math. Quart.* **10** (2002), 375–386.
5. Tadie, On uniqueness conditions for decreasing solutions of semilinear elliptic equations, *Z. Anal. Anwendungen* **18** (1999), 517–523.
6. Tadie, Monotonicity and boundedness of the atomic radius in the Thomas-Fermi theory: mathematical proof, *Canadian Appl. Math. Quart.* **7** (1999), 301–311.

26 A Spectral Method for the Fast Solution of Boundary Integral Formulations of Elliptic Problems

Johannes Tausch

26.1 Introduction

Discretizations of boundary integral equations lead to dense linear systems. If an iterative method, such as conjugate gradients or GMRES, is used to solve such a system, the matrix-vector product has $O(n^2)$ complexity, where n is the number of degrees of freedom in the discretization. The rapid growth of the quadratic term severely limits the size of tractable problems.

In the past two decades, a variety of methods have been developed to reduce the complexity of the matrix-vector product. These methods exploit the fact that the Green's function can be approximated by truncated series expansions when the source and the field point are sufficiently well separated. Typical examples of such methods are the fast multipole method (FMM) and wavelet-based discretizations.

The additional error introduced by the series approximation must be controlled; ideally, this error should be of the same order as the discretization error. Wavelets and the FMM have been shown to be asymptotically optimal in many situations. That is, the complexity of a matrix-vector multiplication is order n, while the convergence rate of the discretization scheme is preserved (see, for example, [1]–[3]).

In this paper, we explore a spectral method to reduce the complexity of the matrix-vector product. Here, the Green's function is replaced by a trigonometric expansion, which is valid globally for all positions of source- and field-points. Spectral techniques have been applied previously by Greengard and Strain to the heat equation [4]. In this work we consider elliptic equations where the Green's function is singular and the convergence of the Fourier series is slow. To overcome this difficulty, we split the Green's function into a local part, which is evaluated directly, and a smooth part, which will be treated with the Fourier series approach. We will also develop nonequispaced fast Fourier transforms for computing the matrix-vector product efficiently.

Although the methodology can be applied to a large class of Green's functions, we limit the discussion in this paper to the Laplace equation. Our focus in this paper is on the presentation of the algorithm. A more detailed analysis of the error will be discussed elsewhere.

26.2 A Fast Algorithm for Smooth, Periodic Kernels

Consider the fast evaluation of a surface integral operator with a generic smooth and periodic kernel G

$$\Phi(x) := \int_S G(x-y)g(y)\,dS_y, \quad x \in S, \tag{26.1}$$

where S is a surface that is contained in the unit cube $[0,1]^3$ and $G(\cdot)$ is a C^∞-function that has period two in all three variables. The kernel can be approximated by the truncated Fourier series

$$G_N(r) := \sum_{|k|\le N} \hat{G}_k \exp(\pi i\, k^T r), \quad r \in [-1,1]^3, \tag{26.2}$$

where the summation index k is in $\mathbb{Z}^3$ and $|k| := \max\{k_1,k_2,k_3\}$. Under the assumptions on the kernel, the convergence of (26.2) is superalgebraic in N. The resulting approximate potential is given by

$$\Phi_N(x) = \int_S G_N(x-y)g(y)dS_y = \sum_{|k|<N} \exp(\pi i k^T x)\hat{d}_k, \tag{26.3}$$

where $\hat{d}_k = \hat{G}_k \hat{g}_k$ and

$$\hat{g}_k = \int_S \exp(-\pi i k^T y)g(y)dS_y. \tag{26.4}$$

This simple computation suggests the following approach to evaluate the potential in (26.1).

1. Compute the Fourier coefficients $\hat{g}_k$ in (26.4).
2. Multiply $\hat{d}_k := \hat{G}_k \hat{g}_k$ for $|k| \le N$.
3. Evaluate the Fourier series (26.3) for $x \in S$.

The choice of the truncation parameter N depends on the approximation properties of the Fourier series (26.2) and can be selected to be much smaller than the size of the linear system n. Stage 2 obviously involves $O(N^3)$ operations; the other two stages can be executed efficiently using nonequispaced fast Fourier transforms (FFTs). This will be discussed next.

26.2.1 Computation of the $\hat{g}_k$

In this section, we describe how FFTs can be used to efficiently compute the Fourier coefficients of the function g. To that end, the three-space is divided into small cubes C_l, $l = (l_1,l_2,l_3) \in \mathbb{Z}^3, 0 \le l_j < N$. These cubes have centers $x_l = l/N$ and side length $1/N$. Note that N is the same as in (26.3) and therefore the cubes get smaller if more terms in the Fourier series expansion of the Green's function are retained. We assume that S is contained in the union of all cubes, and set $S_l = C_l \cap S$ to denote the piece of the surface that intersects with the lth cube (see Fig. 1).

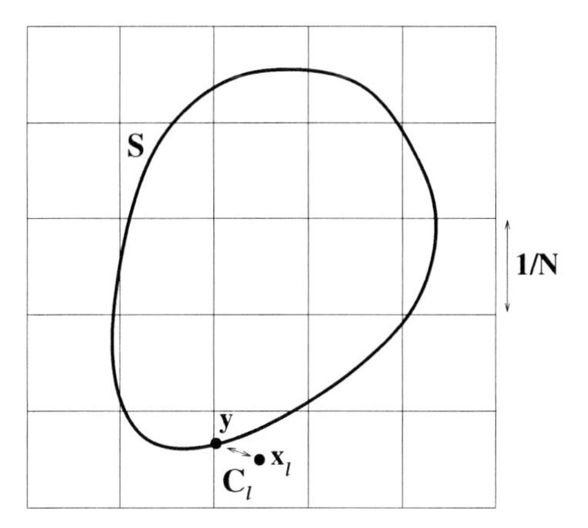

Fig. 1. Two-dimensional illustration of the geometry.

From (26.4) it follows that the Fourier coefficients of g can be written as

$$\hat{g}_k = \sum_l \exp\left(\frac{-\pi i k^T l}{N}\right) \int_{S_l} \exp(-\pi i k^T (y - x_l)) g(y)\, dS_y. \qquad (26.5)$$

The frequency and the spatial variable in the integral can be separated using the Jacobi–Anger expansion

$$\exp(-i\xi t) = \sum_{\nu=0}^{\infty} (-i)^\nu (2\nu + 1) j_\nu(\xi) P_\nu(t), \quad -1 \le t \le 1$$

(see, for example, [5]). Here, $j_\nu(\cdot)$ is the spherical Bessel function of order ν and $P_\nu(\cdot)$ is the Legendre polynomial of degree ν. This formula can be easily applied to the integrand in (26.5); this leads to

$$\exp(-\pi i k^T (y - x_l)) \approx \sum_{|\alpha| \le p} (-i)^{|\alpha|} (2\alpha + 1) j_\alpha(\pi k H) P_\alpha\left(\frac{y - x_l}{H}\right), \quad (26.6)$$

where p is the expansion order, $H = 1/(2N)$, $\alpha = (\alpha_1, \alpha_2, \alpha_3)$ is a multi-index, $|\alpha| = \alpha_1 + \alpha_2 + \alpha_3$, $j_\alpha(x) = j_{\alpha_1}(x_1) j_{\alpha_2}(x_2) j_{\alpha_3}(x_3)$, and $P_\alpha(x)$ is defined similarly. Substitution of (26.6) into (26.5) leads to the approximation

$$\hat{g}_k \approx \sum_{|\alpha| \le p} (-i)^\alpha (2\alpha + 1) j_\alpha(\pi k H) \sum_l \exp\left(\frac{-\pi i k^T l}{N}\right) m_l^\alpha(g),$$

where

$$m_l^\alpha(g) = \int_{S_l} P_\alpha\left(\frac{y - x_l}{H}\right) g(y)\, dS_y$$

is a moment for which exact formulas can be derived if the function and the surface are discretized. In particular, if g is piecewise polynomial, then the moments are linear transformations of the coefficients of g corresponding to the nodal basis. The matrix that maps the coefficients to the αth moments is denoted by M_α. The number of nonzero entries in M_α is n.

In matrix form, the (approximate) coefficient vector $\hat{g}$ is given by

$$\hat{g} = \sum_{|\alpha| \leq p} K_\alpha F M_\alpha \vec{g}, \tag{26.7}$$

where F is the 3D FFT, $\vec{g}$ the vector of coefficients of g, and K_α is a diagonal matrix with the factors $(-i)^{|\alpha|}(2\alpha+1) j_\alpha(\pi k H)$. The computation of $\hat{g}$ involves $(p+1)(p+2)(p+3)/6$ FFTs. Since C_l is smaller than a wavelength of the highest Fourier mode, it suffices to use a small value of p.

26.2.2 Evaluation of the Fourier Series

In the Galerkin discretization, the ith component of the matrix product Φ_i is the inner product of the potential Φ in (26.1) with the ith nodal basis function φ_i. For the fast method, the potential is replaced with the approximated potential Φ_N in (26.3). In order to evaluate the potential efficiently, the Jacobi–Anger approximation (26.6) is used again, just as in the previous section. This is shown in the following computation:

$$\begin{aligned}
\Phi_i &= \int_S \varphi_i(x) \Phi_N(x)\, dS_x \\
&= \sum_k \exp\left(\frac{\pi i\, k^T l}{N}\right) \int_S \exp(\pi i k^T (y - x_l)) \varphi_i(x)\, dS_x\, d_k \\
&\approx \sum_{|\alpha| \leq p} \sum_k \exp\left(\frac{\pi i\, k^T l}{N}\right) i^{|\alpha|}(2\alpha + 1) j_\alpha(\pi k H) m_l^\alpha(\varphi_i)\, d_k.
\end{aligned}$$

In matrix notation, the above can be written as

$$\vec{\Phi} = \sum_{|\alpha| \leq p} M_\alpha^T F K_\alpha^* \vec{d}. \tag{26.8}$$

Hence, $(p+1)(p+2)(p+3)/6$ FFTs are necessary to compute the vector $\vec{\Phi}$. Furthermore, it is evident that the operation (26.8) is the adjoint of operation (26.7).

26.3 Extension to Singular Kernels

If the spectral method is applied to the Green's function of an elliptic equation, then the convergence with respect to the truncation parameter N will be slow due to the singularity at $r = 0$. We will therefore split the Green's function into a smooth and a local part.

26.3.1 Smooth Part

The following discussion is directed at the Laplace kernel, whose Fourier transform is $1/|\xi|^2$. That is, for $r \in \mathbb{R}^3$,

$$G(r) = \frac{1}{4\pi|r|} = \frac{1}{(2\pi)^3} \int_{\mathbb{R}^3} \frac{1}{|\xi|^2} \exp(i\,\xi^T r)\, d^3\xi.$$

A smooth approximation of this function can be obtained by multiplying the Fourier transform by the exponential $\exp(-\delta|\xi|^2)$, where $\delta > 0$ is the mollification parameter which is at our disposal. The smooth Green's function is given by

$$\tilde{G}_\delta(r) = \frac{1}{(2\pi)^3} \int_{\mathbb{R}^3} \frac{\exp(-\delta|\xi|^2)}{|\xi|^2} \exp(i\,\xi^T r)\, d^3\xi.$$

This kernel can be expressed in closed form, and we find that

$$\tilde{G}_\delta(r) = \frac{1}{4\pi|r|} \operatorname{erf}\left(\frac{|r|}{2\sqrt{\delta}}\right), \tag{26.9}$$

where $\operatorname{erf}(\cdot)$ is the error function. The kernel $\tilde{G}_\delta$ is in $C^\infty(\mathbb{R}^3)$, but not periodic. Therefore, we introduce an offset parameter $0 < \mu \ll 1$, re-scale (26.1) so that S is contained in the cube $[0, 1-\mu]^3$, and define a smooth cut-off function

$$\chi_\mu(r) = \begin{cases} 1 & \text{for } r \text{ in } [-1+\mu, 1-\mu]^3, \\ 0 & \text{for } r \text{ outside } [-1,1]^3, \\ \geq 0 & \text{otherwise.} \end{cases} \tag{26.10}$$

Thus, the kernel $G_\delta := \chi_\mu \tilde{G}_\delta$ is smooth and periodic and generates the same potential in (26.1) as the kernel G_δ.

In stage 2 of the spectral method, the Fourier coefficients of the function $\hat{g}_k$ are multiplied by the Fourier coefficients of the kernel $\hat{G}_k$. Since there are no analytic expressions of $\hat{G}_k$ available, these coefficients must be computed numerically, using FFTs and the Jacobi–Anger series. Since this algorithm is completely analogous to the computation of the function coefficients described in Section 26.2.1, we omit the details.

26.3.2 Local Part

Due to the behavior of the error function, the smooth part is a good approximation of the actual Green's function if δ is small and r is large. In the neighborhood of the origin, the two functions are very different and therefore the contribution of this local part must be accounted for. The potential in (26.1) can be decomposed as $\Phi = \Phi_\delta + \Psi_\delta$, where

$$\Phi_\delta(x) := \int_S G_\delta(x-y)g(y)\,dS_y, \quad x \in S,$$

is the smooth part and

$$\Psi_\delta(x) := \int_S E_\delta(x-y)g(y)\,dS_y, \quad x \in S,$$

is the local part. Here, $E_\delta = G - G_\delta$. In what follows we show that the local part has an expansion with respect to the mollification parameter $\sqrt{\delta}$ and indicate how to compute the expansion coefficients.

Because of (26.5), the function E_δ decays exponentially away from the origin. We introduce another cut-off function χ_ν for some $0 < \nu < 1$, which is small enough so that the surface has a parameterization of the form $y(t) = x + At + nh(t)$ in the ν-neighborhood of x. Here, n is the normal to the surface at the point x, $A \in \mathbb{R}^{3\times 2}$ has two orthogonal columns that span the tangent plane at x, and $h(t) = O(|t|^2)$ is some scalar function in $t \in \mathbb{R}^2$. The local potential $\Psi_\delta(x)$ can be written in the form

$$\begin{aligned}\Psi_\delta(x) &= \int_S E_\delta(x-y)g(y)\,dS_y \\ &= \int_S E_\delta(x-y)\chi_\nu(x-y)g(y)\,dS_y + O\left(\exp\left(-\frac{\nu}{\delta}\right)\right) \\ &= \int_{\mathbb{R}^2} E_\delta(t)\tilde{g}(t)d^2t + O\left(\exp\left(-\frac{\nu}{\delta}\right)\right).\end{aligned} \tag{26.11}$$

Here, $E_\delta(t) = E_\delta(x - y(t))$, $\tilde{g}(t) = \chi_\nu(x - y(t))g(t)J(t)$, and $J(t)$ is the Jacobian of the parameterization. For simplicity, we assume that the function $h(t)$ in the parameterization of the surface is analytic, that is,

$$h(t) = \sum_{|\alpha|\geq 2} h_\alpha t^\alpha. \tag{26.12}$$

Thus, there are C^∞-functions H_n such that

$$r(t) := |x - y(t)| = |t| \sum_{n=0}^{\infty} |t|^n H_n(\hat{t}), \tag{26.13}$$

where $\hat{t} := t/|t|$, $H_0(\hat{t}) = 1$, and $H_1(\hat{t}) = 0$. In the neighborhood of the point x, the kernel has the form

$$E_\delta(t) = \frac{1}{\sqrt{\delta}} E\left(\frac{r(t)}{\sqrt{\delta}}\right) \quad \text{where} \quad E(z) = \frac{1}{4\pi z}\left(1 - \operatorname{erf}\left(\frac{z}{2}\right)\right).$$

Note that $E(z)$ is singular at $z = 0$ and decays exponentially as $z \to \infty$. Substituting (26.13) in (26.11) results in

$$\begin{aligned}\Psi_\delta(x) &= \frac{1}{\sqrt{\delta}} \int_{\mathbb{R}^2} E\left(\frac{r(t)}{\sqrt{\delta}}\right) \tilde{g}(t)\, d^2t + O\left(\exp\left(-\frac{\nu}{\delta}\right)\right) \\ &= \sqrt{\delta} \int_{\mathbb{R}^2} E\left(|t| \sum_{n=0}^{\infty} (\sqrt{\delta}|t|)^n H_n(\hat{t})\right) \tilde{g}(\sqrt{\delta}t)\, d^2t + O\left(\exp\left(-\frac{\nu}{\delta}\right)\right),\end{aligned}$$

where the second integral is the result of the change of variables $t \mapsto t/\sqrt{\delta}$. It is easy to see that $H_n(-\hat{t}) = (-1)^n H_n(\hat{t})$, which implies that the integral in the last expression is an even function of $\sqrt{\delta}$. Furthermore, the integral as a function of $\sqrt{\delta}$ is C^∞, and can be expanded in a Taylor series. Since the exponential term does not contribute to the expansion, we obtain

$$\Psi_\delta(x) = \delta^{\frac{1}{2}} \Psi_0 + \delta^{\frac{3}{2}} \Psi_1 + \delta^{\frac{5}{2}} \Psi_2 + \cdots. \tag{26.14}$$

A more detailed analysis shows that the first two expansion coefficients Ψ_k are given by

$$\begin{aligned}\Psi_0 &= \frac{1}{\sqrt{\pi}} \tilde{g}(0), \\ \Psi_1 &= \frac{1}{3\sqrt{\pi}} \left[\Delta\tilde{g}(0) - \left(3h_{20}^2 + 3h_{02}^2 + 2h_{20}h_{02} + h_{11}^2\right) \tilde{g}(0)\right],\end{aligned}$$

where h_{ij} are the coefficients in the expansion (26.12).

26.4 Numerical Example and Conclusions

We present numerical results pertaining to the single-layer equation

$$\int_S \frac{1}{4\pi} \frac{1}{|x-y|} g(y)\, dS_y = f(x),$$

where S is the ellipsoid

$$\left(\frac{x}{2}\right)^2 + \left(\frac{y}{1}\right)^2 + \left(\frac{z}{3}\right)^2 = 1$$

and the right-hand side is $f(x) = 1$. This problem has an analytic solution in closed form, and we compute the $L^2(S)$-error of the numerical solution

$\|e_h\|$ for various values of the meshwidth and the parameters N and δ. To compute the local potential Ψ_δ, the expansion (26.14) is truncated after the first term. Thus, the local potential is replaced by

$$\Psi_\delta(x) \approx \sqrt{\frac{\delta}{\pi}} g(x). \tag{26.15}$$

This approximation is of order $\delta^{3/2}$. The initial triangulation of the ellipsoid consists of 320 panels, which is several times uniformly refined. The finite element space is the piecewise constant functions on this triangulation. Standard convergence analysis implies that the discretization error of the direct Galerkin method is order h, that is, the error is halved in every refinement step. Our goal is to choose the parameters N and δ in such a way that the spectral scheme exhibits the same convergence behavior when refining the mesh. At the same time, the scheme should be efficient, with complexity that is linear or almost linear in the number of panels n. Since the complexity of the FFT is order $N^3 \log N$, we set $N \sim n^{1/3}$ to obtain almost linear complexity.

The parameter δ affects the accuracy in two ways. If δ is small, then the truncation error in (26.15) is small, but on the other hand, the Green's function of the smooth part will be peaked at the origin, which increases the error of the Fourier series approximation. Table 1 displays the behavior of the error when the mesh is refined. In this table, the parameter δ has been determined experimentally to minimize the error. Table 2 displays the effect of δ on the error for the finest mesh.

Table 1. Errors when refining the meshwidth.

n	320	1220	5120	20,480	81,920	327,680
N	8	12	18	32	48	72
δ	1.6(-3)	8.0(-4)	4.0(-4)	2.0(-4)	1.0(-4)	5.0(-5)
$\|e_h\|$	0.3015	0.1549	0.08819	0.04151	0.02036	0.01004

Table 2. Errors for the finest mesh (n =327,680, $N = 72$) for different values of the mollification parameter.

δ	1.0(-3)	1.0(-4)	2.5(-5)	1.3(-5)	6.0(-6)	3.0(-6)
$\|e_h\|$	0.08540	0.01168	0.01214	0.02276	0.04004	0.06116

The hardest problem (n =327,680, $\delta = 3 \cdot 10^{-6}$) took 22 GMRES iterations to converge, the overall time was about 23 minutes on an AMD Athlon64-3200 processor, and the memory allocation was about 550MB. The package FFTW [6] was used for the computation of the FFTs. The order p in (26.7) and (26.8) was set to 4 in all experiments.

The numerical results presented suggest that the parameters in the spectral method can be selected so that the resulting scheme is nearly asymptotically optimal, that is, optimal up to logarithmic factors. Currently we are working on error estimates, trying to confirm this assertion.

References

1. R. Schneider, *Multiskalen- und Wavelet-Matrixkompression: Analysis-basierte Methoden zur effizienten Loesung grosser vollbesetzter Gleichungssysteme,* Teubner, Stuttgart, 1998.
2. S. Sauter, Variable order panel clustering, *Computing* **64** (2000), 223–277.
3. J. Tausch, The variable order fast multipole method for boundary integral equations of the second kind, *Computing* **72** (2004), 267–291.
4. L. Greengard and J. Strain, A fast algorithm for the evaluation of heat potentials, *Comm. Pure Appl. Math.* **43** (1990), 949–963.
5. J.C. Nédélec, *Acoustic and Electromagnetic Equations,* Springer-Verlag, New York, 2001.
6. M. Frigo and S.G. Johnson, The design and implementation of FFTW3, *Proc. IEEE* **93** (2005), 216–231.

27 The GILTT Pollutant Simulation in a Stable Atmosphere

Sergio Wortmann, Marco T. Vilhena, Haroldo F. Campos Velho, and Cynthia F. Segatto

27.1 Introduction

The generalized integral transformation technique (GITT) belongs to the class of spectral methods. This technique has been effective in many applications [1], such as transport phenomena (heat, mass, and momentum transfer). The GITT can be expressed as a truncated series, having as base functions the eigenfunctions of the Sturm–Liouville problem associated with the original mathematical model. The transformed equation is obtained from the eigenfunction orthogonality properties by computing the moments, that is, multiplying by the eigenfunctions and integrating over the whole domain. In general, this strategy produces an algebraic equation system or ordinary differential equations (ODEs) of first order or second order. For the latter case, the solution of the ODEs in the standard GITT is obtained by means of an ODE-solver.

The novelty here is the use of the Laplace transformation (LT) for solving the ODE generated by the GITT. For this specific application, the inverse LT is analytically calculated. After the application of the LT, the resulting system matrix is nondefective, so it can be diagonalized. Therefore, having computed the eigenvalues and eigenvectors, the inversion of the LT is obtained analytically. This new formulation, combining GITT and the Laplace transformation, will be called GILTT (*generalized integral Laplace transformation technique*). In the case of degenerate eigenvalues, a technique is adapted dealing with Schur's decomposition, which demands a greater computational effort.

Here, the GILTT is illustrated in application to the pollutant dispersion problem in the atmosphere under a turbulent flow.

The paper is outlined as follows. In Section 27.2 the GILTT is presented, three cases being described: (i) ODE with constant coefficients, (ii) ODE with variable coefficients, and (iii) nonlinear ODE. A simple pollutant diffusion problem under stable stratification of the atmosphere is presented in Section 27.3. Some conclusions are drawn in Section 27.4.

The authors are grateful to the CNPq (Conselho Nacional de Desenvolvimento Científico e Tecnológico) for partial financial support of this work.

27.2 GILTT Formulation

Some ideas of the GITT are summarized below. Consider the equation

$$A\,v(x,t) = S, \quad x \in (a,b), \;\; t > 0, \tag{27.1}$$

subject to the boundary conditions

$$a_1 \frac{\partial v(x,t)}{\partial x} + a_2 v(x,t) = 0 \quad \text{at } x = a, \tag{27.1a}$$

$$b_1 \frac{\partial v(x,t)}{\partial x} + b_2 v(x,t) = 0 \quad \text{at } x = b, \tag{27.1b}$$

where A is a differential operator, S is the source term, and a_1, a_2, b_1, and b_2 are constants depending on the physical properties. The goal is to expand the function $v(x,t)$ in an appropriate basis. To determine such a basis, the operator A is written as

$$Av(x,t) = Bv(x,t) + Lv(x,t),$$

where L is an operator associated with a Sturm–Liouville problem and B is the operator linked with the remaining terms. Therefore, the operator L is given by

$$L\psi(\lambda,x) \equiv \nabla \cdot [p(x)\,\nabla\psi(\lambda,x)] + q(x)\,\psi(\lambda,x).$$

The functions $p(x)$ and $q(x)$ are real and continuous, and $p(x) > 0$ on the interval (a,b), defining the associated Sturm–Liouville problem

$$L\psi(\lambda,x) + \lambda^2\psi(\lambda,x) = 0, \quad x \in (a,b), \tag{27.2}$$

$$a_1 \frac{\partial \psi(x,t)}{\partial x} + a_2 \psi(x,t) = 0 \quad \text{at } x = a, \tag{27.2a}$$

$$b_1 \frac{\partial \psi(x,t)}{\partial x} + b_2 \psi(x,t) = 0 \quad \text{at } x = b. \tag{27.2b}$$

The Sturm–Liouville problem (27.2) is the general form of the auxiliary problems in the GITT theory [1]. The constants a_1, a_2, b_1, and b_2 are the same as in the original problem (27.1). The solution of the eigenvalue problem (27.2) is used to expand the function $v(x,t)$ in the (orthogonal) eigenfunctions as

$$v(x,t) = \sum_{k=1}^{\infty} \frac{u_k(t)\,\psi_k(x)}{N_k^{1/2}}, \tag{27.3}$$

where $N_k \equiv \int_a^b \psi_k^2(x)\,dx$ is the *norm* of the eigenfunction $\psi_k(x)$. Substituting (27.3) into (27.1), multiplying by $\psi_k(x)/N_k$, and integrating over the entire domain, we arrive at a set of ODEs. The truncated ODE system is the GITT; under certain conditions (not discussed here), the resulting transformed system could be an algebraic one, or even a partial differential equation system. The standard approach in the GITT is to apply an ODE solver to find an approximate solution.

27.2.1 Solving the ODE System by Means of the Laplace Transformation

From the truncated solution of the mathematical model (27.1), the vector $Y(t)$ is defined as the set of N functions

$$Y(t) = [u_1(t) \;\; u_2(t) \;\; \ldots \;\; u_N(t)]^T,$$

and the resulting matrix equation is part of the initial value problem

$$\begin{aligned} EY'(t) + FY(t) &= 0, \quad t > 0, \\ Y(0) &= Y_0. \end{aligned} \tag{27.4}$$

One important issue is linked to the order of the expansion. A higher-order expansion demands expansive computational effort, and the convergence of the iterative process could be hard. However, a low number of eigenvalues could present a low accuracy solution. A strategy for determining the order of the expansion will be commented on later.

Three types of ODE systems can be anticipated, namely (i) type-1: linear ODE systems with constant coefficients, (ii) type-2: linear ODE systems with time-dependent coefficients, and (iii) type-3: nonlinear ODE systems.

27.2.2.1 Type-1 ODE System

The scheme for solving equation (27.4) is the LTS_N method. This method has emerged in the early nineties in the context of transport theory (see [2]–[4]), and its convergence has been proved using C_0-semigroup theory (see [5] and [6]). First, equation (27.4) is multiplied by E^{-1}, then the Laplace transformation is applied to the resulting system, yielding

$$s\hat{Y}(s) + G\hat{Y}(s) = Y_0, \tag{27.5}$$

where $G = E^{-1}F$ and $\hat{Y}(s) \equiv \mathcal{L}\{Y(t)\} = \int_0^\infty Y(t)e^{-st}\,dt$. Next, we factor the matrix G as

$$G = XD_{av}X^{-1}, \tag{27.6}$$

where X is the eigenvector matrix and D_{ac} is the diagonal matrix of the eigenvalues of G. Using (27.6) in (27.5) results in the solution

$$Y(t) = XH_{av}X^{-1}Y_0,$$

where the matrix H_{av} is given by

$$H_{av}(t) = \mathcal{L}^{-1}\{(sI + D_{av})^{-1}\} = \text{diag}\,[e^{-td_1} \quad e^{-td_2} \quad \ldots \quad e^{-td_N}].$$

27.2.2.2 Type-2 ODE System

Now the ODE system is expressed in the form

$$\begin{aligned} E(t)Y'(t) + F(t)Y(t) = 0, \quad t > 0, \\ Y(0) = Y_0. \end{aligned} \tag{27.7}$$

In order to solve equation (27.7), the expression $E_mY'(t)+F_mY(t)$ is added to both sides, where $E_m = \int_0^{\mathcal{T}} E(t)\,dt$ and $F_m = \int_0^{\mathcal{T}} F(t)\,dt$. After some algebraic manipulation, we arrive at

$$\begin{aligned} E_mY(t) + F_mY(t) = S(Y'(t), Y(t), t), \\ Y(0) = Y_0, \end{aligned} \tag{27.8}$$

where the heterogeneous function S is given by

$$S(Y'(t), Y(t), t) = [E_m - E(t)]Y'(t) + [F_m - F(t)]Y(t).$$

A solution of (27.8) is obtained in terms of a convolution, namely

$$Y(t) = XH_{av}(t)X^{-1}Y_0 + XH_{av}(t)X^{-1} * S(Y'(t), Y(t), t). \tag{27.9}$$

The above equation is an implicit expression. To overcome this drawback, Adomian's decomposition method [7] is employed. Initially, the vector $Y(t)$ is written as

$$Y(t) = \sum_{k=1}^{\infty} w_k(t). \tag{27.10}$$

Substituting (27.10) in (27.9) yields

$$\begin{aligned} \sum_{k=1}^{\infty} w_k(t) = XH_{av}(t)X^{-1}Y_0 \\ + XH_{av}(t)X^{-1} * S\Big(\sum_{k=1}^{\infty} w_k'(t), \sum_{k=1}^{\infty} w_k(t), t\Big). \end{aligned}$$

Finally, the first term on the left-hand side is identified with the first term on the right-hand side, and so on; therefore,

$$\begin{aligned} w_1(t) = XH_{av}(t)X^{-1}Y_0, \\ w_k(t) = XH_{av}(t)X^{-1} * S(w_{k-1}'(t), w_{k-1}(t), t), \quad k = 2, 3, \ldots. \end{aligned}$$

27.2.2.3 Type-3 (Nonlinear) ODE System

Here, the differential equations produced by the GITT methodology are nonlinear ODEs of the form

$$E(Y,t)Y'(t) + F(Y,t)Y(t) = 0, \quad t > 0, \qquad (27.11)$$
$$Y(0) = Y_0.$$

An iterative procedure is adopted to look for a solution. This is a simple feature, where system (27.11) is written as

$$E(Y^{(m)},t)Y'(t) + F(Y^{(m)},t)Y(t) = 0, \quad t > 0, \qquad (27.12)$$
$$Y(0) = Y_0.$$

System (27.12) is formally identical to that in equation (27.7). Therefore, the method described in Section 27.2.2.2 can be applied. The procedure is repeated until there is convergence, with the first guess $Y^{(0)} = Y_0$.

27.3 GILTT in Atmospheric Pollutant Dispersion

Consider a pollutant puff released from an area source (an urban region, for example), in the evening (characterizing a stable atmospheric stratification), under a weak-wind condition, which means that vertical transport will be the main process involved. This situation can be modeled by [8]

$$\frac{\partial}{\partial z}\left[K_{zz}(z)\frac{\partial c(z,t)}{\partial z}\right] = \frac{\partial c(z,t)}{\partial t}, \quad z \in (0,h), \; t > 0, \qquad (27.13)$$

with the initial and boundary conditions

$$c(z,t) = Q\delta(z - h_f) \quad \text{at } t = 0, \qquad (27.13a)$$
$$K_{zz}\frac{\partial c(z,t)}{\partial z} = 0 \quad \text{at } z = 0 \text{ and } z = h. \qquad (27.13b)$$

In this model, $c(z,t)$ denotes the average pollutant concentration as a function of the level z and the time t, Q is the source strength, h is the boundary layer height, $\delta(z)$ is the delta function, and h_f is the level where the pollutant is released.

The turbulent eddy diffusivity tensor is represented in orthotropic form, with the horizontal diffusion considered negligible. Under the assumption of a stable boundary layer (SBL), in [9] an expression has been derived for the turbulent exchange coefficient, with its vertical component given by

$$\frac{K_{zz}(z)}{u_* h} = \frac{0.33(1 - z/h)^{\alpha_1/2}(z/h)}{1 + 3.7(z/h)(h/\Lambda)},$$

where u_* is the friction velocity and Λ the local Monin–Obukhov length [9]

$$\frac{\Lambda}{L} = (1 - z/h)^{3\alpha_1/2-\alpha_2};$$

here L is the Monin–Obukhov length for the entire SBL and α_1 and α_2 are experimental constants depending on several parameters such as evolution time of the SBL, topography, and heat flux. Numerical values of the constants α_1 and α_2 for the Minnesota (SBL in transition) and Cabauw (more steady state SBL) experiments are given in Table 1.

Table 1. Values of the experimental constants of the SBL [8].

Experiment	Minnesota (Exp-M)	Cabauw (Exp-C)
α_1	2	3/2
α_2	3	1

In the GITT formulation, the first step is to identify the associated Sturm–Liouville operator. Two approaches can be used, namely

$$L_1 \equiv \frac{d}{dz}\left(K_{zz}\frac{d}{dz}\right) + \lambda I,$$
$$L_2 \equiv \frac{d^2}{dz^2} + \lambda I, \tag{27.14}$$

where I is the identity operator. Both operators have the same boundary conditions as the original problem (27.13). Operator L_1 implies a more difficult Sturm–Liouville problem. On the other hand, L_2 is a simpler operator, but it also conveys less information about the basic problem. The price to pay for this simplicity is to get more terms in the expansion, in order to maintain an appropriate accuracy for the computed solution. Our option is to work with operator L_2.

The Sturm–Liouville problem defined by the operator L_2 has the eigenfunctions [10]

$$\psi_k(x) = \cos(\lambda_k z), \tag{27.15}$$

where the eigenvalues λ_k are the positive roots of the equation

$$\sin(\lambda_k z) = 0.$$

The next step is to represent the concentration $c(z,t)$ as an expansion of the form (27.3), that is,

$$c(z,t) = \sum_{k=0}^{\infty} u_k(t)\, \psi_k(z), \tag{27.16}$$

where $\psi_k(z)$ is the eigenfunction given by (27.15). Substituting (27.16) into (27.13), multiplying by $\psi_j(z)$, integrating over the domain, and taking into account the auxiliary Sturm–Liouville problem (27.14), we arrive at

$$\sum_{j=0}^{\infty}\left[u_k(t)\left(\int_0^h K_{zz}\psi_k(z)(-\lambda_k^2\,\psi_j(z))\,dz+\int_0^h \frac{dK_{zz}}{dz}\psi_k(z)\frac{d\psi_j(z)}{dz}\,dz\right)\right.$$
$$\left.-\frac{du_k}{dt}\int_0^h \psi_k(z)\psi_j(z)\,dz\right]=0. \qquad (27.17)$$

A similar procedure is applied to the initial condition, which yields

$$\int_0^h \sum_{j=0}^{\infty} u_k(t)\,\psi_k(z)\,\psi_j(z)\,dz=\int_0^h Q\,\delta(z-h_f)\,\psi_j(z)\,dz,$$

leading to

$$u_k(0)=a_r\,\frac{\psi_k(h_f)}{h},\qquad a_r=\begin{cases}1 & \text{for } k=0,\\ 1/2 & \text{for } k\geq 1.\end{cases} \qquad (27.18)$$

In matrix notation, equations (27.17) and (27.18) are written as

$$\begin{aligned} EY'(t)+FY(t)&=0, \\ Y(0)&=Y_0, \end{aligned} \qquad (27.19)$$

where the matrices E and F and the vectors $Y(t)$ and Y_0 are given by

$$E_{kj}=-\lambda_k^2\int_0^h K_{zz}\psi_k(z)\psi_j(z)\,dz+\int_0^h \frac{dK_{zz}}{dz}\psi_k(z)\,\frac{d\psi_j(z)}{dz}\,dz, \qquad (27.19a)$$

$$F_{kj}=-\int_0^h \psi_k(z)\,\psi_j(z)\,dz, \qquad (27.19b)$$

$$Y(t)=[u_0(t)\ \ u_1(t)\ \ \ldots\ \ ]^T, \qquad (27.19c)$$

$$Y(0)=[u_0(0)\ \ u_1(0)\ \ \ldots\ \ ]^T. \qquad (27.19d)$$

Equation (27.19) was derived using the GITT procedure [1], which is typically implemented by numerical solvers. An analytic solution for the time integration is presented here, and the approach described in Section 27.2.2.1 can be applied.

For a numerical example, some parameters must be known. For the simulation performed, these parameters are shown in Table 2.

Table 2. Parameters for simulations.

Parameter	L	Q	h	u_*
Value:	116 m	400 g m^{-2}	400 m	0.31 m s^{-1}

Tables 3 and 4 show that enhancing the order of the expansion of the GITT leads to a more accurate solution. The result is numerical evidence of the convergence.

Table 3. Computed pollutant concentration for different evolution times and several expansion degrees at $z/h = 0.2$.

	Number of eigenvalues							
t	5	10	15	20	25	30	40	50
$1h$	216.29	216.74	216.81	216.83	216.78	216.74	216.67	216.64
$2h$	175.95	176.22	176.36	176.38	176.36	176.35	176.31	176.30
$3h$	152.57	152.71	152.84	152.85	152.84	152.84	152.82	152.81
$4h$	137.78	137.88	137.98	138.00	137.99	137.99	137.98	137.97

Table 4. Computed pollutant concentration at different levels (z/h) for several expansion degrees at $t = 6700$ s.

	Number of eigenvalues							
z/h	5	10	15	20	25	30	40	50
0.20	1.8028	1.8059	1.8073	1.8075	1.8073	1.8071	1.8067	1.8065
0.47	1.0738	1.0647	1.0643	1.0640	1.0641	1.0643	1.0644	1.0644
0.73	0.3452	0.3336	0.3339	0.3337	0.3338	0.3341	0.3344	0.3345
1.00	0.0121	0.0321	0.0223	0.0233	0.0220	0.0215	0.0209	0.0207

Fig. 1 displays the computational effort in terms of the CPU time when the number of eigenvalues increases. The results obtained for several degrees of the expansion show the convergence process of the solution strategy. However, the degree of the expansion can be stipulated according to the tolerance required, as suggested in [1], p. 246.

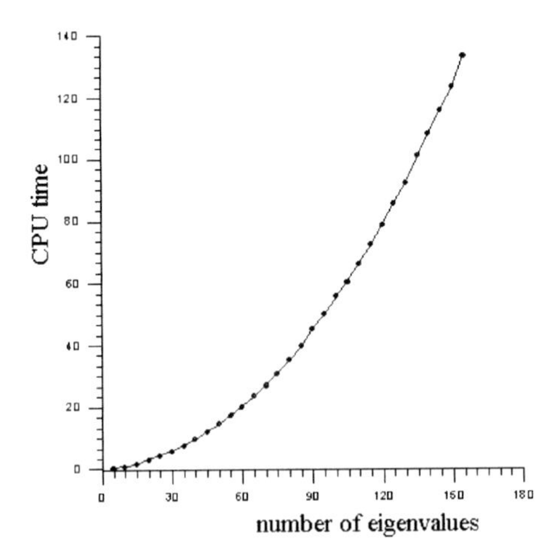

Fig. 1. Computational effort in terms of the order of the expansion.

The evolution times for two different SBLs are depicted in Fig. 2, showing the concentration profiles at different evolution times, and in Fig. 3, which illustrates the evolution times for the concentration at the center of the SBL. In both cases, the diffusion process is slow, in comparison with the vertical transport in the convective boundary layers. These results are in agreement with the expected turbulent diffusion under stable boundary conditions. The results present very good agreement with those obtained using the finite difference numerical method [11].

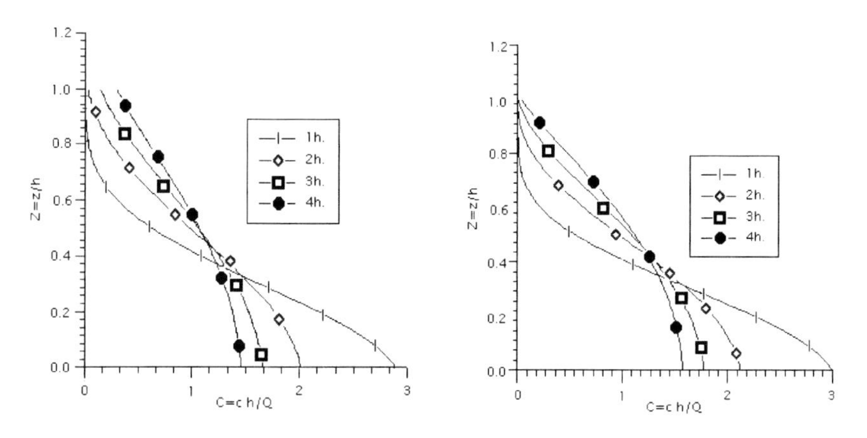

Fig. 2. Concentration profile for (left) the Minnesota experiment and (right) the Cabauw experiment.

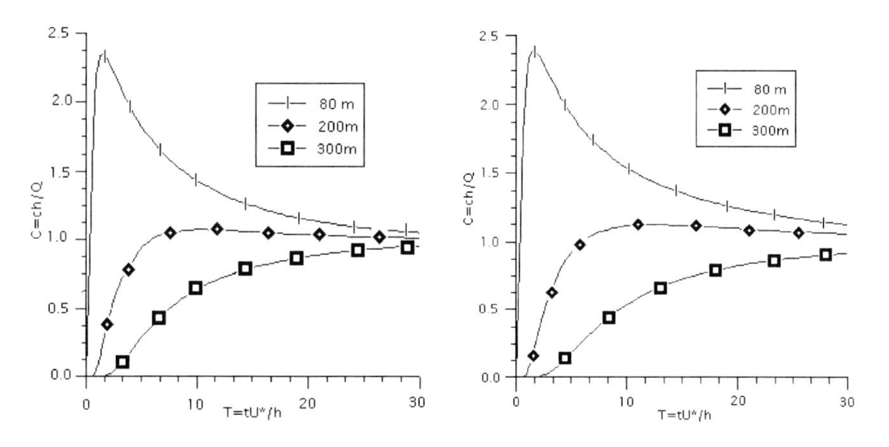

Fig. 3. Concentration evolution at the central point of SBL: (left) the Minnesota experiment and (right) the Cabauw experiment.

27.4 Final Remarks

This paper presents a new approach to time-integration in the GITT (*generalized integral transformation technique*), based on the Laplace transformation. Our formulation, the **GILTT** (*generalized integral Laplace transformation technique*), has been developed in this paper and has been applied to the atmospheric dispersion model.

The results computed using the GILTT indicate accurate solutions. Some advantages can be pointed out in the use of the method. Thus, the solution can be obtained at each moment of time, without marching in time, through a closed mathematical formula; in addition, the GILTT reduces the computational effort required by the GITT when an ODE-solver is employed.

References

1. R.M. Cotta, *Integral Transforms in Computational Heat and Fluid Flow,* CRC Press, Boca Raton, FL, 1993.
2. M.T. Vilhena and L.B. Barichello, A new analytical approach to solve the neutron transport equation, *Kerntechnik* **56** (1991), 334–336.
3. C.F. Segatto and M.T. Vilhena, Extension of the LTS_N formulation for discrete ordinates problem without azimuthal symmetry, *Ann. Nuclear Energy* **21** (1994), 701–710.
4. C.F. Segatto and M.T. Vilhena, State-of-art of the LTS_N method, in *Mathematics and Computation, Reactor Physics and Environmental Analysis in Nuclear Applications,* J.M. Aragonés, C. Ahnert, and O. Cabellos (eds.), Senda Editorial, Madrid, 1999, 1618–1631.
5. M.T. Vilhena and R.P. Pazos, Convergence in transport theory, *Appl. Numer. Math.* **30** (1999), 79–92.
6. M.T. Vilhena and R.P. Pazos, Convergence of the LTS_N method: approach of C_0 semigroups, *Progress Nucl. Energy* **34** (1999), 77–86.
7. G. Adomian, A review of the decomposition method in applied mathematics, *J. Math. Anal. Appl.* **135** (1988), 501–544.
8. F.T.M. Nieuwstadt, The turbulent structure of the stable nocturnal boundary layer, *J. Atmospheric Sci.* **41** (1984), 2202–2216.
9. G.A. Degrazia and O.L.L. Moraes, A model for eddy diffusivity in a stable boundary layer, *Boundary-Layer Meteorology* **58** (1992), 205–214.
10. M.N. Özişik, *Heat Transfer*, Wiley, New York, 1980.
11. G.A. Degrazia, O.L.L. Moraes, H.F. Campos Velho, and M.T. Vilhena, A numerical study of the vertical dispersion in a stable boundary layer, in *VIII Brazilian Congress on Meteorology and II Latin-American and Iberic Congress on Meteorology,* vol. 1, Belo Horizonte, 1994, 32–35.

Index